WORK IS DANGEROUS
TO YOUR HEALTH

WORK
IS DANGEROUS
TO YOUR HEALTH

A Handbook of Health Hazards in the
Workplace and What You Can Do About Them

Jeanne M. Stellman, Ph.D.

Susan M. Daum, M.D.

With Contributions by

James L. Weeks, M.S.
Steven D. Stellman, PH.D.
Michael E. Green, PH.D.

With illustrations by Lyda Pola

PANTHEON BOOKS
A Division of Random House, New York

Library of Congress Cataloging in Publication Data

Stellman, Jeanne M. 1947–
 Work Is Dangerous to Your Health.

Bibliography, pp. 423–25
1. Occupational diseases. 2. Industrial
hygiene. I. Daum, Susan M., 1941–   joint author.
II. Title.
[DNLM: 1. Industrial medicine—Popular works.
2. Occupational diseases—Popular works. WA400 S824w
1973]
RC964.S7 1972b   613.6′2   72-12386
ISBN 0-394-48525-4

Manufactured in the United States of America by
American Book-Stratford Press
Charts and diagrams by Bob Cordes

3 5 7 9 8 6 4 2

*Dedicated to a world in which workers have eradicated
the health hazards of work*

CONTENTS

List of Illustrations	viii
List of Tables	ix
Permissions Acknowledgments	xi
Foreword by George Wald	xiii
Acknowledgments	xix
How to Use This Book	xxi
CHAPTER ONE Occupational Disease: The Silent Killer	3
CHAPTER TWO The Human Body and Occupational Disease	17
CHAPTER THREE Stress	76
CHAPTER FOUR Noise and Vibration	93
CHAPTER FIVE The Effects of Heat and Cold	121
CHAPTER SIX Light, X-rays, and Other Radiation	132
CHAPTER SEVEN Chemical Hazards	154
Threshold Limiting Values (TLVs) for Commonly Used Chemicals	261
CHAPTER EIGHT Welding Hazards	275
CHAPTER NINE Controlling Pollution in the Workplace	290
CHAPTER TEN Measurement and Monitoring	327
CHAPTER ELEVEN Keeping Health Records	345
CHAPTER TWELVE What Is to Be Done	360
Some Exposures to Health Hazards Listed by Occupation	368
Bibliography	423
General Index	427
Index of Substances	439
About the Authors	449

ILLUSTRATIONS

Figure 1 The Lungs 23

Figure 2 The Paths of Blood Circulation 29

Figure 3 How the Kidneys Work 41

Figure 4 The Nervous System 49

Figure 5 Microscopic Cross-section of the Skin 55

Figure 6 The Back 69

Figure 7 What Stress Does to Your Body 77

Figure 8 Causes of Fatigue 81

Figure 9 The Frequencies Corresponding to Sound That Can Be Heard 95

Figure 10 The Ear 98

Figure 11 How the Temporary Threshold Shift Grows with Time 101

Figure 12 Comparison of Effects of Different Levels of Noise 105

Figure 13 Noise Control 110

Figure 14 A Sound-Level Meter 114

Figure 15 The Percentage of Industrial Workers Showing 25 dB Hearing Loss 117

Figure 16 An Audiometer 118

Figure 17 The Energy of Different Types of Radiation 134

Figure 18 The Eye 136

Figure 19 Shielding Needed for Ionizing Radiation 147

Figure 20 Some Welding Hazards 276

Figure 21 A Portable Hood 289

Figure 22 Some Types of Fans 302

Figure 23 Four Respirators 307

Figure 24 A Sampling Train 329

Figure 25 An Impinger Assembly 332

Figure 26 A Paper-Tape Sampler 334

Figure 27 A Universal Tester 337

Figure 28 A Film Badge 344

TABLES

Table 1 The Effects of Cigarette Smoking 31

Table 2 Components of the Blood 34

Table 3 Some Chemicals and Drugs That Cause Liver Damage 39

Table 4 Common Causes of Kidney Disease 44

Table 5 Some Chemicals That Affect the Nervous System 53

Table 6 Important Irritating Industrial Chemicals 57

Table 7 Ten of the Most Common Groups of Chemicals That Cause Contact Dermatitis 59

Table 8 Some Chemicals That Cause Skin Cancer 63

Table 9 Recommended Weights for Loads for Occasional Lifting by Untrained Workers 73

Table 10 Estimated Noise Levels at Various Machines and Locations 96

Table 11 Effectiveness of Various Ear-protective Devices 112

Table 12 Estimated Ideal Temperatures for Various Types of Work 122

Table 13 Sources and Effects of Non-ionizing Radiation 137

Table 14 Recommended Lighting Levels 144

Table 15 Legal Standards for Exposure to Radiation 153

Table 16 Causes of Death Among Asbestos-Insulation Workers   174

Table 17 Effects of Diethylene Glycol Ethers   189

Table 18 Effects of Some Diphenyls   193

Table 19 Health Hazards of the Nitro and Amino Aromatics   199

Table 20 Cancer-Causing Ability of Anthracene Derivatives   202

Table 21 Some Ketones   209

Table 22 Organic Acids   215

Table 23 Effects of Acetates   220

Table 24 Toxicity of Common Pesticides   224

Table 25 Effects of Some Epoxy Compounds   228

Table 26 Toxic Effects of Plasticizers   233

Table 27 Some Nitro Compounds and Their Effects   235

Table 28 Toxic Properties of Some Amines   236

Table 29 Heavy Metals Found in Major Industries   243

Table 30 Hazards of Welding Processes   285

Table 31 Velocity of Particles Generated by Different Processes   295

Table 32 Minimum Air Flow Needed for Dust Collection 297

Table 33 Using the Right Respirator   311

Table 34 Color Assigned to Canister or Cartridge   314

Table 35 Some Gases That Can Be Detected by Continuous Monitoring Devices   336

Table 36 Chemical Tubes for the Universal Tester   338

Table 37 Percentage of Lung Cancer Patients and Controls Who Were Non-smokers, and Who Were Heavy Smokers   349

Table 38 Observed and Expected Number of Deaths Among 632 Asbestos Workers Exposed to Asbestos Dust 20 Years or Longer, 1943–1962   351

PERMISSIONS ACKNOWLEDGMENTS

Figure 14 reprinted from *NOISE! NOISE! NOISE!*, General Radio 1564-B and 1563, by permission of General Radio, Concord, Mass.

Figure 15 adapted from W. L. Baughn, "Noise Control—Percent of the Population Protected," *International Audiology*, vol. 5 (1966).

Figure 16 reprinted by permission from *Beltone D-Series and DW-Series Audiometers*. Copyright © 1971 by Beltone Electronics Corporation, Chicago.

Figure 22 reprinted by permission from American National Standards Institute Publication no. Z9.2-1971, *Fundamentals Governing the Design and Operation of Local Exhaust Systems*. Copyright © 1972 by the American National Standards Institute, New York.

Figure 23: demand-type compressed-air mask by permission of Scott Aviation, a Division of ATO, Inc., Lancaster, New York; M-S-A special hose mask by permission of Mine Safety Appliances, Inc., Pittsburg, Pa.; pocket respirator and single-filter respirator reprinted from *Whatever the Hazard . . . There's an AO Respirator* by permission of American Optical Corporation, Southbridge, Mass.

Figures 25, 26, and 27 reprinted by permission and courtesy of The Bendix Corporation Environmental Science Division, Baltimore.

Figure 28 reprinted by permission of R. S. Landauer Junior and Company, Glenwood, Ill.

Tables 4 and 8 adapted by permission from *Encyclopaedia of Occupational Health and Safety*. Copyright © 1971 by International Labour Office, Geneva. Table 8 courtesy Dr. W. C. Hueper.

Tables 6 and 7 reprinted from *Occupational Contact Dermatitis*, by R. M. Adams, by permission of J. B. Lippincott Company, Philadelphia.

Table 14 adapted from "Practice for Industrial Lighting," *Illuminating Engineering* 1-1965, by permission of the Illuminating Engineering Society, New York.

Tables 16 and 38 adapted from E. C. Hammond, I. J. Selikoff, and J. Churg, "Neoplasia Among Insulation Workers in the United States, with Special Reference to Intra-abdominal Neoplasia," *Annals of the New York Academy of Sciences*, vol. 132 (1965), by permission of the New York Academy of Sciences, Mr. E. C. Hammond, and Dr. Irving J. Selikoff.

Table 29 adapted from *Environmental*, vol. 6, no. 6 (1972), by permission of the copyright owner, the American Chemical Society, Washington, D.C.

Tables 31 and 32 reprinted by permission from American National Standards Institute Publication no. Z9.2-1971, *Fundamentals Governing the Design and Operation of Local Exhaust Systems.* Copyright © 1972 by the American National Standards Institute, New York.

Table 34 adapted from American National Standards Institute Publication no. K13.1-1973, *Identification of Air-Purifying Respirator Canisters and Cartridges.* Copyright © 1973 by the American National Standards Institute, New York.

Table 36 adapted by permission and courtesy of The Bendix Corporation Environmental Science Division, Baltimore.

Foreword

THIS IS A BOOK written to and for workers. It tells them how to stay alive while earning a living. That isn't always easy. Modern industrial processes introduce new hazards all the time. The easy ones to spot are those that kill or maim quickly. Others take a long time, but are just as certain. When those hazards are at work, the signs are all around; one gets used to them: persistent coughing, headaches, dizziness, lassitude, pains that won't go away, stiffening of joints. Then they begin to add up: premature aging, quitting work at age 50, early death. It's a big price to pay, and all needless. Dealing with it just takes attention and concern, and perhaps some special effort. This book tells what to look for, and how to deal with it.

It's a vast problem. Among the 80 million workers in the United States, more than 14,000 deaths on the job are recorded annually, and about 2.2 million disabling injuries. Those are probably minimal figures, for every worker knows the devices by which industry hides or disguises accidents on the job and pads its safety records. A recent report has estimated that the actual numbers may run as high as 25,000 deaths and 20 to 25 million job-related injuries annually.

Such figures take no account of the slowly developing occupational diseases that eventually cripple or kill workers. In 1968 Dr. William H. Stewart, former Surgeon General of the United States, told a congressional committee that

United States Public Health Service studies showed 65 percent of industrial workers to be exposed to toxic materials or harmful physical conditions such as excessive noise or vibrations.

Some of those occupational diseases have become famous. Black lung (pneumoconiosis), for example, afflicts about 100,000 miners, about 4,000 of whom die of it each year. About 17,000 cotton, flax, and hemp workers have brown lung (byssinosis). Over the next twenty years, 6,000 uranium miners are expected to develop cancers. Of about 500,000 workers now or previously employed with asbestos, Dr. Irving Selikoff, of the Mount Sinai Hospital in New York City, estimates that 100,000 will die of lung cancer, 35,000 of abdominal or chest cancers (mesotheliomas), and about 35,000 of asbestosis (scarring of the lungs). A study of 689 asbestos and textile workers during the thirteen years from 1959 to 1971 revealed 72 deaths from all types of cancer compared with an expected 27.8 in the general population; among these, 35 deaths were from cancer of the lungs, pleura, trachea, and bronchi, as compared with an expected 8.4 such deaths in the general population. As this book says, it's like a major war, going on all the time. The casualties, however, aren't fighting, but just trying to make a living.

All kinds of workplaces are involved. In 1968 the Chicago Institute of Medicine sponsored a survey of 803 workplaces employing 260,000 persons in the Chicago area, including not only manufacturing plants but transportation, wholesale and retail trade, hotels, and places of entertainment. In 73 percent of such places, they found that workers were exposed to one or more potentially hazardous materials.

Such hazardous materials include toxic gases and dusts, and increasing numbers of chemicals and their vapors. The *Standard Reference Manual* of the American Conference of Governmental Industrial Hygienists, an unofficial professional society, estimated that in 1970, 6,000 to 12,000 toxic industrial chemicals were in common use; yet it listed standards for only 410 of them. At present about 3,000 new chemicals are introduced into industry every year; yet standards are being developed for only about 100 new chemicals per year.

All of this makes a frightening account. The point, however, is not to be frightened, but to do something about it. None of it is inevitable, none of it necessary. The time has come to make a fresh start, to see to it that workers in our country live healthy lives, and live them out as people do outside the shops: to say it plainly, to see to it that workers live long enough to collect their pensions.

Recent legislation invites that new deal: the Federal Coal Mine Health and Safety Act of 1969, and the Occupational Safety and Health Act of 1970. Yet such legislation in itself can accomplish little without strong and continuing and well-informed pressure and participation by workers.

The legislation itself is beset with frustrations. As this book says, there are fewer than 500 inspectors, and only 50 industrial hygienists to examine 4.1 million workplaces. During the first eight months under the 1970 act, inspectors visited 17,743 workplaces. At that rate it would take them 230 years to visit all the workplaces. Also, an inspector is not permitted to take an air sample; only one of those few industrial hygienists is permitted to do that.

Many employers have a deep and sincere concern for their workers' health and safety; others, unfortunately, seem to be less concerned. All of them are caught between this type of consideration and obvious pressures to keep costs down and to speed production. There is a widespread and understandable tendency, sometimes with the cooperation of plant physicians, to dismiss injuries and diseases as other than work-related. This holds down insurance premiums and workmen's compensation costs. Industry can also exert pressure in these directions upon government.

For example, *Nucleonics Week* of February 12, 1970, reported that the now-defunct Federal Radiation Council had placed a $200,000 contract with Arthur D. Little Incorporated of Cambridge, Massachusetts, funded by the Atomic Energy Commission, to study the economic effects of lowering uranium-mining standards. The article quoted an unnamed government official as explaining, "There actually are two questions to be answered. First, we must determine if the [maximum exposure] standard is too high from the standpoint of the health of the miners. Then we must decide if a lower standard is economically practical, regardless of the hazards involved. It boils down to the old question of benefit versus risk." One might add a further question: Whose benefit? whose risk?

A few months later A. D. Little issued its report, summarized in *Nucleonics Week* for December 24, 1970. It concluded that the maximum limits of exposure could be lowered—as had been ordered in a federal regulation issued in the last days of the Johnson administration—without large expense. The new standards were to go into effect in January 1970; but "interested parties in industry and government" asked for an extension of the old standards for another six months to permit "further study," and this was granted by the White House.

In the same vein, the *New York Times* for January 12, 1973, reported that the United States Bureau of Mines had revised procedures for imposing fines for safety violations on mine operators because of "adverse comments" from the operators. Yet the General Accounting Office—a congressional watchdog agency—found that of $12.5 million in safety-violation fines assessed by the Bureau of Mines since 1970, only $1.4 million had been collected.

If the far-reaching provisions of the new health and safety legislation are to be realized, it will be only because workers insist upon them, organize for them, and participate in their enforcement. That is just what the legislation invites. The Occupational Safety and Health Act of 1970 provides that an authorized representative of the workers must accompany all inspections; also that the inspector must have an opportunity to question employees in private. The workers are to be informed of any violation found by an inspector. Workers or their representative may request a special inspection at any time they feel that an imminent danger or a health or safety violation exists. Employers must monitor and keep records of the exposure of workers to potentially toxic materials or harmful physical conditions, and workers must have opportunity to observe such monitoring and have access to the records. Arrangements are to be provided for the education and training of both management and workers in sound health and safety practices.

The present manual is designed to help make workers ready to assume these functions; to train them to deal effectively with hazardous work situations, to be able to discuss them knowledgeably with inspectors and employers, and to judge the adequacy of inspections and what is done about them.

There will surely be places and times when the workers themselves should monitor their working environment: take air and dust samples, measure the concentrations of toxic fumes, and the like. Many such procedures are described here. Such tests require special equipment, sometimes simple, sometimes complicated and costly. There is nothing in the procedures themselves, however, that need intimidate workers, nothing intrinsically more difficult than the kinds of thing they do in their jobs. What an inspector or industrial hygienist has learned to do, workers can learn to do. It is important to realize that, and act upon it.

A big reorientation is needed also toward worker health and safety. The old emphasis on compensation after injury and disease have taken their toll needs to be replaced by a preventive attitude, by an insistence that killing, maiming, and disease not be part of the work experience. In the past it has too often been taken for granted that work inevitably involves physical injury to the workers. The new attitude should be an insistence that workplaces be kept decent and healthful, that the health and strength and longevity of the workers be a condition of their work. In the long run, that will be advantageous for the employers too.

As I write these words, one union—the Oil, Chemical and Atomic Workers International Union—has taken a major step in this new direction. It has written a series of clauses that implement features of the Occupational Safety and Health Act of 1970 into its new contracts. Three features of the new agreements are: (1) Surveys of plant health and safety conditions are to be made by *independent* health consultants at company expense. (2) In the light of the findings of such independent consultants, physical examinations and medical tests are to be provided to the workers. (3) The companies are to supply the union annually with all available information on the disease and death statistics of its workers. Basic control of health and safety on the job will rest with a joint Labor-Management Health and Safety Committee, made up of equal numbers of persons from both groups.

It would be hard to exaggerate the importance of this development. We may hope that in future such clauses in labor contracts will give full scope to the new workers' health and safety legislation. That could lead the way to a new and better era in the lives of all American workers.

GEORGE WALD

Cambridge, Massachusetts
April 1, 1973

Acknowledgments

WE WISH TO ACKNOWLEDGE Anthony Mazzocchi of the Oil, Chemical and Atomic Workers International Union (OCAW), who introduced us to many of the problems of occupational health that are faced by workers throughout the world. We thank the OCAW under the leadership of its president, A. F. Grospiron, for its support of this effort and its pioneering role in occupational health. We are grateful to the Scientists Committee for Occupational Health, which produced the manual *Industrial Health Hazards,* a predecessor to this work, and to the Rutgers Labor Education Center, their staff, and our many worker-students, from whom we learned much about the daily hazards that workers face on the job, and who freely criticized and helped us attempt to make this book relevant. We wish to thank the following people who read the manuscript and made many useful suggestions: Stanley Weir, A. C. Sabatine, Charles Lienenweber, Barbara Garson, and Drs. Frank Collins, Kurt Deuschle, Ruth Lillis, William Nicholson, and Irving Selikoff. We are indebted to them for their time and effort, and are of course ourselves solely responsible for any errors that may remain.

We appreciate the encouragement and help that we received from all. Of course, all viewpoints expressed in this book are our own and do not reflect upon the Mount Sinai School of Medicine.

How to Use This Book

IN ORDER FOR health hazards of the workplace to be eliminated, workers and the public must understand the nature of these hazards and learn what can be done about them. Therefore, this book is written in language that can be understood by a person who is not trained in technical matters. So that it may be useful as a reference guide, however, it also contains detailed information on many individual chemical and physical hazards.

The introductory chapter and the ones on the human body, stress, noise, and heat and cold (Chapters One through Five) are of general interest. In fact, they apply not just to the workplace but to many of the problems of living in a modern industrial society. Parts of Chapter Six, on light, X-rays, and other radiation, are also of general interest because of the increasing pollution problems that some of these hazards present. We therefore suggest that everyone read these chapters. Chapters Seven and Eight, on chemical hazards and welding, and Chapters Nine and Ten, on eliminating pollution in the workplace and on measurement and monitoring, are more technical and may not be of immediate concern to all readers. They can be skipped without loss of continuity or of points of general interest. However, they will be useful for reference.

One way to use this book as a reference is simply to look up a particular chemical or physical hazard in the indexes and then read what the text has to say about it. Another way is to look up a particular profession in the appendix, "Some Exposures to Health Hazards Listed by Occupation," determine what hazards the workers in that profession are exposed to, and then refer to the indexes. The reader may also want to skim through Chapters Seven to Ten to see what material they may contain that is relevant to his or her occupation. It is not necessary to read these chapters straight through; in fact, reading through the chapter on chemical hazards would be rather like reading an encyclopedia or a dictionary. Preferably, the General Index, the Index of Substances, and the appendix should be used to locate the material in which the reader is really interested.

The authors of this book do not pretend to be unbiased. The subject of occupational health involves too many unknowns and too many controversies to be presented simply as a set of "facts." The facts must be interpreted, and interpretation always introduces bias. Our bias is one that favors the worker. For example, if a particular chemical has been shown to cause disease in animals and its effects have not yet been studied in humans, this book questions its safety for human exposure. It emphasizes that concrete knowledge of the effects of that chemical on humans must be obtained before workers are exposed to it. We hold that industry and government must prove that a chemical is safe, not that workers must prove it is dangerous by developing occupational disease.

We also stress that it is the workers' own interest to gain more control over their working conditions and to strive to change them. Uncontrolled and unpressured, industries and the government will do little or nothing to eliminate pollution either within the workplace or in the general environment. Chapter Eleven, "Keeping Health Records," discusses ways that workers themselves can collect the necessary data for further research on occupational diseases. Chapter

Twelve, "What Is to Be Done," shows how they may work through their unions to bring about the needed changes.

The book often refers to the work of the Oil, Chemical and Atomic Workers International Union (OCAW) in health and safety. We do not mean to stand as partisan supporters of the OCAW or to overlook the contributions of others. However, our own experience in the field of occupational health has been so closely involved with OCAW that we inevitably draw upon the work of that union for examples.

We have not spoken of the many people who have dedicated their lives toward achieving decent working conditions for those who labor in the factories, mines, and other workplaces throughout the world. Dr. Alice Hamilton is one of many who come to mind as pioneers in the field. However, the purpose of this book is not to retell history, but to provide material for the carving out of a future— a future in which men and women will not have to lose one minute of life or health while earning a living.

WORK IS DANGEROUS
TO YOUR HEALTH

CHAPTER ONE

OCCUPATIONAL DISEASE: THE SILENT KILLER

EACH DAY millions of workers in America enter a battlefield, but they fight no foreign enemy and conquer no lands. No borders are in dispute. The war they are fighting is against the poisonous chemicals they work with and the working conditions that place serious mental and physical stress upon them. The battlefield is the American workplace, and the casualties of this war are higher than those of any other in the nation's history.

Official estimates of the extent of occupational illness and injuries are startling. Government figures state that over 14,000 workers are killed each year in work accidents, and over 2.2 million more are either permanently or temporarily disabled as a result of work accidents and occupational disease. Quite separately from accidents, almost a half million workers develop "official" occupational diseases each year.[1] Horrifying as these statistics are, they *underestimate* the real numbers involved. A recent study commissioned by the United States Department of Labor stated that at least 25 *million* serious injuries and deaths go uncounted each year and that the department's biased counting methods incorporate at least an 8 to 10 percent underre-

---

[1] U.S. Department of Health, Education, and Welfare, *Environmental Health Problems* (Washington, D.C.: Government Printing Office, 1970).

porting rate for injuries.[2] For example, it is common practice in some industries to bring injured men to work on
stretchers or in casts and bandages so that they can check
in and not be counted as part of the on-the-job injuries.[3]

What is more important is that these statistics and numbers are largely concerned with accidents and injuries and
not with occupational disease that develops from long-term
exposure to noisy, dirty, hot, or cold working conditions and
to various toxic chemicals and physical hazards. The hun-
dreds of thousands of cases that are counted are only those
which are "recognized," those which result from such extremely dangerous or unhealthy conditions that the development of disease is obvious and provable. Uncounted in these
already shocking statistics are the millions of workers who die
prematurely of some common illnesses, such as heart or
lung disease or cancer or a multitude of other illnesses that
are also caused or aggravated by exposure to the environmental insults of the workplace.

If present trends continue, it is almost certain that conditions will get worse rather than better. Fifteen thousand
chemical and physical agents are already in use in industry.
With expanding technology more than 3,000 new chemicals
are introduced into industry each year.[4] The introduction of
new chemicals is not accompanied by health and safety programs that ensure their safe use.

The primary concern of businessmen and engineers with
American workers lies in increasing their productivity. What
does increasing productivity mean? It means that maintenance crews are cut down to the minimum number and
machinery is not kept in good working order. It means there
are fewer shutdowns for preventive maintenance. Increasing

2 Jerome B. Gordon et al., Industrial Safety Statistics: A Re-examination; A
   Critical Report Prepared for the Department of Labor (New York:
   Praeger Publishers, 1971).
3 Ray B. Davidson, Peril on the Job: A Study of Hazards in the Chemical
   Industries (Washington, D.C.: Public Affairs Press, 1970).
4 Department of Health, Education, and Welfare, Environmental Health
   Problems.

productivity means that the speed of production lines is kept at a maximum, which physically drains and injures workers and produces shoddier merchandise for the American consumer. Greater productivity often means less investment in proper ventilation, air-pollution control, and other devices that make the work environment safer.[5] In essence, greater productivity may demand that the American people sacrifice their lives and well-being for more production and profit.

The Great Unknown

Why are the statistics for occupational injury and disease so unfamiliar to the general public, and why do occupational diseases often go unrecognized? The answers to these questions are complex and varied. One reason is the nature of the dangers of the workplace. Only accidents or extreme exposures to noise, toxic substances, or physical hazards result in acute, easily identified illnesses. But most workers are exposed to low levels of these insults, which may be just as deadly in the long run, though less apparent in the short run. Low-level insults cause chronic illnesses. The onset of these illnesses is often not noticed. The illnesses themselves, such as lung cancer and heart disease, are attributed to non-occupational causes by industry and the medical profession. They therefore go unrecognized, uncounted, and uncompensated.

Oddly enough, in this advanced technological society, practically no one is trained to study, recognize, or treat occupational disease. The average medical student spends as much time studying tropical diseases such as malaria as she or he does learning about occupational medicine. A person who visits a doctor is not usually asked about his or her occupation or what kind of work is performed. The relationship between work and disease is rarely brought out.

---

[5] If these assertions seem controversial, almost any issue of any business publication will support them; for example, "Pollution controls . . . will add millions of dollars to industry's cost of production . . . without increasing productivity" (*Harvard Business Review*, January 1971).

It calls for inspection of the workplace without prior notice
to a company, and directs that a representative of the workers accompany the inspector on the tour. OSHA allows any
worker to register a complaint and call for an inspection,
while protecting the worker from discrimination for using
the act. The act carries fines for violations and forces the
employers to make and keep records of work-related deaths,
injuries, and illnesses, as well as of exposure to toxic materials or harmful agents. The records must be available to
the government and to the workers.

Thus the Occupational Safety and Health Act of 1970 at
last recognizes the rights of workers. But unfortunately,
recognition is meaningless unless the act is enforced and
utilized, and the trends thus far are disheartening indeed.
First, there are only about 500 inspectors for 4.1 million
workplaces. As of January 1, 1973, there are only 50 industrial hygienists employed by the government. Inspectors
are not permitted to take the air samples necessary for ascer-
taining whether a toxic substance in the air exceeds the concentration limit set by law. Legally, this type of sample can
be taken only by one of the 50 industrial hygienists.

During the first eight months of the act, inspectors visited
17,743 workplaces. At this rate, it would take them 230 years
to visit all the workplaces in our country. Even if the number of inspectors were increased to the projected level for the
year 2000, it would still take 46 years to inspect all workplaces once.[8]

If the law may seem weak to workers in uninspected factories, it does not seem so to the business community. They
have been quite busy, lobbying Congress and the Department of Labor to further weaken the law. First they per-
suaded the House of Representatives to eliminate small
workplaces from the protection of the law entirely. In a
House bill, five out of every six workplaces—or 15 million
workers—were eliminated from coverage by the act, since

8 *Health-PAC Bulletin,* September 1972.

they employed less than 15 workers each. A somewhat more favorable compromise version of this bill was later vetoed by the president, not because of concern for workers' health, but only because it was also associated with appropriations for social welfare.

Another attack on the act's effectiveness came from the Department of Labor, which has been trying to give over its enforcement powers to the state governments. This can mean no less than disaster for workers, since most of the state governments are notoriously lax in their enforcement procedures and capabilities. The power of the federal government is necessary for enforcement of the act.

Finally, we must consider the standards used in the act itself. Whereas standards set by the Environmental Protection Agency for the protection of communities are based purely on health considerations, standards set by the Department of Labor take into account "economic feasibility." The Secretary of Labor derives these standards from the recommendations of semiofficial bodies like the American National Standards Institute, business representatives, and a bare sprinkling of labor representatives. Thus, rather than considering only the health of the workers, these *consensus* standards rely on the criteria of economic feasibility in setting a maximum level of poison or physical hazard to which workers can be exposed in the context of economic feasibility to the employers.

The guideline for noise exposure provides a good example of how standards are set. Government officials wanted a standard of 85 decibels, but industry objected and the standard was set at 90 decibels, *an increase of practically seven times in noise intensity.* The government has now come out with recommended criteria for setting a new standard of 85 decibels.[9] Even this level is a compromise, as

[9] U.S. Department of Health, Education, and Welfare, *Criteria for a Recommended Standard: Occupational Exposure to Noise,* HSM no. 73-11001.

it still guarantees a hearing loss to a large percentage of the population. But the government feels that this loss is not really an "impairment," because a worker would still be able to communicate. Of course, he or she wouldn't have normal hearing ability, but as the government puts it, "The question of how much hearing should be protected and in what percentage of the people hearing losses of certain magnitudes should be permitted has long been an issue of much controversy." This controversy, however, does not extend to the workers whose hearing has been and continues to be so blithely sacrificed.

The case of the asbestos standard is also illuminating. Asbestos is a substance encountered in almost every industry, since it is used in insulating most industrial plants. It is also used in over 3,000 products made in this country. For the past seventy years, scientists have known that inhalation of asbestos dust causes *asbestosis,* a severe, even fatal scarring of the lungs. More recently, it has been found that even small exposures to asbestos dust result in cancer in a large fraction of the people exposed, cancer that first appears as long as 30 to 40 years after the exposure. Because it is so hazardous, the Department of Labor declared the control of asbestos to be a prime goal of the new Occupational Safety and Health Administration (OSHA). The first step in controlling asbestos hazards was the publication of an emergency standard in December 1971.

Labor representatives had proposed a standard of 2 fibers per cubic centimeter of air. This level was adopted as a temporary emergency standard for a three-month period in 1972. A permanent standard became law three months after the emergency standard was announced. The asbestos industry forced the government to compromise; the 2-fiber level will be the permanent standard, but the government will not put it into effect *until 1976*. Moreover, the standards are aimed only at preventing asbestosis, not cancer. In fact, there is new evidence that cancer and even asbestosis have occurred among British workers who have been exposed to no more than 2 fibers per cubic centimeter throughout their working lives. If health were really the criterion, no asbestos exposure would be permissible, and specific industrial hygiene techniques that eliminate exposure would be included in the law. But these techniques cost money, while the cost of cancer is not calculated in the profit-loss columns of industry.

What Can Be Done

It does no good simply to criticize a law or to throw up one's hands in despair. In the same way that business interests can cripple a law, workers and organized labor can force it to be administered more effectively. Some unions have been fighting for the proper interpretation of every clause in the Occupational Safety and Health Act, so that the workers' rights provided in the law are preserved.

The "imminent danger" clause of the act furnishes an example of the government's willingness to weaken it, and of the necessity to fight to preserve and strengthen it. This clause was the strongest section of the original version of the law that was proposed in the House of Representatives. Originally, the clause allowed compliance officers to close down an industrial operation in which a danger existed that could be expected to cause death or serious injury to workers in the area—including "danger from substances or conditions which would cause irreversible harm even though the resulting physical disability might not manifest itself at once."

The final version of the law requires, even in cases of emergency (imminent danger), that the compliance officer go to federal court to stop the process causing danger. Because of this requirement, the clause has become meaningless and is rarely applied. You can't save lives from an explosion by discussing it in court. In fact, when 22 workers were blown up in the Port Huron Tunnel in Michigan on De-

cember 13, 1971, it took the government until December 23
—eleven days—to post the notice of imminent danger at the
tunnel site.

Another problem comes in defining "imminent danger."
This problem was specifically raised by the Oil, Chemical,
and Atomic Workers International Union in a test case in
May 1971 on the clause. OCAW asked if an obviously hazardous condition that will cause irreversible damage or illness to
workers is an imminent danger. The union claimed that
pools of mercury on the floor of the Allied Chemical factory
in Moundsville, West Virginia, represented an imminent
danger—that is, a situation in which gradual toxic buildup
of mercury must occur in the workers in the plant.

The union showed through its own investigation that at
least 25 workers in the cell area of the plant had symptoms
—tingling in the hands and feet, tremors, irritability, drowsiness, loss of memory, and sore gums—or chemical evidence
of mercury poisoning. The Labor Department, however,
stated that "imminent danger" applied only when there was
a "risk of sudden great physical harm . . . like a boiler explosion." They refused to consider that gradual illness, even
from what is generally recognized as an obvious and serious
hazard, was an imminent danger.

Allied Chemical, whose annual profits are around $40 mil-
lion, was fined $1,000 for a "serious hazard" and told to
abate the hazard by June 2, two days after the citation was
issued. A repeat inspection on June 18 still showed excessive
mercury exposure in the plant. But OCAW is not relenting,
and using the bargaining table as a tool to change plant
conditions, it has been making steady progress.

It is important that workers insist that their unions use
their lobbying strength to realize the rights guaranteed by
the Occupational Safety and Health Act of 1970. It is obvious that the problems of the workplace will not be solved
by the government. No essential rights now enjoyed by working men and women were simply bestowed by the govern-

ment; rather, they were won through struggle. Indeed, every
improvement and benefit that workers know has come from
long years of struggle and negotiation. The same is true of
occupational health. Lobbying, concerted action, and all the
tools of trade unions will be needed to make the workplace
safe.

There is no doubt that a strong concern exists. A 1969
Department of Labor survey showed that 71 percent of all
workers felt occupational safety and health to be more important than wages. Workers are beginning to come, with
enthusiasm and interest, to courses and seminars on health
and safety. But the tremendous problem remains, that the
workers themselves know very little about the effects of
hazardous working conditions. They know that many people
are getting sick and dying, but no one has been able to pinpoint cause and effect for them.

Medical Mysteries

Workers cannot go to the medical literature or turn to
most doctors for help. The major reason is that medicine
and science are written in terms so obscure that even a
scientist in one branch of specialization will have trouble
understanding a scientist in another. Medicine is the worst.
Doctors insist on using words no one else understands. For
example, people with a ringing in the ears only think they
have a ringing in the ears. According to the doctor, what
they really have is *tinnitis*. Nobody suffers a loss of appetite
—he only gets *anorexia*. And so on. No wonder people are
confused about what is happening to them.

Furthermore, physicians and scientists often feel that only
they can understand the problem, take samples of air or
measurements of noise, and administer medical tests. Actually, *anyone* can take a representative sample of air if he or
she is only shown how to do it. Yet the professional community refuses to recognize this. If workers can keep the
wheels and offices of America running, then they certainly

can handle equipment that is quite simple to operate, and understand material that is not really complex, if it is written in language that can be understood.

The purpose of this book is to translate the relevant medicine, biology, and industrial hygiene into the English that people use every day. This does not mean that the subject matter is oversimplified or inaccurate, but only, we hope, that many of the obscure technical aspects are made less obscure. With this knowledge, workers can then use the grievance procedure, contract negotiations, the law, and all the other means at their disposal for instituting changes. What changes to demand and suggest, and how to achieve these changes, are discussed throughout the rest of the book.

Where Does Ecology Fit In?

In our now pollution-conscious society, workers often feel threatened by the ecology movement, fearing shutdown of their plants and loss of their jobs. Often, this fear is extended to health and safety improvements. In a recent article [10] the *New York Times* showed that not only were plants not shutting down, but that in fact pollution control could be used to make money. Those plants which were shutting down were marginal (unproductive) operations in the first place, and many of them were more than a century old and would soon have closed anyhow. According to the latest federal data, unemployment due to pollution control numbered fewer than 1,500 people in all. There is no reason to think that improving the conditions on the inside of the plant will cause any more shutdowns or job loss than cleaning up the external environment.

Moreover, it is almost never realized that the outside environment cannot be cleaned up if the industrial workplace itself remains polluted and deadly. Unfortunately, what has happened time and time again in the ecology movement is

[10] "Cost of Cleanup, or, a Myth of Factory Closings is Exploded," *New York Times,* Business and Finance Section, June 4, 1972.

that consumers and community groups have forced workers into the company's corner by demanding plant shutdowns or abandonment of processes that would cost many jobs. Faced with a choice between unemployment and pollution, workers must choose the job in order to live. But workers and the community do not have to be at odds. When they act together, real change can be accomplished, because it is the workers who can really pressure the company. After all, without the labor of working people, nothing can be produced.

The power of the worker-community coalition was shown recently in Alloy, West Virginia. The Union Carbide plant, called the dirtiest plant in America by *Business Week* magazine, was under attack for pollution cleanup by community groups, Ralph Nader aides, and the government. Immediately, the company threatened that it would have to lay off 625 employees, expecting this to still the movement. Instead A. F. Grospiron, president of the Oil, Chemical, and Atomic Workers International Union, stated that he would not let the people in OCAW be used as pawns in this game of "environmental blackmail" of workers. The union strongly supported the community activists. Carbide retreated, and no one lost a job. In fact, in May 1972 the *New York Times* ran a piece, "Cleaning Up Carbide," talking of the Alloy plant and showing a smokestack with flowers. The Environmental Protection Agency says that Carbide is actually ahead of schedule in its proposed cleanup campaign. United, workers and the community, who battle the same dirty air, can achieve a clean environment for all.

A Difficult Path

For the workers in the Kanawha valley area, where Alloy, West Virginia, is located, this ecology struggle was not a new experience. In 1967 ten Union Carbide plants were on strike for three months to force the company to pay for medical insurance. Other workers have waged similar battles over insurance, pensions, and working hours in plants across the nation. But what good is the battle over pensions if no one lives to collect them? How much of that insurance premium is really paying for the cost of occupational illness? (That payment is not a negotiated benefit, but an obligation of the industry. It should not be figured into a worker's fringe-benefit package.)

It took a three-month strike by workers from ten plants coordinated by the AFL-CIO Union Carbide Council to break company policy on insurance. It will take similar unity and will to make the workplace safe and to see that workers do not have to sacrifice one minute of their lifetimes or their good health in order to have the dignity and sense of achievement that comes from earning a living. Workers have long suspected that job conditions were injuring their health. This book seeks to eliminate the guesswork and document the suspicions so that formal protest and battle can be made—to stop the killing. There are many things that can be done to change the workplace, and many specific suggestions are made in the following chapters on how to achieve this. We have no fear that when armed with the facts, those who are well enough organized to achieve job protection will act to make the job a safer and healthier one.

CHAPTER TWO
THE HUMAN BODY AND OCCUPATIONAL DISEASE

HUMAN BEINGS are complicated, and their bodies are a huge complex of parts and systems. Many of these parts can be compared to mechanical things we are familiar with, as for example, a car. Like a car, the body has fluid, the blood, which travels around supplying fuel for energy. It has a pump, the heart, to keep it going and many filters for cleansing out wastes and poisons. But one thing the body has that no machine has is the ability to repair itself. If it is cut, it can mend itself. If it becomes infected, it can very often combat the foreign germs without the help of medicine. If you bleed, your body forms a blood clot to stop the bleeding. But if the cut is very bad or the infection extremely serious, then the body needs help in order to repair itself, and this help may be antibiotics or a few stitches or even a surgical procedure to remove a sick part.

The body's ability to heal and protect itself is its natural "defense system." This defense system helps keep us alive from day to day even though we live in a world of germs, dirt, and other medical dangers. The system took millions of years to develop, and the ways it did develop depended on the environment that people lived in. For example, people who evolved in hot, sunny climates have a different skin texture than people who evolved in cold, darker areas.

There are differences in the makeup of the blood of people whose ancestors developed in areas where malaria was present, as compared with those whose ancestors didn't. But one thing all people have in common is that *no one* evolved in order to work in a noisy, dirty factory surrounded by poisonous fumes and dusts, and usually at temperatures that are either too hot or too cold. The modern work environment constantly challenges the body's defenses. It keeps them working all the time, and unfortunately it defeats them in many cases.

Acute Versus Chronic Effects of Toxic Substances

One reason it is sometimes hard to recognize an occupational disease—that is, a sickness that comes from exposure to industrial hazards—is that the symptoms may come on so gradually that they can be ignored by the worker. The symptoms that do occur are often common to many non-occupational diseases: for example, cough, increasing fatigue, loss of appetite, indigestion, and pain. In some cases, a worker may not actually develop the disease for as much as twenty to thirty years after exposure, as with cancer caused by asbestos, benzene, or radiation exposure.

This slow onset of symptoms is typical of a *chronic* disease and is different from an *acute* reaction. *Acute reactions* are reactions or illnesses that occur immediately upon exposure. These are usually severe. *Fulminant reactions* are acute reactions that are so severe and occur so quickly that medical help of any sort is too late to prevent permanent disability or death. Most acute reactions, however, are usually over quickly, lasting only minutes, hours, days, or at most a few weeks. If a person survives, he or she may either completely recover or else be permanently disabled. State laws today almost always accept acute reactions and injuries, rather than chronic disease, as grounds for compensation.

When you become acutely ill, you know it. If a chemical

knocks you out or stops you from breathing or seeing, you know it. This is very different from the chronic effects spoken of earlier. Chronic effects of industrial exposures are much more subtle and insidious—they creep up on you. A chronically ill person often gets used to having mild symptoms, such as a headache or ringing in the ears. You may ignore these symptoms and be unaware that you are becoming chronically ill.

Chronic diseases can come from repeatedly being exposed to low doses of irritating substances. For example, look at exposure to chlorine. Each exposure causes severe irritation of the large, medium, and small airways of the lungs (*acute bronchitis* and *acute bronchiolitis*). Daily irritation eventually damages the small and medium airways of the lungs and probably leads to *chronic bronchitis*. How this happens is discussed in more detail later, but the important point to note is that the exposed person is not knocked out or killed by the chlorine, but just irritated a little each day until he or she finally gets a chronic lung disease.

This brings us to another problem. The chronic lung disease that can arise from chlorine exposure is the same disease that can result from long-term exposure to things like nitrogen dioxide, cigarette smoke, or cotton dust. It all looks the same to the doctor, no matter what the cause. This development of similar symptoms from many different causes makes chronic disease especially hard to diagnose, and even harder to win compensation for, because cause-effect relationships are difficult to establish. If the symptoms are hard to put your finger on, the cause is hard to pinpoint. If many things *may* cause the same effect, it is difficult to prove that a specific one *did*. While the controversy goes on, workers will continue to be exposed to dangerous substances that are *most probably* causing chronic illness.

Means of Contact and Entry of Toxic Substances

It is important to understand how the harmful substances that industrial workers are exposed to pass through the body's defense barriers and enter the system. That is, we should learn how these substances enter the body.

Skin contact is a common means of industrial poisoning or illness and occurs when the body is directly exposed to toxic—that is, poisonous or harmful—substances. The skin has a protective coating of oils and proteins that helps prevent injury or penetration by dangerous chemicals. But some chemicals, like phenol or carbolic acid, can penetrate the skin without being felt, or else, like trifluoroacetic acid, burn their way through the skin barrier. Sometimes a substance may react with the proteins of the skin and cause an allergy. When a chemical passes right through the skin, it can be absorbed into the bloodstream and carried throughout the body. For example, aniline, a common dye, can be absorbed through a worker's skin (sometimes coming right through shoes and clothes), and after reacting with the blood, can cause the whites of the eyes, the lips, and the fingers to turn blue, as well as seriously damaging other organs of the body.

Another route of entry into the body is through the respiratory tract. The majority of poisons that affect the internal organs enter the body by being breathed in. Substances like chlorine and ammonia can cause local irritation of the airways and lungs. Other substances may not do their damage locally, but may be absorbed from the lungs into the blood and cause damage to other organs. These are called *soluble* substances because they dissolve in the blood. Finally, *insoluble* substances, substances that do not dissolve in the blood or dissolve only in small amounts, tend to remain in the lungs for long periods of time. There they may cause serious local reactions either immediately or many years after the initial exposure. Among these substances are dusts like silica, asbestos, and beryllium.

The behavior in the lungs of dusts, whether soluble or insoluble, depends on the size of the particles breathed in. In general, the large particles are captured more easily by the mucus of the nose and the upper breathing passages. The hairs that line these passages sweep the mucus up to the throat, where it is swallowed or spit out. The smaller particles are not trapped and go right on to the lungs, where they either remain or are carried away by the blood.

The third way things enter the body is through the mouth and digestive tract. This is a less common route of entry than the respiratory tract or skin, but mouth contact with contaminated hands, food, and cigarettes does occur and is dangerous to the worker who has been handling extremely toxic substances like lead or arsenic. Once swallowed, the substances enter the digestive tract. Not everything that is eaten enters the blood system right away. One of the body's natural defenses is that it chooses what materials to absorb from the digestive tract, and also dilutes them. Those materials that are harmful are absorbed slowly and in small amounts. However, once absorbed into the bloodstream, the toxic substance goes directly to the liver, which then attempts to change the substance chemically to make it harmless to the body. The liver does not always succeed, however, and in fact, it may be damaged by trying to handle too many toxic substances.

In order to understand the effects of industrial hazards, it is necessary to know how the body works, and what kinds of diseases the different organs and systems of the body can develop. The rest of this chapter will be devoted to a discussion of the individual parts of the body, and will try to tell you what you need to know about your body in order to understand how it works and why it stops working properly. We will explain how the various industrial chemicals and processes affect the body, and discuss some of the tests that can be taken to show these effects.

LUNGS AND BREATHING

The respiratory system is one of the most important routes of entry for polluting or toxic substances. Many occupational diseases result from accumulation of toxic chemicals in the respiratory system itself, and other diseases are caused by passage of harmful substances through the lungs to the rest of the body.

The purpose of the respiratory system is to absorb oxygen from the air and transfer it to the blood. It also removes carbon dioxide, the waste gas produced by the body's processes, from the blood and transfers it to the exhaled air. This process is carried out in the lungs.

The lungs contain millions of tiny *air sacs*. Blood flows all around the air sacs and is separated from the air in the sacs by a membrane only one-millionth of an inch thick. This membrane is so thin that gases can pass through it: oxygen from the air into the blood, and carbon dioxide from the blood into the air. With each breath inhaled, fresh air penetrates all the way down to the air sacs. The blood then absorbs the oxygen from the air through the thin air-sac wall, while it discharges carbon dioxide into the air. These processes take place very rapidly, and within a few seconds, the air in the sacs is exhaled. This is the process of breathing.

The other part of the respiratory system is the *upper respiratory* tract, consisting of the conducting airways: the nose, the throat, the windpipe or *trachea,* and the smaller, branching air tubes called *bronchi.* Because the air sacs are extremely delicate, to avoid injury the air must be filtered and humidified before reaching them. The large hairs and moist mucous membrane of the nose are the first step in the filtration process. The windpipe and the bronchi also

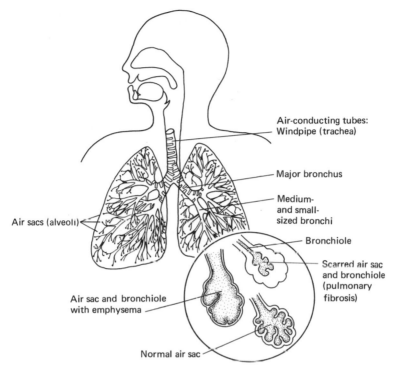

Air-conducting tubes:
Windpipe (trachea)

Major bronchus

Medium-
and small-
sized bronchi

Air sacs (alveoli)

Bronchiole

Scarred air sac
and bronchiole
(pulmonary
fibrosis)

Air sac and bronchiole
with emphysema

Normal air sac

Figure 1. THE LUNGS

have a carpet of hairs, *cilia,* but these are very small (*micro-scopic*). These cilia sweep particles up the respiratory tract just as if the particles were on an escalator. The bronchi and even smaller tubes called *bronchioles* are lined with a coating of mucus, which also aids in trapping dusts and chemicals. The bronchioles have tiny muscles that can close them off completely, preventing air from reaching the sacs. This happens when you are exposed to an extremely irritating gas like sulfur dioxide. Your body refuses to breathe in the gas, and you are forced to flee the area, gasping for breath.

The sense of smell also has an important function in the respiratory system. The cells sensitive to odor are located in the nose. Air passes these sensitive cells long before it reaches the lungs. If you smell a harmful substance, you can stop breathing. This automatic response, or *reflex,* is part of the body's defense system which prevents damage to the lungs. The problem is that these cells can lose their ability to detect odor if they become accustomed to a smell. Anyone working in a bad-smelling environment knows how easy it is to get used to a smell. This is known as *olfactory fatigue* and is a major reason why the sense of smell cannot be used as a warning system in the industrial environment, where workers are constantly surrounded with unpleasant and irritating odors.

In spite of the sense of smell, the muscles of the small tubes, and the mucous membranes and hairs of the respiratory system, the lungs are often injured by substances that enter the air and are breathed in. The amount and type of damage to the lungs depend on what the chemical substances are, and how much and how long the exposure is.

Acute Lung Reactions

Substances such as caustic acids and bases, if inhaled in large enough doses, can cause severe burns in the lungs. When the skin is burned, there is pain, redness, and swelling, and large amounts of tissue fluid accumulate to form a blister. This same reaction can occur in the lungs when corrosive or reactive chemicals are breathed in. Chemicals such as ammonia, nitrogen oxides, sulfur dioxide, phosgene, and chlorine burn the lungs in the following way. First the gas dissolves in the mucus of the lungs. The length of time needed for the gas to dissolve may be minutes—or hours, as in the case of nitrogen dioxide. The strong solution that results is what actually burns the air sacs. Just like the skin, the sacs then fill with fluid, which makes them useless for the transfer of oxygen. The medical name for this condition is *pulmonary edema.* It is actually just like drowning in water: the lungs fill with fluid so that you can't breathe. Pulmonary edema can therefore be compared to dry-land drowning. *Pneumonia,* or infection of the lungs, is a frequent complication of these burns. In order to recover, the victim needs intensive medical care. There is often permanent damage to the lungs from such exposure.

Chronic Lung Diseases

Two very common forms of lung disease in this country are *chronic bronchitis* and *emphysema.* They are the final result of a long series of small injuries to the lungs. Many chemicals, such as those in cigarette smoke, urban air pollution, and industrial air pollution, irritate the large and medium-sized air-conducting tubes. Each time the air tubes are irritated, their membrane linings increase the amount of mucus they produce, in order to dissolve, dilute, and remove the irritating substances. The affected person then coughs to remove the increased mucus.

As the irritation is repeated day after day, the mucus-producing glands in the air tubes begin to produce excessive mucus regularly even if the irritating substance is not present. Infections thrive in the excess mucus, so that chronic irritation may easily lead to increased susceptibility to respiratory infections. The infections then lead to more mucus production, and the cycle goes on and on. Repeated infections are believed to damage the small air tubes and probably also the air sacs.

Other elements in this chain of events lead to *emphysema,* a serious disease that results when the delicate air-sac walls are destroyed. As already mentioned, repeated infections may damage the air sacs. The sacs may also be damaged when they are exposed to increased air pressure. Such pressure increases occur when the small and medium-sized air tubes are filled with mucus and breathing out is blocked, or when the small tubes are narrowed by the contraction of the muscles that encircle them, as in asthmatic attacks. The air trapped behind such blocked or narrow tubes may stretch and eventually break the air-sac walls, which leads to larger, empty spaces in the lungs. These enlarged, balloonlike spaces are not close to the blood vessels in the remaining air-sac walls, and are therefore not useful to the lungs for air exchange. They are like a swamp. They do not receive fresh air containing oxygen, and do not give up the waste carbon dioxide with each breath. The air they contain is stale and does not move.

This disease process may come on slowly after many years of repeated exposure to such irritating substances as nitrogen dioxide and hydrogen sulfide. Of course, many people are also exposed to cigarette smoke and polluted urban air. All these substances have an additional (*cumulative*) effect.

People who develop chronic bronchitis and emphysema gradually notice increasing cough and mucus production over the years. They may tend to get frequent and increasingly severe respiratory infections, especially during the winter months. They very gradually begin to suffer from shortness of breath, at first only when they exercise vigorously, later with ordinary activity such as walking. Finally, the lungs cannot supply enough oxygen and remove enough waste carbon dioxide from the body to allow other organs, such as the heart, to work normally.

Another form of chronic lung disease is *pulmonary fibrosis,* also called *interstitial fibrosis.* This disease is not very common among the general population, but it does occur more frequently in groups of people exposed to certain dusts and chemicals found in industry. *Pulmonary fibrosis* means "lung scarring." A typical example of lung scarring is caused by asbestos dust. In response to this dust, which the lungs can neither remove nor destroy, the lungs produce a large amount of scar tissue, a process called *fibrosis.* The scar tissue is located mainly in the air-sac walls and causes them to become thickened. These thickened walls do not permit adequate transfer of oxygen from the air sac to the bloodstream. As a result, the oxygen content of the blood is decreased, and again the organs of the body do not get enough oxygen to function properly. In addition, the scar tissue is very hard and inelastic, unlike the normal spongy lung tissue it has replaced. The scarred lungs are much harder to expand with each breath. The simple act of breathing has become hard work.

The Diagnosis of Lung Disease

From the preceding paragraphs it is apparent that there are different types of lung disease, and that they cause different degrees of disability. Doctors can usually figure out what kind of lung disease you have and how severe it is by listening to the story of your symptoms—whether you cough, wheeze, have chest pain, are short of breath, or have other symptoms that are characteristic of particular types of lung disease. Doctors also use X-rays for diagnosis. X-rays show the appearance and location of the air in the lungs. Air shows up black on the X-ray film, while other parts of the chest—the heart, the bones, the muscles, the lung lining (*pleura*)—show up white, as do the lung scars. In addition to X-rays, there are tests of the ability of the lungs to do their job of bringing oxygen into the body and exchanging it for carbon dioxide. These are called lung function tests, and there are several different kinds of them. Some are used to find out how freely air flows into and out of the lungs. Others examine how well air or oxygen crosses the thin air-sac membranes into the blood. There is also a blood test that measures the amount of oxygen and carbon dioxide in the blood. Unfortunately, lung tests often do not reveal any abnormality in the lungs while there is still time to prevent or cure the disease that is causing the abnormality. In most cases, when the lung tests are abnormal, the damage to the lungs is done and cannot be reversed, that is, repaired. The only hope then is to prevent further damage.

Using some or all of these tests, doctors can determine what is wrong with your lungs and how well (or poorly) the damaged lungs are functioning to keep you active or alive. Many of the lung tests can be done by a local union office and the records kept on file. This establishes a health base-

line for the union members, in addition to possibly pin-
pointing areas or processes in the plant that are producing
disease.

THE HEART AND BLOOD VESSELS

The heart, blood vessels, and circulation are all known as
the *cardiovascular system*. The heart is a special muscle that
constantly pumps blood to all parts of the body. There are
two separate circuits of blood: one from the right side of
the heart (*pulmonary*) and one from the left side (*systemic*).
The lungs receive all the blood from the right side of the
heart. In the lungs the blood gets a new supply of oxygen
and gives up the carbon dioxide it has picked up in the
tissues. It then travels to the left side of the heart and out
to the other organs of the body: the brain, the liver, the
kidneys, the intestines, the skin, and so forth. In other words,
the blood passes first through the pulmonary system, then
through the systemic system, and so on back and forth.

The blood vessels form a branching series of pipes for
the blood. The smallest blood vessels, the *capillaries,* have
very thin walls that allow substances in the blood, such as
oxygen or sugar, to enter the tissues. In return they pick up
the body's waste products and transport them to the lungs
or kidneys for removal from the body.

The small blood vessels and small arteries are under very
careful control by the nerves and by chemical substances
in the bloodstream such as *hormones*. The nervous impulses
and chemical messages received by the muscle in the blood-
vessel wall tell the muscle to contract or relax. When the
muscle contracts, the blood vessel becomes narrower and
its diameter smaller, permitting less blood to flow through
it to the tissues it supplies. A stimulus, or message, that re-
laxes the muscle of the blood-vessel wall increases the diameter of the vessel and allows more blood to flow through
it to the tissues. Dilation, or widening, of the blood vessels
occurs under many conditions: for example, when waste

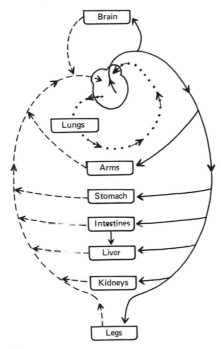

Figure 2. THE PATHS OF BLOOD CIRCULATION

products such as carbon dioxide or acids such as lactic acid build up in the tissues, or when there is a lack of oxygen. Blood vessels also dilate in a hot environment. Blood vessels constrict, or narrow, in the presence of the hormone *adrenaline*. Nicotine from cigarettes and noise cause generalized narrowing of blood vessels. In a cold environment, the blood vessels of the skin are narrowed so that heat will not be lost from the body through the skin.

Since the blood vessels themselves are composed of living muscle tissue, they also need oxygen to survive. Large blood vessels such as the *aorta*, the main vessel of the body carrying blood away from the heart, actually have small blood vessels that supply oxygen to their walls. Smaller blood vessels get their oxygen directly from the bloodstream.

When blood vessels are deprived of oxygen, they can be damaged. This happens most commonly when a person inhales cigarette smoke. The carbon monoxide in the smoke combines with the *hemoglobin,* the oxygen-carrying substance in the red blood cells, and makes much of it unavailable for bringing oxygen to the tissues. The medium-sized blood vessels, as well as other body tissues, are therefore repeatedly deprived of the amount of oxygen they need. Scientists think this lack of oxygen leads to damage of the blood-vessel wall.

The heart itself is a modified blood vessel. It has a very thick muscular wall that contracts rhythmically, 60 to 100 times a minute, pumping blood through all the blood vessels. Because it can never stop pumping, the heart needs a constant supply of oxygen. Certain substances are known to cause it to pump more rapidly and forcefully. When this happens, the heart is working harder and needs even more oxygen than usual. The nicotine in cigarette smoke and the hormone adrenaline both increase the rate and force of the heart. However, the carbon monoxide in cigarette smoke at the same time prevents oxygen from reaching body tissues.

The heart muscle has a blood-supply system that consists of numerous branching arteries, called *coronary arteries,* and capillaries. *Coronary arteriosclerosis,* which means "hardening of the arteries of the heart," is the most common cause of heart attacks. Over a period of 20 to 30 years during adult life, the small blood vessels supplying the heart muscle become lined with patches called *plaques,* which contain *cholesterol.* Cholesterol is a type of fat found in eggs, liver, and fish roe. It is also made by the body and becomes part of certain body hormones. Increased levels of blood cholesterol result from a diet high in animal fat, cream, and butter, which contain "saturated fat." Cholesterol deposits in the small blood vessels of the heart muscle are believed to be increased by certain substances. For example, low levels of carbon disulfide and lead may raise the amount of cholesterol in the bloodstream, and this may cause more cholesterol plaques to be deposited in the blood

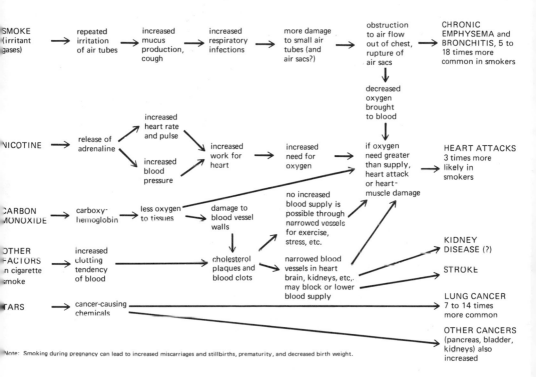

Table 1. THE EFFECTS OF CIGARETTE SMOKING

vessels. Carbon monoxide found in industrial sites and in cigarette smoke is believed to speed up the formation of cholesterol plaques in the medium-sized arteries.

Arteries that are lined with cholesterol plaques become stiff and narrowed, and may eventually be totally blocked off. When this happens, the part of the heart muscle to which they circulated oxygen-containing blood is damaged or dies from lack of oxygen. This is a heart attack. People may have mild heart attacks or severe ones, depending on the amount of muscle damaged or destroyed and the ability of the rest of the heart muscle to take over for the damaged part.

If a sufficient amount of heart muscle is weakened but

not killed by a heart attack, or if the remaining muscle is not able to take over completely for the lost muscle, then the heart does not pump the blood to other parts of the body adequately. This is called *congestive heart failure* and is often debilitating, occasionally fatal.

The heart muscle can also be damaged by having to pump through narrowed blood vessels either in the lungs or in the rest of the body. This occurs when scar tissue in the lungs is formed as a result of chronic lung disease. People with high blood pressure or those who are exposed to noise have a narrowing of the small blood vessels of the rest of the body, which causes increased resistance to blood flow. This results in a bigger workload for the heart, and over long periods of time probably causes the heart to wear out more quickly than is normal.

Disease from narrowing of the small arteries of the heart and brain caused 44 percent of all deaths in America in 1965. It accounted for 28 percent of all deaths below the age of 65. About 5 million Americans between the ages of 18 and 79 have heart disease as a result of narrowing of the heart-muscle arteries. Of Americans under the age of 65, 3.4 million suffer from coronary heart disease. About 700,000 people died from coronary heart disease in 1968, and of these almost 200,000 were younger than 65.

Heart disease due to stiffening and narrowing of the blood vessels by cholesterol plaques is obviously very common. Because it is so common, scientists have been doing research to find out which factors in the environment and which illnesses increase a person's chances of developing the disease. Cigarette smoking has been found to increase the risk of dying of coronary heart disease from one-and-a-half to three times the risk of nonsmokers. Other factors that increase the risk of coronary heart disease are high blood pressure, overweight, diabetes, and emotional stress. There is not much scientific data on the effects of the work environment on heart disease because there have been very few studies of the risks of heart disease in different occupational groups.

Rayon viscose workers were found to have a greater incidence of coronary heart disease. Workers exposed to carbon disulfide had twice the risk of getting coronary heart disease compared with workers in an industry with no carbon disulfide exposure. This finding will need to be confirmed by other studies before it is widely accepted by most scientists and doctors. There are also some studies which suggest that the occurrence of heart and blood-vessel diseases are increased among workers exposed to high levels of noise. Tunnel workers, exposed to high levels of carbon monoxide, are currently being studied to determine the effects of this exposure upon their health.

Medical Tests for Heart Disease

Chest pain, fatigue, and breathlessness are frequent symptoms of heart disease, although they may also occur in other diseases. Often a chest X-ray may be useful in determining if the heart is enlarged. A specific medical test, the *electrocardiogram,* measures the electrical activity of the heart muscle, that is, the rate and rhythm of heartbeats. These may be abnormal if the heart muscle is enlarged or damaged. People with any symptoms of heart disease should have a complete medical examination immediately.

THE BLOOD

The blood consists of a complicated solution of all the substances that the body organs and tissues use and dispose of, including salts and numerous proteins. Some of the proteins are clotting molecules. Some carry vitamins, fats, and minerals. In addition, the blood contains *white blood cells* to fight infection, *red blood cells* to carry oxygen, and *platelets* to begin the blood-clotting process.

The blood is pumped by the heart through the lungs,

where it picks up oxygen and gives off waste carbon dioxide. It is then pumped out to the rest of the body, where it gives up its oxygen to cells that need it, and picks up their waste carbon dioxide. The blood passes through the digestive system, where it picks up dissolved, digested food to carry to all the cells of the body. It passes through the kidneys, where it is filtered to remove waste products.

The transport of oxygen by the blood is of particular importance. The red blood cells are filled with the special iron-containing protein, *hemoglobin,* which has the ability to pick up large quantities of oxygen from the air in the lungs, carry it to the rest of the body, and release it where it is needed. The organs, especially the brain, heart, and kidneys, cannot survive and function when they are not supplied with oxygen.

Table 2. COMPONENTS OF THE BLOOD

COMPONENT	FUNCTION
Red blood cells and hemoglobin	Carry oxygen to the tissues
White blood cells	Make antibodies and fight infections
Platelets	Begin the blood-clotting process
Plasma	Contains many different proteins: Clotting factors
	Proteins which transport vitamins, minerals, and fats
	Antibodies

Many chemicals encountered in industrial processes interfere with the ability of hemoglobin to pick up and release oxygen. Carbon monoxide poisoning is actually hemoglobin poisoning. Carbon monoxide resembles oxygen chemically, and is picked up by hemoglobin in the lungs in the same way as oxygen. But unlike oxygen, carbon monoxide becomes very tightly bound to hemoglobin and cannot easily be released from it. Thus the hemoglobin that is bound to carbon monoxide is unavailable to the body to carry oxygen. If a high concentration of carbon monoxide is present in the air, all the hemoglobin becomes bound to it, and none is left to carry oxygen. The tissues of the body are deprived of oxygen. A person will die if this condition lasts for more than a few minutes. Carbon monoxide hemoglobin has a cherry-red color, and this is the color of the face and nails of a person with carbon monoxide poisoning.

Another compound that affects the hemoglobin is aniline. Aniline and other chemicals like it react with the hemoglobin and change its form and color. The altered hemoglobin is blue and cannot carry oxygen. A person with this kind of poisoning will turn blue in the face and fingertips. There are simple blood tests that will show if a person is suffering from this condition. Workers handling these chemicals should have the proper tests.

Aniline, as well as acetanilide, toluene, trinitrotoluene, benzene, methyl chloride, arsine, lead, and other chemicals, damages the red-blood-cell membrane and causes the cell to rupture. The membrane of a red blood cell is like an envelope. It holds the contents of the cell, mainly hemoglobin, inside and maintains the cell's shape and function in the transport of oxygen. When the cell ruptures, the hemoglobin, which has large molecules, is released into the bloodstream and damages the kidneys.

Rupture of the red blood cells can occur rapidly or slowly, depending on the chemical cause and the amount of exposure. When it occurs slowly, the symptoms are the same as those of anemia: weakness, fatigue, palpitations, headache, and a pale complexion. Under some circumstances, a person's skin will have a yellowish color, caused by jaundice. When rupture occurs rapidly, severe anemia develops. Because there may not be enough red blood cells left to carry oxygen to the brain, heart, and kidneys, damage to these organs may complicate the illness.

Other components of the blood are the white blood cells

and platelets. The white blood cells are important in controlling infection. They make antibodies and contribute to the body's defense against bacteria and viruses. Platelets are important in initiating blood clotting. They contain prepackaged chemicals all ready to react to a specific stimulus that is released by injured tissues.

A steady supply of red blood cells, white blood cells, and platelets is produced in the bone-marrow cavities of the flat bones of adults. Many drugs and chemicals and exposure to X-rays and ionizing radiation can cause damage to the bone marrow. When the marrow is damaged, it doesn't produce enough blood cells, and the results are severe anemia from lack of red blood cells, repeated, serious infections from lack of white blood cells, and uncontrollable hemorrhage from lack of platelets. Benzene is one of the most hazardous chemicals. Chronic exposure to small doses of benzene, though it may produce no symptoms at the time of exposure, may cause bone-marrow failure years later. In some victims of this disease, the dried-up bone marrow may later develop *leukemia,* in which cancerous cells multiply rapidly and replace the entire bloodstream. Most types of leukemia can be treated and controlled for varying periods of time, but ultimately they are fatal. Leukemia may also develop suddenly, without a previous period of bone-marrow failure.

THE LIVER

The liver is a large organ located in the upper right corner of the abdomen, just below the rib cage. It is the "chemical factory" of the body. The liver has a special circulation, which receives all the blood from the intestines. This blood supply, called the *portal system,* brings to the liver all the digested foods and drugs and chemicals absorbed from the intestines. The liver may break the foods down to the simple chemicals necessary for the energy supply of the body, or it may build new chemicals, such as protein clotting factors, enzymes, hormones, storage forms of energy, and other necessary chemicals. In the case of harmful drugs, the liver may destroy or change the chemical into a harmless or inactive form. In addition to foreign chemicals, the liver also destroys body chemicals such as hormones when they are no longer needed. The liver has one additional function: it makes *bile,* a thick green mixture containing green *bilirubin* and bile salts. Bilirubin is the chemical that results from the breakdown of hemoglobin, the red oxygen-carrying part of the blood. Bile salts are natural detergents that allow the intestines to dissolve fats in a water solution within the intestines. Just as a dishwashing detergent cuts grease and fats, bile cuts the fats in food so they can be absorbed through the cells of the intestinal wall. The bile is stored in the gall bladder, a small pouch near the liver. The gall bladder commonly causes illness when it becomes inflamed or when the bile forms stones, which block the tubes leading from the gall bladder to the intestines.

When the liver is not functioning, any or all of these many chemical processes are interrupted. The symptoms of liver disease may be quite vague. There may be tenderness or swelling of the liver, nausea, or loss of appetite. *Jaundice,* a yellow or greenish coloring of the skin, is a common symptom of liver disease. Severe liver disease can result in tremors, confusion, personality changes, or coma due to the action of toxic chemicals on the brain. These toxic chemicals may be normal body chemicals that the liver, when it is functioning properly, usually renders harmless by breaking them down or combining them with other chemicals. When the liver is severely damaged, these chemicals are not broken down and can build up in the body.

Hepatitis means inflammation of the liver. This disease may be caused by a virus or by chemicals such as alcohol, carbon tetrachloride, tetrachloroethane, and other chlorinated hydrocarbons (see Table 3 for examples). Various drugs, such as anesthetics and testosterone, the male hormone, also can cause this inflammation. Some substances, such as Thorazine (chlorpromazine), cause hepatitis in only a few people—those who become allergic, or hypersensitive, to it. As part of the reaction to the drugs, the liver becomes inflamed. This reaction is in many ways similar to the reaction that causes irritation and inflammation of the skin, described on pages 58–61 of this chapter.

Most types of hepatitis heal without any lasting effects on or damage to the liver, or any disability. However, severe hepatitis, from either chemicals, drugs, or a virus infection, may leave scars. Repeated episodes of hepatitis, even if not severe, may also lead to liver scarring, called *cirrhosis of the liver*. Scientists think that liver scarring can result from exposure to combinations of two or more chemicals, particularly carbon tetrachloride and alcohol. Occasionally scarring seems to occur from such chemicals without the victim being aware of having had hepatitis before.

Scarring of the liver causes many changes in body functions. If it is so severe that a large number of the liver cells are replaced by scar tissue, there is not enough normal liver left to do the necessary chemical work. Toxic substances may build up, the person may get jaundice, and a lack of clotting factors in the blood may lead to bleeding. Cirrhosis of the liver has other effects on the body as well. The scarring presses on the blood vessels of the liver and blocks them. The normal circulation to the liver from the intestines is blocked and backs up. The abdominal cavity fills with fluid that has leaked from the blocked blood vessels. This condition is called *ascites*.

The diagnosis of liver abnormalities is usually made by a thorough examination of the chemical content of the blood. Routine blood tests should check for elevation of the bilirubin, which can appear as jaundice. Increased bilirubin and a decrease in blood proteins and clotting factors indi-

Table 3. SOME CHEMICALS AND DRUGS THAT CAUSE LIVER DAMAGE

METALS AND INORGANIC CHEMICALS

Antimony (acute)
Arsine (acute)
Beryllium
Bismuth
Cadmium
Manganese
Selenium
Yellow phosphorus

ORGANIC CHEMICALS

Acetonitrile
Acrylonitrile
Benzene
Bromoform
2-butyl-3-one
Carbon tetrabromide
Carbon tetrachloride
Chlorinated benzenes (including **DDT**)
Chlorinated diphenyls and naphthalenes
Chlorobutadiene
Chloroform
Cresol
Dimethyl sulfate
Dimethyl formamide
Dinitrophenol
Dioxane
Epichlorohydrin
Ethyl alcohol
Ethyl bromide
Ethyl silicate

Ethylene chlorohydrin (beta-chloroethanol)
Ethylene dibromide (if swallowed)
Ethylene dichloride (in animals)
Ethylene glycol monomethyl ether
Ethylidene dichloride
Hydrazine
Mesityl oxide
Methyl alcohol
Methyl chloride
Naphthalene
Nitrobenzene
Phenol
Phenylhydrazine
Propylene dichloride
Pyridine
Styrene
Tetrachloroethane
Tetrachloroethylene
Toluene
Trichloroethane
Trichloroethylene

DRUGS

Ferrous sulfate
Fluothane
Hydrazine derivatives (INH)
Methoxyfluorane
Nilovar
Oral contraceptives
Steroids
Tetracyclines

cate a poorly functioning liver. Higher than normal amounts of certain enzymes indicate liver irritation. Many industrial chemicals that cause liver disease can usually be found by chemical analysis of the blood or urine.

THE KIDNEYS

The bloodstream carries proteins, sugars, salts, and water. It also carries various waste products that result from the body's chemical processes. For good health it is necessary to keep all these parts of the blood balanced. This is the job of the kidneys, which are a delicate, complicated filtering system. Since the kidneys act as a filter to all substances in the blood, they can be seriously injured by many toxic chemicals that pass through the body. They are also affected by diseases in other parts of the body. A good urinalysis will often pinpoint many diseases, and an understanding of kidney function is important for understanding occupational health.

The kidneys, which are located in the back just below the rib cage, are about 5 inches long and 2 inches wide. They each contain over a million microscopic filter tufts through which blood circulates and is filtered. These capillary filter tufts, called *glomeruli,* separate all the water and salts from the proteins and cellular components of the bloodstream, and bring the water and salt into the kidney. The proteins and cells stay in the blood. This water and salt mixture is called *urine.*

After this separation, the urine passes from each tuft into little tubes, or *tubules,* which are long coiled structures. At this point it still contains sugar. The job of the cells of the tubules is to put the sugar back into the blood. These cells actually operate like a chemical pump to remove the sugar from the urine. If the amount of water and salt in the urine is not right, the tubules correct the condition by a feedback mechanism. For example, if there is too much acid in the body, the tubules reabsorb alkaline sodium bicarbonate from the urine to neutralize the acid in the blood. If there is too much water in the body, they absorb more from the blood, and so on, keeping the body balanced. All these tubules lead into a single tube to the bladder, which is the outlet for the urine.

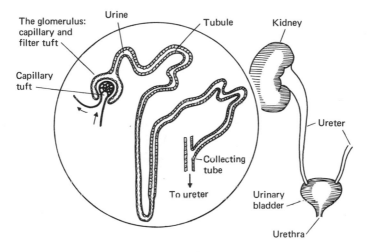

Figure 3. HOW THE KIDNEYS WORK

Symptoms of Kidney Disease

When the filters or tubes or the kidney as a whole is damaged, the abnormalities may show up in various ways. But the symptoms of kidney disease are often vague. They are called *non-specific,* which means that from the patient's description of the symptoms, the doctor is unable to make a definite diagnosis of the cause of the illness. The volume of liquid may not be maintained properly, and this can result in either high blood pressure or low blood pressure. A urinalysis or blood test can show too much or too little sodium, potassium, magnesium, calcium, or chloride. If there is too much nitrogen from the body proteins, a person will have *uremia,* and if there is too much acid, she or he will develop *acidosis.*

If the kidneys are causing an increase in body salt and water accumulation, then there may be high blood pressure.

Increased salt may also cause *edema,* or swelling due to accumulation of fluid in the tissues, or *pulmonary edema,* accumulation of fluid in the lungs. High blood pressure in itself may cause heart strain and lead to heart failure. The heart failure in turn may cause congested circulation and fluid accumulation. When the salt composition is abnormal, muscle weakness can develop; heartbeat irregularities may also result, which will be felt as palpitations or dizziness. A person suffering from acid imbalance and increased blood nitrogen content will have a loss of appetite and generally feel weak and fatigued. Moderately severe kidney disease is usually accompanied by anemia. A victim of kidney disease may have any one of these symptoms or a combination of them. In severe kidney disease, many of these abnormalities may be present together.

Acute Kidney Disease

Industrial chemicals and industrial accidents are two causes, among a long list of causes, of kidney disease. Each type of kidney disease may result in loss of any or all of the functions of the kidneys. One of the important causes of acute kidney damage is loss of circulation to the kidneys. Like the heart and the brain, the kidneys are always "on" and must have a constant supply of blood nutrients and oxygen. Lack of circulation or lack of oxygen may result from major injuries that lead to blood loss, low blood pressure, and poor circulation. Carbon monoxide is an example of an industrial chemical that prevents circulation of oxygen and may cause kidney damage. The changes in circulation that occur during heat exposure may cause more blood to go to the skin and less blood to go to the kidneys. If this shift in the blood is great, kidney damage can result. A shift of blood circulation to the muscles occurs during heavy exercise. This shift is usually not harmful in healthy young working people, but in older workers or in those with mild kidney damage, it causes further damage to the kidneys.

The kidney tubes may be damaged directly by many chemicals such as those listed in Table 4. These chemicals must be removed by the tubules, which are thus exposed to high concentrations of them and may be poisoned. A poisoned person usually is nauseated and has stomach irritation and liver damage. With a toxic chemical like carbon tetrachloride, kidney poisoning may not appear for a few days, or it may come on only a few hours after exposure. A poisoned person may cease urinating entirely, or pass only a few drops a day. All the functions that the kidneys normally perform are affected. The accumulation of fluids leads to congestion of the circulation. Accumulation of the body's waste chemicals, or *uremia,* may end in coma and death. This condition must be treated in a hospital. Body wastes and the poison have to be removed from the body by use of an external artificial kidney. If the victim survives, recovery begins about 4 to 10 days after exposure. One of the first signs is increased urine production. However, it may take weeks or months before all of the kidney's functions are fully restored.

Another type of acute damage to the kidney tubes is caused by chemicals and injuries that break up the red blood cells and release the red oxygen-carrying pigment, hemoglobin, into the bloodstream. Normally, because it is enclosed in the membrane envelope of the red blood cells, hemoglobin does not enter the kidney tubes. However, when the cells are ruptured and the hemoglobin is free in the bloodstream, it passes through the filter tufts of the kidneys and enters the kidney tubules. The hemoglobin molecules, which are very large, form a gelatinous clump within the tubes and block further formation and excretion of urine. Muscle cells contain a substance called *myoglobin,* which is similar to hemoglobin. Kidney damage may also result from clumps of myoglobin, which are formed in the tubes when muscles have been injured by crushing accidents or high-voltage electrical shocks.

Table 4. COMMON CAUSES OF KIDNEY DISEASE

POOR OXYGEN OR BLOOD SUPPLY TO KIDNEYS	DAMAGE TO TUBE CELLS	RELEASE OF HEMOGLOBIN OR MYOGLOBIN INTO URINE
Shock from injury causing hemorrhage	Mercury	Arsine
	Chromium	Crushing injuries
	Arsenic	Lightning or other high-voltage electricity
Allergic reactions causing low blood pressure ("anaphylitic shock")	Oxalic acid	
	Tartrates	
	Ethylene glycol	
Acute carbon monoxide poisoning	Carbon tetrachloride	
Heat stroke	Tetrachloroethane	
Vibration	Cadmium	
Chronic exposure to heat	Bismuth	
Carbon disulfide	Uranium	
Lead		

Chronic Kidney Diseases

There are three types of abnormalities that may result in chronic kidney disease. One is injury to the large and medium-sized blood vessels, which may decrease circulation to the kidneys. A second type is damage to the capillary filter tufts or the tubes, which may be caused by injury or illness. High blood pressure, which may result from kidney disease, is in its turn a cause of damage to the small and medium-sized blood vessels of the kidneys. When these vessels become thickened and narrowed due to injury from high blood pressure, the kidney is damaged from chronically decreased circulation. Hardening of the arteries may occur in any part of the body. When it occurs in the kidneys, circulation to the kidneys may be decreased and their functions affected.

When the body has lost blood, or when there is low blood pressure for other reasons, the kidneys, as well as all other organs, receive too little blood. When this happens, the kidneys change their function and increase the volume of blood for the whole body by increasing the salt and water content of the bloodstream. The kidney tubules make this correction in body blood volume by reabsorbing the right amount of salt and water from the urine back into the bloodstream. Such salt and water reabsorption is the body's normal response to a decrease in blood pressure or blood volume. By this mechanism the kidneys keep blood pressure and volume normal. However, in kidney disease, where only the kidneys have a "low blood pressure" because of narrowed blood vessels, and where the blood pressure and blood volume in the rest of the body are normal, the kidneys will still use this mechanism of increasing blood volume. The result is an abnormal reabsorption of salt and water, and the person with such kidney damage will have high blood pressure. In most forms of blood-vessel abnormality in the kidneys, high blood pressure is an important and often serious complication.

Chronic lead poisoning, as well as repeated episodes of acute lead poisoning, may result in blood-vessel changes and kidney abnormalities that closely resemble this hardening of the kidney blood vessels. Kidney-function damage and high blood pressure are typical symptoms of the disease. Chronic carbon disulfide poisoning may cause a similar disease of the kidney blood vessels and have the same signs.

Disease of the filter tufts is not a common form of industrial kidney disease. It does occur, though rarely, from chronic inhalation of mercury over a period of months or years. The disease that results affects the filter as well as the tiny blood vessels that bring blood past the filter surface. This disease is called *glomerulitis*. The damage to the filter makes it lose its selectivity, or ability to separate the salts and water from the proteins. The filter tuft actually develops a leak, so that protein and red blood cells enter the urine. If large amounts of protein are lost, a person will develop swelling of the face and the legs. High blood pressure may result from damage to the capillary circulation of the filter tufts, just as in other types of kidney damage.

Chronic disease of the kidney tubules most commonly results from kidney and urine infection. It is not clear at present whether occupational factors increase the risks of kidney infection. Studies are under way to determine if there is a relationship between body posture, vibration, and the risk of kidney infection. Damage to the tubes from infection or other causes results in abnormal tubule function. Loss of the ability to reabsorb salt and regulate the amount of body fluid may result in *low* blood pressure. The body may also lose its ability to rid itself of acids and keep the concentration of salts at the correct level. These abnormalities may be mild or severe. If they are severe, the illness may be disabling or fatal, as with the other forms of severe chronic kidney disease previously discussed. The kidney tubules may also be damaged by long-term exposure to certain chemicals, including cadmium, bismuth, uranium, and perhaps chromium.

In addition to chemically caused disease of the different structures of the kidney, kidney cancers are now known to be associated with exposure to several industrial chemicals. Coke-oven workers have a high incidence of kidney cancers as compared with other steelworkers. Cancers of the urinary bladder, where urine is stored, are also a result of exposure to certain chemicals, especially those associated with the manufacture of aniline dyes.

Medical Tests

Tests for kidney function include examination of the urine for total daily volume, concentration, and the presence of cells and protein. Specific industrial chemicals can also be looked for. Changes in blood pressure, so often present in kidney disease, should be looked for. Blood chemicals—especially the "blood electrolytes," potassium, sodium chloride, calcium, and blood acids—should be checked. These substances do not require special tests and can all be analyzed in routine blood tests. Further information about kidney function can be obtained from an X-ray examination of the kidneys while they are excreting a dye that shows up on X-ray film. This examination is called an *intravenous pyelogram,* or IVP. More detailed examinations of kidney circulation are made by injecting dye into the arteries supplying the kidneys. This examination is always performed in a hospital.

THE NERVOUS SYSTEM

In order for you to exist from minute to minute, your body must be constantly working. You have to keep breathing, your heart must keep pumping, many of your muscles must remain active, and all the other organs and systems of your body must keep functioning. In addition to the mechanical functions your body performs to stay alive, you as a human being think, have emotions, and respond physically to the outside world. All these complicated jobs that your body and mind carry out are controlled by the nervous system, and many of them go on occurring without your being aware of them. Because the industrial environment affects so many functions of the nervous system, and because this system is so vital, we will discuss it in some detail.

The nervous system is classified into different parts. First there is the controlling center, the *central nervous system,* which is made up of the brain and spinal cord. The way the brain works and how you think are hardly understood at all. What is known, to a large extent, is which parts of the brain are responsible for what functions. There are 12 pairs of cranial nerves that come from the brain. These nerves control your vision and the way your eyes work, your senses of smell and hearing, and several other functions. They also branch out and connect with other nerves that extend throughout the body. The spinal cord is really an extension of the brain. It too has nerves, 31 pairs, that extend and divide out through the body, going to the limbs, organs, muscles, and skin.

The nerves that leave and return to the central nervous system extend to every part of the body. Since this part of the nervous system reaches the outer, or peripheral, areas, it is called the *peripheral nervous system.* When the nerves leave the spinal cord, they are assembled in bundles called *plexuses.* From these plexuses, peripheral nerves branch out to the back and sides or to the front of the body, in a definite pattern. Since the activity of each muscle in the body is controlled by nervous impulses, the nerves must extend throughout the body to every muscle. For example, the peripheral nerves that come from the bundle of nerves in the shoulder area, the *brachial plexus,* supply the skin and muscles of the shoulder and arm and allow them to move. Peripheral nerves supply the muscles that turn your head from side to side, or move your little finger up and down.

If the brain or spinal cord is cut or damaged, it will never repair itself. Damage to peripheral nerves can sometimes be reversed, but any injury to the central nervous system is permanent. Nerve functions can be altered or destroyed by some of the chemicals that people work with. Pesticides and metals such as mercury and lead, for example, interfere with the chemical transfer of information necessary for an impulse to be sent through the nervous system. This interfer-

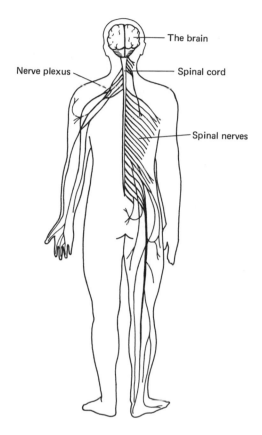

Figure 4. THE NERVOUS SYSTEM

ence may result in no impulses passing through, which means paralysis, or in too much or too little stimulation, which may cause tremors, loss of reflexes, and loss of feeling. These effects can be reversed in many cases if exposure is discontinued, but long-term exposure may lead to permanent damage.

The brain is probably the most delicate and most vital part of the body. It must have a constant supply of oxygen. When a worker is exposed to a toxic substance, the first response he or she may feel is dizziness or drowsiness. Although the chemical may not be causing permanent brain damage, it *is* interfering with the proper functioning of the central nervous system. For example, the worker's reflexes may be dulled. When a worker begins to feel dizzy or drowsy, the brain is acting as a warning system, telling him that things which do not belong in his body are entering it and affecting the way it works. It pays to heed such warnings.

The brain and spinal cord are protected by a hard, bony exterior, the skull and the bones that make up the spine. The brain and cord are covered by a thick outer membrane for protection, as well as by a delicate inner membrane containing many blood vessels that supply some of the blood to the system. These protective layers make up the *meninges,* and *meningitis* is an inflammation of these layers. The brain and spinal cord float in a liquid that contains many substances necessary for the nervous system to transmit messages. If a person is thought to have an infection of the central nervous system, or another neurological disease, a doctor will often perform a spinal tap. This means that some of the liquid from the spinal column is actually removed and examined for abnormalities.

Besides sending messages out through the peripheral nervous system, the brain and spinal cord receive messages back. If you touch something hot or stub your toe, the peripheral nerves transmit these sensations back to the brain and spinal cord, which then tell the body how to respond. If you are burned, you immediately pull back. The peripheral nerves that carry messages from the central nervous system are called *motor nerves,* and those that bring messages back are called *sensory nerves.*

Besides the nerves that stimulate the muscles of your limbs and those that transmit sensation, there are nerves that control your blood vessels, the movement of your intestines, the temperature of your body, the hormones you secrete, and so on. Making your stomach move to digest food is not controlled in the same way that you move your finger up and down. These nerves normally operate independently of your conscious mind. That is, they display autonomy, or independent government; hence they are appropriately called part of the *autonomic nervous system.*

The messages of the autonomic nervous system are carried by chemicals in the blood. An interesting experiment has been done to show this. Scientists used electricity to stimulate the heart muscle of a frog and then transferred the blood of this frog into the heart of a second frog, whose heart had not been stimulated. The first frog's blood caused the second frog's heart to be stimulated, even though it was not exposed to the electricity. This meant that some substance in the blood was the initiator of the stimulation. Therefore, when the body receives or sends a nervous impulse through the autonomic nervous system, it transfers this impulse with chemicals.

The autonomic nervous system is subdivided into two parts. One part is the *sympathetic nervous system.* Just as the name implies, this system is particularly concerned with responses to stress. That is, it is the part of your body that responds "in sympathy" with your surroundings and your feelings. For example, the nerves of the sympathetic nervous system activate the muscles that dilate or contract your blood vessels. If you get angry, you may get red in the face. In medical terms, your sympathetic nervous system has caused the blood vessels in your face to dilate so that more blood passes through them, making your face red. When a girl blushes at a flirtatious comment, her blood vessels are dilating in response to an emotional input. Accompanying these changes, the blood pressure may rise and the rate of heartbeat increase. If you have ever experienced stage fright, or been afraid to take an examination, you have felt the sympathetic nervous system causing you to respond physically to an emotional or stressful situation.

The smooth muscles of the skin, the muscles that make the hair on your body stand up, and the sweat glands are all stimulated by the sympathetic nervous system. If you break into a sweat during a trying situation, this is again your body's neurological response to stress. A serious response to stress occurs when the heart muscle is stimulated, increasing its rate. When a person is under stress —exposed to intensely loud noise, for example—the sympathetic nervous system increases the rate of heartbeat and therefore the amount of work the heart must do.

The other part of the autonomic nervous system is called the *parasympathetic nervous system*. This system has fairly specific functions, such as stimulating digestion, emptying the bladder, and causing mucus secretion. Again, stimulation is brought about by release of chemicals that activate the particular muscles involved, just as in the experiment with the two frogs.

Just as the peripheral nervous system functions by receiving and sending messages to the brain and spinal cord, so the functions of the autonomic nervous system are also controlled by the central nervous system. The lowermost part of the brain, the *brain stem,* is the area that controls breathing, blood circulation, and heart action, all of which are reflexive or automatic actions. These responses are involuntary; you don't have to think about continually making your heart beat. Cats that have had their entire brains removed except for the brain stem can still breathe and maintain blood pressure, although they can't perform other functions like eating or keeping their body temperature up.

From this short discussion, you can see that the operations of the nervous system are very complicated. The brain and spinal cord act as control centers, and literally millions of nerves throughout the body work to keep it functioning. The nervous system is a delicately balanced operation, and many of the conditions and substances that workers are exposed to can upset this balance. The effects of various chemicals and conditions on the nervous system are discussed

Table 5. Some Chemicals That Affect the Nervous System

DEPRESSION OF CENTRAL NERVOUS SYSTEM	BRAIN POISONING	BRAIN DAMAGE BY OXYGEN DEPRIVATION	NERVE-FUNCTION DISORDERS
Acetates	Carbon disulfide	Asphyxiating gases	Organophosphate pesticides
Alcohols	Hydrogen cyanide	Carbon monoxide	Organophosphate plasticizers
Brominated chemicals	Hydrogen sulfide		
Chlorinated chemicals	Stilbene		Heavy metals
Ethers	Arsine		Mercury
Ketones			Lead
			Manganese
			Arsenic

throughout this book. Of particular importance are the chapters on noise, stress, and chemical hazards. Table 5 lists some of the chemicals that affect the nervous system.

THE SKIN

In spite of its remarkable protective properties, the skin is the organ most commonly injured in industry. About 65 percent of all reported occupational diseases are skin diseases. About 800,000 cases of skin disease related to occupation are reported each year. However, many cases fail to be reported or are not recognized as being caused by occupational exposure. In 1945, the United States Public Health Service estimated that the cost of skin diseases to working people, industry, and insurance companies was about $100 million. Today that cost must be much higher. Since compensation covers, on the average, less than half of a worker's earnings and does not count human suffering, the real social cost of skin disease is not measured.

The skin has two layers, the *dermis* and the *epidermis*. The epidermis is the outer layer. It consists of a densely packed layer of cells, which are constantly dividing and replacing themselves with young cells. As these cells grow older, they move toward the outer surface of the skin. During this process the cells make *keratin,* a tough protein that is resistant to many chemicals and to dust and germs. The outermost layer of the skin consists of this protein entwined with layers of dead epidermis cells. Skin color is also made in the epidermal outer layer, and is important in protecting the skin against damage from sunlight. The epidermis contains sweat glands, hair follicles, and blood vessels, which are important in regulating body temperature. It also contains glands that make a substance similar to ear wax which acts as a protective film over the outer layers of the epidermis. It is the outer layer of dead cells plus the tough

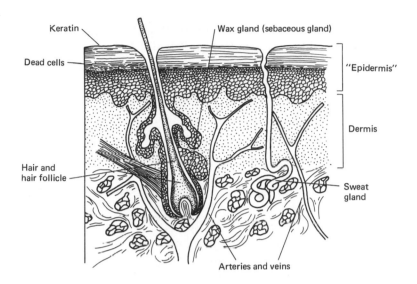

Figure 5. MICROSCOPIC CROSS-SECTION OF THE SKIN

keratin and the waxy coating that act as a barrier to chemical and physical injury.

The inner skin layer, or *dermis,* consists of fat and connective tissue. It holds the outer layer to the body and provides insulation for the body.

In spite of all these protective mechanisms, the skin can be injured in many industrial processes. Caustic chemicals, soaps, and caustic alkalies may dissolve the keratin protein layer. Organic solvents and detergents can dissolve the waxy coating. Inorganic acids, anhydrides, salts, and some alkalies take water from the skin in a process called *dehydration.* Some acids, bleaches, and peroxides react with the skin chemically. Coal tar, petroleum, and arsenic increase its tough keratin coating. Still other chemicals damage the skin in ways that are not understood by scientists. When the damaged skin becomes reddened, itchy, blistered, thickened, hardened, or flaky, a doctor calls this condition *dermatitis,* which means "skin inflammation."

Direct Irritation of Skin by Chemicals

Chemicals that damage the skin directly are called *primary irritants*. These substances penetrate the normal outer barriers and injure the lower skin layers. Strong acids and alkalies, certain metallic salts or simple metals, many organic chemicals, and many oils are strong primary irritants for everyone. That is, *all* people exposed to the same concentration of these substances will have an identical skin irritation.

Other chemicals, usually milder acids and alkalies and many solvents, are less severe direct irritants. These milder chemicals frequently damage the skin, but only after prolonged or repeated exposure, or when the skin has previously been damaged by physical injury such as friction, pressure, cuts, dryness, heat, or cold. If areas of the skin are enclosed in plastic gloves or under tight rings, they may develop injury that sets the stage for further damage by these milder direct irritants.

The skin irritation caused by direct contact with chemicals can look like any other dermatitis; it can consist of sores, pimples, scales, or flakes. But it *always* appears where direct contact with the irritant has occurred, and it *usually* goes away when contact with the irritant is ended. Such dermatitis does not spread to other parts of the body. It is likely to recur after another exposure to the chemical that caused the first irritation. Usually, a second case of irritant dermatitis is caused by exactly the same conditions of contact, the same concentration and length of exposure, that caused the earlier episode. This type of dermatitis generally occurs among *all* the workers who handle the same amounts of an irritating chemical. A list of irritating chemicals is provided in Table 6.

Table 6. IMPORTANT IRRITATING INDUSTRIAL CHEMICALS

Acetaldehyde
Acetic acid
Acetic anhydride
Acetone
Acetonitrile
Acridine
Acrolein
Acrylonitrile
Allyl alcohol
Aluminum salts
Ammonia
Amyl acetate
Amyl alcohol
Aniline
Antimony compounds
Arsenic compounds
Arsine

Barium compounds
Benzene
Benzidine
Benzyl chloride
Beryllium compounds
Bismuth subnitrate
Boron compounds
Bromine
1,3-butadiene
n-butyl acetate
n-butyl alcohol
Butyl cellosolve
 (2-butoxyethanol)

Cadmium compounds
Calcium oxide
Carbon disulfide
Carbon tetrachloride
Carbonyl chloride
Cellosolve
Cellosolve acetate
Chlorinated lime
Chlorine
Chlorobenzene
Chloroform
1-chloro-1-nitro-
 propane
Chromic acid
Chromium compounds
Coal tar compounds

Cobalt compounds
Copper compounds
Cresol
Cyclohexane
Cyclohexanol
Cyclohexanone
Cyclohexene

ortho-dichlorobenzene
1,2-dichloroethylene
1,1-dichloro-1-
 nitroethane
Dimethyl aniline
Dimethyl formamide
uns-dimethylhydrazine
Dimethyl sulfate
Dinitrobenzene
Dinitrophenol
Dinitrotoluene
Dioxane

Epichlorohydrin
Ethyl acetate
Ethyl benzene
Ethyl bromide
Ethyl chloride
Ethylene chlorohydrin
Ethylenediamine
Ethylene dibromide
Ethylene dichloride
Ethyl formate

Fluorine compounds
Formaldehyde
Formic acid
Freon
Furfural

Gasoline

Hexamethylene-
 tetramine
Hydrazine
Hydrocyanic acid
Hydrogen chloride
Hydrogen fluoride
Hydrogen peroxide
Hydrogen selenide

Hydrogen sulfide
Hydroquinone

Iodine

Kerosene
Ketones

Mercaptans
Mercury compounds
Methyl bromide
Methyl cellosolve
Methyl cellosolve
 acetate
Methyl chloride
Methyl formate

Naphtha (coal tar)
Naphtha (petroleum)
beta-naphthylamine
Nickel compounds
Nitric acid
Nitrobenzene
Nitroglycerine
Nitromethane

Oxalic acid

Pentachlorophenol
Perchloroethylene
Phenol
Phenylhydrazine
Phosgene
Phosphine
Phosphorus
 trichloride
Phthalic anhydride
Picric acid
Platinum salts
Propyl acetate
Propylene dichloride
Pyridine

Quinone

Selenium compounds
Silver compounds

Table 6. IMPORTANT IRRITATING INDUSTRIAL CHEMICALS
(CONTINUED)

Sodium and potassium hydroxides	Tetramethylthiuram disulfide	Turpentine
Stoddard solvent	Thallium compounds	Vanadium tetrachloride
Styrene (monomer)	Thorium compounds	
Sulfur dioxide	Tin compounds	Vinyl chloride
Sulfuric acid	Titanium tetrachloride	
Sulfur monochloride	Toluene	Xylene
	Toluene diisocyanate	
Tetrachloroethane	Trichloroethylene	Zinc compounds
Tetrachloroethylene	Tricresyl phosphate	Zirconium compounds

Contact Dermatitis

Substances that are not normally irritating may cause dermatitis if the skin becomes allergic to them. Many substances cause this type of reaction between a chemical, the skin, and the body's normal system for handling foreign substances—its *immune system*. First the chemical attaches to the skin, chemically changing some protein that is normally found there. This turns the normal protein into a foreign protein. Then the immune system, the body's protective mechanism against foreign substances, reacts against this altered protein, causing skin inflammation, or dermatitis. Normally the immune system reacts only to proteins that are not part of the body's own tissues. In this type of reaction, however, the immune system reacts against the combination of the altered skin protein and the chemical just as if the protein were not part of the body. It is this process that causes skin inflammation in contact dermatitis. Once the immune system has "learned" to react against this combination of skin and chemical, it becomes so sensitive that it can react against even tiny quantities of the combination. In some cases, even when the chemical is no longer present, the reaction continues against the already inflamed skin. In other cases, the reaction occurs again when the skin comes in contact with substances that are similar to the original chemical. The immune system can remember the skin-chemical combination for years, and may react against it whenever it occurs again, even years after the exposure that caused the first reaction. This "memory" is the same type of immune memory that prevents us from getting the measles more than once because the body has built up a lasting defense against the foreign measles virus. Thus the immune system can help us, as with the measles, or hurt us, as in allergies.

This complicated reaction is often called a *hypersensitivity reaction* or *contact dermatitis*. A very common example of contact dermatitis is caused by poison oak or poison ivy. A chemical in the poison ivy plant causes an itchy, blistered rash in about 70 percent of the people exposed to it. Other substances are weaker sensitizers and do not cause an allergic reaction in such a great proportion of those exposed.

Table 7. TEN OF THE MOST COMMON GROUPS OF
CHEMICALS THAT CAUSE CONTACT DERMATITIS

Bichromate salts of sodium, potassium, and ammonium

Plastics, especially epoxy resins and catalysts

Rubber accelerators and antioxidants

Germicidal agents used in soaps and other cleaners, especially hexachlorophene, bithionol, and halogenated salicylanilides

para-substituted aromatic amines and their derivatives, such as para-phenylenediamine, azo dyes, "-caine"-type topical anesthetics, and topical antihistamines

Formalin

Nickel compounds

Mercury compounds, both organic and inorganic

Poison ivy, poison oak, sumac, and related plants

Cobalt compounds

Contact dermatitis takes time to develop. It *never* occurs on the first exposure to a new chemical. The earliest reaction to a new chemical takes at least 5 days to develop. Usually it takes 2 or 3 weeks to appear, and it may occur only after years of repeated exposure to a chemical. A person may suddenly develop an allergic reaction to a substance he or she has handled repeatedly for 20 years. The length of time depends on two factors: how allergic the individual worker is, and the ability of the chemical to sensitize people—that is, to cause the reaction between the skin-chemical combination and the immune system.

Once the skin is sensitized to a specific chemical, it still takes about 48 to 72 hours for the allergic reaction to occur after a repeat exposure. Everyone is familiar with the reaction to poison ivy which develops at least 2 days after the picnic!

A person who is sensitive to a chemical generally reacts only to that chemical, though if the allergic reaction is very strong, it may be prompted by closely related substances. Skin sensitivity to one chemical, however, does not usually increase the chances of becoming sensitive to another, unrelated chemical.

Certain skin conditions lead more readily to the development of contact dermatitis. The most common condition is skin irritation from direct primary irritants, discussed previously, or from friction, pressure, sweating, or prolonged exposure to water. All of these factors damage the skin and make it more likely that the immune system will react against a skin-chemical combination.

The diagnosis of contact dermatitis is based on the story of how the skin eruption occurred and on the results of patch testing. Contact dermatitis usually occurs in areas of the skin that were in direct contact with the chemical. The skin becomes itchy, reddened, and blistered, and may ooze fluid. Crusts form from dried fluid and peeling skin. There must be contact for at least 5 days with the suspected new chemical for it to be considered the cause of the dermatitis. Contact dermatitis will usually improve after vacations from work, but weekend interruptions in exposure may be too short for any noticeable improvement to occur. Of course, some contact dermatitis results from handling household or garden chemicals, and some household chemicals may be similar to or the same as those causing sensitivity at work. In these situations there will be no improvement in dermatitis during periods away from work.

The patch test is used by dermatologists to discover the exact cause of contact dermatitis. A standard, diluted amount of the suspected chemical is placed on a patch, which is applied to the skin. After 48 hours, the patch is removed and the skin examined for a reaction. The skin is examined again after 72 hours. Patch testing may have false positive results if other irritating substances are applied, if strong pressure is applied, or if severe, widespread dermatitis is already present. False negative results, in which there is no reaction to the actual sensitizing substance, occur when the substance is applied in a way that is not the same as the usual manner of contact. False negative reactions may also occur when the concentration of the test substance is too weak, or when light happens to be a necessary part of the allergic reaction. Sometimes the test substance may have deteriorated because it is old or has been stored improperly.

Patch testing is not without dangers. People who are given a battery of patch tests may become newly sensitive to one of the test chemicals. Severe reactions to patch-test chemicals may cause lasting scars. Because of these dangers, pre-employment screening with patch tests is not a good idea. Moreover, patch testing does not predict a possible "tendency" to develop skin allergy, because a person who is sensitive to one substance is *not* more likely to become sensitive to another.

Other Types of Skin Disease

Oils, waxes, tars, and chlorinated hydrocarbons cause blockage of the *hair follicles* (the depressions in the skin where hair grows), oil glands, and sweat pores. In addition, oils and waxes act as direct irritants by washing away the skin's own protective oils so that it becomes dry and cracked. Contact with oils, tars, waxes, and chlorinated hydrocarbons may cause a variety of rashes, such as acne and blackheads, which result from blocked glands. Blocked hair follicles may become infected, causing *folliculitis,* and skin abscesses may result.

Frequent use of soap and water is usually an excellent way of preventing this kind of dermatitis. Waterless cleaning preparations are also satisfactory. Abrasive cleansers, however, or mineral spirits such as kerosene harm the skin. Barrier creams are of no help in preventing skin exposure to oils, since they dissolve in the oils and are washed away.

Areas of darkened skin may result from the combination of tars and sunlight. Tars are difficult to remove from the skin because they stick within the hair follicles. Coloration from tar is prevented by the use of protective clothing, sun screens, and conveniently placed, frequently used soap and water. Greasy ointments used as protective coatings on the skin actually *increase* the reaction between the tar, the skin, and the ultraviolet rays of sunlight. Water-soluble barrier lotions offer some protection by preventing the entry of tars into the hair follicles. To be effective, barrier lotions must be reapplied every time the hands are washed.

Cancers of the skin are caused by excessive exposure to sunlight and by some chemical substances. Chemical or physical substances that cause cancer are called *carcinogens* or *carcinogenic substances.* Most carcinogens can probably cause several types of cancers. For example, mineral oil or arsenic, which can cause skin cancer after prolonged or re-

peated contact with the skin, may cause lung, kidney, or
bladder cancer if mists, dusts, or fumes containing these substances are absorbed into the body over long periods of time.
Skin tumors commonly occur after 20 or 30 years of exposure
to carcinogens. Skin cancer may also result from shorter exposures in susceptible people, but it more commonly occurs after repeated or intermittent exposure over many
years. Several skin cancers can occur at the same time, and
these may be multiple tumors. The skin areas most likely
to be affected are those of greatest contact: the hands and
arms, the upper legs, and the scrotum in men. Cancers of
the scrotum result from passage of the cancer-causing substances through the clothing. The chemicals may be splashed
on the clothes or get on them when the workman wipes his
oily hands on them or touches the oiled moving parts of
machines with his body.

The most important group of chemicals that cause skin
cancers are listed in Table 8. Some of them are present

Table 8. SOME CHEMICALS THAT CAUSE SKIN CANCER

Coal tar and derivative fractions	*Petroleum oils and derivative fractions*
Pitch	Tar
Asphalt	Asphalt
Tar oil	Carbon black
Creosote oil	Fuel oil
Anthracene oil	Diesel oil
Soot	Grease
Lamp black	Cutting oil, machine oil
Lignite	Wax paraffin oil
Tar	Methylated naphthalene
Oil	Paraffins
Wax	*Arsenic*
Shale oil and wax	*Ultraviolet light*
Hydrogenated coal oil and tar	*Sunlight*
	X-rays

naturally in parts of oil. They are also produced by any process that burns carbon-containing substances incompletely. These chemicals are found in soot, carbon black, mineral oils such as cutting oils, paraffin, pitch, tar, bitumen resins, wood fats, vegetable and animal oils, and combustible gases. They are used in many different industrial operations.

Another group of chemicals associated with skin cancers are compounds of arsenic. It is suspected that arsenic compounds can also cause lung cancer when people are exposed to arsenic dusts over a period of time. Radiation-induced cancers of the skin occur from exposure to X-rays and radioactive chemicals such as radium.

Occupational skin cancers can be completely prevented, and general preventive measures will be discussed next. Special measures for the prevention of occupational cancers are mentioned here. For example, exposure to carcinogens can be avoided by substituting less dangerous substances. Purified oils from which the carcinogens have been removed are available for some operations. Barrier creams are of little help in the prevention of occupational cancers. Frequent and thorough medical examinations must accompany any prevention program. Skin cancers, if detected early, are not fatal and can be easily removed.

Prevention of Occupational Skin Diseases

More than 90 percent of occupational dermatitis is preventable. All of the usual principles of industrial hygiene must be practiced (see Chapter Nine). Machines and operations must be enclosed to prevent contamination of the worker and the workplace with dangerous chemicals. Chemicals must be handled in closed containers. Shields, guards, and automatic equipment should be used where possible to separate workers from harmful materials. Effective, properly designed ventilation equipment should collect toxic fumes, vapors, and dusts that are liberated in the industrial process.

Cleanliness of the workplace and of the workers themselves is a very important part of the prevention of industrial dermatitis. General housekeeping and maintenance of production and ventilation equipment should be assigned to work crews specially trained in the skills necessary for maintenance. These work crews generally should not be brought in by outside contractors, but should be made up of the workers who work with the equipment and processes day in and day out and are most familiar with them. Such crews should work closely with the union safety committee and make sure that the maintenance is sufficient. Housekeeping means proper storage of all materials used or produced in the factory. It means prompt cleanup of spilled chemicals. It also means general maintenance of equipment and surrounding areas. Dust, dirt, drippings, splatters, and overflows must be cleaned up promptly, and containers or shields that collect unwanted waste should be emptied and cleaned regularly. Maintenance of ventilation equipment is discussed in Chapter Nine.

Personal cleanliness is another important factor in preventing dermatitis from industrial chemicals. Obviously, thorough, frequent washing removes the chemicals from the skin and prevents a reaction with it. Different skin cleansers have been designed for different types of chemicals and dirt. They must also not harm, irritate, or sensitize the skin. Abrasives are frequently added to help in the removal of dirt; however, frequent use of strong abrasives can damage the skin. Lanolin is often added to skin cleansers to help restore natural oils to the skin.

A leading medical specialist in occupational dermatitis has stated that if a worker must walk more than 100 yards to wash, she or he will probably remain dirty until the next break, and risk dermatitis. Where washing facilities are far away from the work area, it is sometimes the practice to use "waterless" hand cleaners. These are not as good as soap and water, but they are better than nothing at all. There are three varieties of waterless cleaners. One type contains a solvent (kerosene), soap, and lanolin. This type may be quite irritating because of the combination of the solvent and the alkaline soap. The combination may even cause a primary-irritant dermatitis (see p. 56 above). Another type of waterless skin cleaner acts as a soap. The third type, the detergent cleansers, are the least irritating. Waterless skin cleaners should be removed from the skin with *clean* towels or rags.

Waterless cleaners should be used as a temporary measure. Workers must demand and obtain adequate wash-up facilities and paid wash-up time so that the risk of dermatitis is decreased. Of course, wash-up facilities are no substitute for the proper industrial hygiene and preventive maintenance necessary for avoiding all contact with chemicals.

Certain jobs require a daily shower before leaving work. Washrooms and showers must be conveniently located, sanitary, and supplied with soap and towels. The dressing area should have double lockers, one for dirty, contaminated work clothing and the other for street clothes. Work clothing, which should be supplied by the company, and any other clothing that has become contaminated with dangerous chemicals should be cleaned at the company's expense. The laundry workers should also be warned of the presence of hazardous chemicals in the clothes, and should be protected from them. Disposable work clothes can be used when regular contamination with hazardous chemicals occurs. Personal cleansing is part of the job. It should not be left to the individual worker, whose worktime is taken up fully by production tasks. Paid wash-up times must be provided.

Protective clothing is necessary for some operations. Aprons should be full, extending below the knees in order to protect properly. Gloves should be properly designed;

they should extend at least one-third of the way up the arm and fit snugly. They should be lined with cotton so that the hands don't sweat too much. Rest periods must be provided during which the gloves can be removed to allow sweat to evaporate and permit the skin to breathe. Holes in the gloves can be very dangerous; chemicals may get into the gloves and be held in close contact with the skin. Protective sleeves should be worn out over the cuffs of the gloves. The sleeves should fit comfortably and have perforations to allow for air circulation. Sleeve material must be designed so that it *tears* immediately if it gets caught in machinery.

Common skin conditions such as psoriasis and acne and skin allergies such as eczema are thought to increase the risk of industrial dermatitis. People with these conditions especially should have little or no contact with irritating or sensitizing chemicals on their job. When a person with a skin condition gets industrial dermatitis, it may be more severe and more difficult to treat than usual.

BONES AND MUSCLES

The bones and muscles make up another system of the body, called the *musculo-skeletal system*. Its purpose is to hold the body up and together in the way a steel framework supports a building. The system is made up of several parts with different functions. The bones give the body structure. In between the bones are joints, which are smooth surfaces like ball bearings; these allow motion between the bones to occur without friction. The bones are held together by *ligaments*. The muscles, which contract to cause active motion of the bones around the joints, are anchored firmly to the bones by the *tendons*.

The Back and Back Injuries

Back injuries are among the most common disabilities caused by work. Approximately one half of all back injuries occur at work and the other half at home. Because they are so common, it is difficult to get an exact estimate of how important back injuries are. The National Safety Council estimates that in 1970, injuries to the trunk represented 26 percent of all compensation cases, and 36 percent of all compensation dollars spent. These percentages accounted for 570,000 injuries to the trunk of the body.[1] These half-million injuries do not account for all the back injuries in the country, because the National Safety Council statistics are based only on injuries that are first reported by doctors as being occupational, and that win compensation awards in the compensation courts. Using these criteria, many industrial injuries are unreported.

A British survey suggested that in any single year 4 to 5 percent of the population has "disc" injuries. Disc injuries, however, represent less than 1 percent of all back injuries that occur in industry. Most of the back injuries, as described below, are injuries to the muscles and ligaments. These figures are used only to give an estimate of how important and common back injuries are.

The back is made up of the bony spinal column and the ligaments and muscles that support and move the spine. The bones of the spine, called *vertebrae,* support the weight of the body. There are 24 of these blocks of spongy bone. Each of the blocks has winglike projections that surround and protect the spinal cord and the nerves. The joints between the vertebrae allow flexible movements. In the joint between each pair of vertebrae are cushions called *discs,* which are kept tense by their special structure. The discs consist of a jellylike center encased in a criss-cross network of tough ligament fibers. They act as pivots and shock absorbers for the blocks of bone of the spinal column (see Figure 6). The bones, joints, and discs are held together by strong, elastic ligaments, which bind the winglike projections to the vertebrae, as well as binding the bones and discs together. These

1 Injuries to the trunk include more than just back injuries.

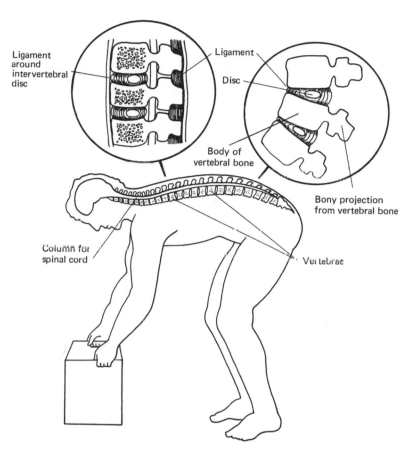

Figure 6. THE BACK

connections keep the spinal column stable. The back muscles surround and are attached to the bones and their projections. The muscles allow the trunk of the body to make active movements. When the muscles on one side of the spine contract, the body can bend sideways. When the muscles in front or in back of the spine contract, the body can bend forward or backward. The muscles not contracting steady the spine so that the movements are smooth and balanced.

Another function of the muscles is to fix the joints of the spine when the back needs to be braced against movements of the hips, arms, or legs. This automatic cooperation or coordination between all the spinal muscles protects the joints and ligaments from strains. The back muscles also maintain the body's posture so that it can keep an upright position against the force of gravity.

The spine is divided into three sections: the neck, chest area, and lower back (or *cervical spine, thoracic spine,* and *lumbar spine*). The neck is the most movable section, while the lower back carries the weight. During lifting, the weight of the body and any additional weights being carried behave as if they were on a lever, with the center point of the lever at the hips. The lower back is important for keeping this lever action. The chest area is very important in posture. Adolescents and children can injure the midsection of the spine by improper posture or by carrying too much weight while their bones are still forming. An injury in this section can lead to a humpbacked, stooped posture.

Many backaches are incorrectly blamed on the cushion, or disc, between the vertebral bones. Disc injuries do cause some back troubles, and may come simply from growing old. Beginning at ages 30 to 40, the ligaments that encase the disc start to dry out and become less elastic. When they are dried out, a person can snap or break them by suddenly rotating or bending the spine. If the ligament mesh surrounding the disc has been torn, the pulpy core of the disc may squeeze through it. This is called a "ruptured disc," and it may cause pain by pressing on the nerves that enter and leave the spinal cord.

A person who has a back strain has suffered injuries to the joints and ligaments. Many times it is hard to pinpoint what exactly has been injured, because only the bones show up on X-ray film. When too much force is applied to the ligaments of the spine, some of the ligament fibers can tear, which causes severe back pain, stiffness, and cramping of the muscles. The tear in the ligament will heal, but a scar will be formed that is weaker than the original ligament. If a person has repeated back strains, the ligaments lose their elasticity and generally become loosened. Loosened ligaments mean that the joints are no longer held in a stable position and are more likely to be injured. If a joint is injured, there may be a buildup of new bone around its edges. If an X-ray shows joint injury and bony spurs around the joint, this disease is called *degenerative arthritis,* or *osteoarthritis.* When one joint in the spine is damaged, the other joints must take the load that the damaged joint cannot handle. The greater load may put the rest of the spinal joints and ligaments under stress and lead to further strains.

Lifting and carrying put great stress on the spine. In spite of tremendous technological advances in many industries, manual lifting and carrying are still the cause of almost one-quarter of all compensated industrial injuries in the United States. There is controversy among doctors about the way the back functions in lifting and carrying. It is designed to handle loads most efficiently when it is in the erect position, because the discs absorb compression best in this position. When loads are carried at an angle with respect to the body's pivot at the hips, there is stress on the spinal ligaments and discs, which may lead to back strain. In order to avoid this stress, many doctors and safety experts recommend the "straight-back, bent-knees" method of lifting.

This straight-back, bent-knees method may not always be best, one reason being that the ligaments do not lift alone but are supported and balanced by the muscles. Proper lifting also depends on strong, well-coordinated muscles that protect the ligaments from strain. If your muscles are tired, they can be less well coordinated. Muscles become fatigued from many causes. In industry, abnormal or unnatural work posture may be an important cause of muscle fatigue. Good work posture must be designed into the industrial process. Placement of workbenches, chairs, stools, lighting, platforms, and storage shelves must be designed to allow correct natural posture. Muscles that have maintained an abnormal posture may be too tired to perform even a light task safely. Another cause of muscle fatigue is standing still while handling loads. When lifting, the rhythm of movement should help in moving the load. Standing still prevents this.

The back muscles use energy when they are involved in lifting and carrying, and the more energy they use, the greater the fatigue that may result. Canadian scientists have tried to measure the energy used in various styles of lifting. They compared straight-back, bent-knees lifting with the bent-back, straight-knees method and with a free-style method. For light loads, up to 30 pounds, the "cheapest" style of lifting, in terms of using up energy, was the style that *felt most comfortable* to the volunteer, the free-style. For heavier loads, the volunteer chose the straight-back, bent-knees position. These experiments suggest that the straight-back method of lifting is necessary only for very heavy loads, and may be impractical and unnecessary for many lighter tasks. However, these observations should be confirmed by further research.

Certain movements of the spine are more likely to cause stress and may result in back strains. Sudden twisting or rotation off the body's center of balance is particularly dangerous. Overreaching to the side or overextending to the rear are also dangerous movements. Adjusting a load while carrying it, or other sudden moves from loss of balance, slipping, tripping, fright, laughing, or sneezing, may add further stress that ligaments and muscles cannot absorb.

Prevention of Back Injuries

What is most important is to prevent the *first* back injury. Once the first injury has occurred, a second is more likely.

Mechanization and mechanical lifting are obviously the most important means of preventing injuries. In a society that is capable of manned space travel, the use of mechanical lifting devices such as fork-lift trucks, hand trucks, conveyors, lifting tackle, and cranes should be routine.

Table 9. RECOMMENDED WEIGHTS FOR LOADS FOR
OCCASIONAL LIFTING BY UNTRAINED WORKERS

AGE	MEN	WOMEN
16–18	44 lb.	26 lb.
18–20	51 lb.	30 lb.
20–35	55 lb.	33 lb.
35–50	46 lb.	28 lb.
over 50	35 lb.	22 lb.

Limits on weights to be lifted have been recommended. These limits apply to workers who are untrained in special lifting techniques and who lift only occasionally during work hours. For regular or frequent lifting, a fatigue factor must be taken into account. Then the figures in Table 9 should be reduced by 25 percent. Specially trained young workers—aged 25 to 30—may lift heavier weights. A maximum for all lifting, recommended by the International Labor Organization, is 55 kilograms, or about 120 pounds. Older workers or those with heart disease should do little if any lifting, as it places an additional physical stress on the body as a whole. Since everyone gets old and many people develop back injuries, unions should negotiate jobs with duties that do not involve heavy lifting. These jobs should, of course, involve no loss in pay or in equal-pay progression to other jobs. One should not be penalized simply for growing old, and certainly those people who have been injured from their work should suffer no economic loss.

Because it is important to know what type of movement caused a back injury, a careful history of the injury should be taken by the doctor. The function of nerves in the leg and the leg movements must be examined. A person will have abnormal leg movements or nerve function only when a disc has been dislocated. Most back injuries, luckily, are strains, which heal in one or two weeks. Pain relievers are usually helpful. Other treatments, such as heat, diathermy, short-wave, or manipulation, probably do not improve the rate of healing very much, but they may make you feel more comfortable in the first few days after an injury. If a back strain does not heal in two weeks, X-rays of the spine should be taken to look for evidence of a slipped disc or of damage to the bones of the spine.

If a person has already had one back injury, every effort should be made to prevent another. First of all, the person should avoid lifting; however, further injury can result from a sudden movement of the wrong kind, even without a load. The same movement that caused the first injury is often responsible for a second. For example, if a twisting motion to the right caused the first injury, a similar motion may cause further damage. It is important to determine which motion triggered the strain and to avoid it. A careful program of physical exercise may be prescribed to strengthen the muscles of the back.

The Hands, Arms, and Legs

Other parts of the muscular and skeletal system can also be injured in industrial work. All these other parts, such as the hands, the arms, and the legs, are made up in a way similar to the back: they consist of bones, joints, cartilage cushions between joints, ligaments, and tendons. Each joint of the body is designed to do work best in a natural position. For example, the hand is strongest and most comfortable when it is extended on a straight line from the arm. The fingers also are strongest when the hand is in this position. The shoulder is most comfortable and strongest when the arms are close to the body. The legs are most comfortable when they are straight, or when the knees are bent to a right angle. If you do repetitive or heavy work in an unnatural position, your muscles, ligaments, and joints will become fatigued and may be damaged.

Repetitive use of a part of the body such as the hand in an unnatural position causes injury to the tendons. Tendons are similar to ligaments; they are strong elastic connections between the muscles and the bones. When the tendons become inflamed, the hand becomes swollen and sore: this is called *tendonitis*. Such injuries, which result from unnatural posture, can be prevented by using properly designed equipment. Workers know the position that is comfortable for the tasks they have to do. Comfortable positions are usually the "right" positions for the anatomy of the body. The designers of machines, trucks, and workbenches should consult the workers who use them to find out what is most comfortable for them. In this way they could avoid making harmful machines that require unnatural work positions. Any truck driver, for example, can probably suggest at least a dozen ways to redesign the cab to make it a more comfortable and healthful place.

CHAPTER THREE
STRESS

THROUGHOUT THIS BOOK we discuss the effects of chemical and physical hazards on the body. We illustrate how the body reacts to various injuries and the long-lasting damage they can do. Not all workers, however, are exposed to these things, and yet they too have a good chance of developing disease as a result of their working conditions. This is because there are very few working people who are not under stress on the job, and stress can and does have serious medical effects.

Most people have heard that the stress of being an executive can lead to ulcers, heart attacks, and other problems. What most people unfortunately don't recognize is that workers experience as much and often more stress. It is not recorded, because workers do not enjoy the social prestige of executives and are therefore not the subject of television programs, movies, magazine articles, or scientific research. An understanding of the kinds of stress that most workers encounter and what it can do to them is important.

This chapter will describe the body's natural response to stress, some of the causes of stress, and some ideas on how to overcome it. It is directed to both white-collar and blue-collar workers. It will probably also be of interest to women who are raising families and do not have outside jobs, because stress has become part of the way of life for everyone in America.

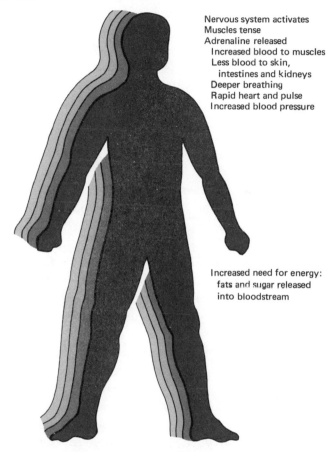

Nervous system activates
Muscles tense
Adrenaline released
  Increased blood to muscles
  Less blood to skin,
    intestines and kidneys
  Deeper breathing
  Rapid heart and pulse
Increased blood pressure

Increased need for energy:
  fats and sugar released
  into bloodstream

Figure 7. WHAT STRESS DOES TO YOUR BODY

The Stress Reaction

Stress is difficult to define because it is common, has many causes, and is almost impossible to measure. One definition is that it is the response of the body to any demand made upon it.

When the body is subjected to stress, a whole series of

biological changes occur. Many of the changes are basic to man and similar to an animal's "fight or flight" reaction. When a human being or an animal senses danger, there is an instantaneous response. Adrenaline hormone is released from the adrenal glands into the bloodstream, causing an increase in pulse rate, blood pressure, and rate of breathing and stimulating the release of energy-supplying body chemicals—glucose and fats—into the blood. The energy will be used by the muscles, which are supplied with an increased flow of blood through their dilated blood vessels.

In all organs except the muscles and the brain, and especially in the intestines, the small blood vessels constrict, so that no energy is diverted from meeting the threat that caused the stress. The complicated blood-clotting system is also partially activated to provide for the rapid clotting of wounds.

Some stress is normal to life, and life would be dull without it; however, if stress is repeated, prolonged, or continuous, the body becomes worn out and various illnesses can result. Stress may actually wear the body out and increase aging. One theory of wear and tear says that each time the body undergoes a stress reaction, it uses up life energy. Some scientists think that this body energy is not replaceable once it's used up, and that each individual is born with a set supply. This reserve has been called "adaptation energy." (It can be compared to the supply of coal in a mine: once the coal is mined and burned, it is gone.) At any rate, people should not have stress imposed upon them at work, but should have the option of choosing their form of stress.[1]

The early symptoms of a chronic stress reaction may not be symptoms of a specific disease. They can be indecision, reduced appetite, loss of weight, irregular bowel movements, headache, backache, skin rashes, insomnia, nervousness, tremors, poor memory, and irritability. On the other hand,

1 Hans Seyle, *Stresses of Life* (New York: McGraw-Hill Book Co., 1956).

these symptoms may never appear, and chronic stress can lead directly to actual disease.

Numerous diseases are associated with or aggravated by stress: ulcers, migraine, asthma, ulcerative colitis, and especially coronary heart disease. Coronary artery disease (see Chapter Two, pp. 30–33) is a particularly important stress disease. It is the most common cause of death in this country. One out of every five men under 60 in the United States will have a heart attack. Scientists think that the increase in blood fats, the increased efficiency of the clotting system, and the rise in blood pressure, all of which happen under stress, speed up the formation of cholesterol plaques in the walls of the small arteries. The narrowing of the heart-muscle arteries by cholesterol plaques (*coronary arteriosclerosis*) leads to a decrease in blood supply and a decreased oxygen supply to the heart muscle. A heart attack—the death of the heart muscle from lack of oxygen—can occur when there is complete blockage of an artery to the heart muscle. An attack can also happen when the heart is pumping rapidly—during a stress reaction, for example—and the demand for blood oxygen is greater than can be supplied through a narrowed coronary artery. Cigarette smoking also decreases the blood supply to the heart muscle, and a rise in blood pressure increases the amount of work the heart must do. Both of these factors contribute to the development of coronary heart disease.

Causes of Stress in Industry

Stress has a variety of causes. One cause—**noise**—is dealt with extensively in Chapter Four. The body stress reaction occurs continuously in the presence of noise. The fatigue caused by a noisy environment is due to continuous energy loss. Other physical hazards in industry probably also use up the body's "adaptation energy."

The rate of work for repetitive tasks can be a source of stress. A very fast work rate can cause strain, feelings of fear, anxiety, fatigue, nervousness, depression, isolation, loneliness, and loss of identity or individuality. Aside from these emotional stresses, fast, repetitive motions may also cause inflammation of the joints and tendons of the hands.

There can be a somewhat slower rate where there is no strain and the worker is fully absorbed in the task. If the task itself has several steps, or requires skill to perform, this pace of work may be satisfying. However, this rate may be emotionally upsetting if it requires full attention and yet is totally boring, monotonous, and meaningless. Work may also be performed at a still slower rate where the worker becomes bored and unwilling to continue working. Under these conditions, workers will become fatigued and make an increased number of mistakes.

Finally, an even slower rate of work may be the most tolerable. At this very slow rate, the worker spends most of the time doing something other than the main task. He or she may read a book or talk to co-workers.[2]

The level of speed of these four work rates varies with individuals and the type of work. The rate of work can be set at a tolerable level by a silent agreement among the group of workers who perform the tasks together, in an informal work group. Many processes, however, are totally out of control of the workers. In this case, workers can change the pace through concerted action and effective union representation, because it is the right of the union to bargain over their working conditions.

Fatigue is closely related to stress. The experience of fatigue is familiar to all. Two types can be recognized. One is muscle fatigue, which is the sensation of pain when specific muscles are moved. It is the result of exhaustion of the supply of chemical energy available to the muscles and of the buildup of waste products of the body's reactions in the muscle tissue. General fatigue is a more psychological form of fatigue. It leads to a decreased willingness to work. Gen-

2 A. T. Welford, *British Journal of Industrial Medicine*, vol. 15 (1958), p. 99.

eral fatigue is caused by accumulation of all the various stresses that a person experiences in the course of a day. Examples of the stresses that cause fatigue are shown in Figure 8 and include monotony; long hours of work; mental and physical exertion at work; environmental conditions, climate, light, and noise; disease, pain, and nutritional deficiency; and mental causes such as responsibility, worries, and conflict. All of these stresses are additive. They all consume the body's energy, and energy consumed must be supplied through recreation, food, and rest. The greater the amount of any single stress, the less energy will be available to handle some other stress. For example, a person who is subject to a great deal of noise on the job may be worn out by the end of the work day and unable to deal effectively with the worries and responsibilities of home and family.

Some scientists have suggested that the sensations of fatigue come from certain areas of consciousness in the brain,

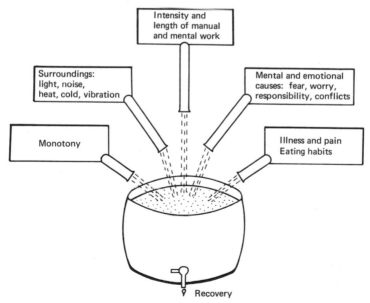

Figure 8. CAUSES OF FATIGUE

which are turned "on" or "off" by two opposing systems of nerves located in the lower brain structures. The "on" or "activating" system tells the brain to be alert, active, and responsive to the environment. The "off" or inhibiting system tells the brain to rest and sleep. Each of these systems in turn receives unconscious signals from the environment and the body, which govern its degree of activity. If there is danger or stress, the activating system will be turned on. But when the body energy level is lessened because of active response to the stress, the inhibiting system will become dominant. If surroundings are monotonous, the inhibiting system will be "on" and the activating system will be "off." Any interruption in the monotonous routine, however, will turn the activating system on and diminish the signals from the inhibiting system. The resulting behavior, whether active or quiet, will depend on the balance between the inhibiting and activating systems. The balance of signals between these two systems explains why fatigue results from monotony, even though there is no physical work load.[3]

The various feelings associated with fatigue, such as boredom, weariness, depression, lassitude, anger, and exhaustion, can cause diminished skill in performing tasks and changes in the heart rate and the brain waves. Fatigued workers pay less attention, receive new visual information more slowly, need more reaction time, take more time to think, and show decreased motivation and performance. Because of these factors, they have more accidents and lower output.

Fatigue is a sensation that is supposed to protect the body, as hunger, thirst, and fear protect the body. It tells the body to change activity, or to rest. If a person disregards the feeling of fatigue, the sensation may become almost painful, and eventually become overwhelming and force a collapse. Re-

3 E. P. Grandjean, "Fatigue," *American Industrial Hygiene Association Journal*, 1970, pp. 401–11.

peated daily fatigue leads eventually to chronic fatigue. A person in this state feels tired much of the time, before the day's work is done or even at the beginning of the day. Chronic fatigue is accompanied by changes in mood such as depression, anger, nervousness, and irritability. There may also be physical symptoms: general ill-feeling, loss of appetite, headaches, dizziness, sleeplessness, and indigestion. There is probably increased susceptibility to many diseases during this state.

Fatigue and its serious consequences on the job are prevented by sensible work hours, rest periods, job rotations to relieve monotony, and pleasant surroundings. The elimination of noise is especially important in preventing fatigue in the workplace.

The effect of the hours of work is another aspect of stress that must be considered. Work assignments during hours other than the daytime have, in the past, been common to service industries—for example, hospitals. In the past ten or fifteen years, they have become much more common in industrial operations. The increase in shift work results from the changing nature of industrial production. Some processes where highly automated machinery is used, such as iron- and steelmaking, computer operations, and oil refining, are so expensive that according to management they must be operated continuously to be efficient. In management's terms, this means that the machines and the workers operating them must pay for the cost of the equipment with increased earnings from increased production. Therefore, a portion of the work force must work during times other than the day.

Man is a daytime animal. Almost all body functions are periodic. Hormone levels, body temperature, the chemical reactions through which the body produces energy, pulse rate, and blood pressure, all fluctuate on a 24-hour schedule, with peak activity in the midday period. These internal rhythms, plus the external rhythms of daylight and dark, temperature, and social activities, regulate a wide range of body functions and thereby affect fatigue, sleepiness, rest, activity, and alertness. The biological fluctuations continue with only slow changes or adjustments to time changes, as in shift work or in long-distance travel. For example, President Nixon's trip to Peking was carefully scheduled to allow rest periods and adjustments for the time changes between the United States and China. The purpose of these rest periods was to prevent the fatigue and irritability that ordinary travelers experience on such a journey. The day-night rhythm of the body is very slow to change. It has been estimated that at least one month of an uninterrupted routine of being active and awake at night and of sleeping during the day is necessary before the body can completely reverse all of its around-the-clock functions.[4] Such an adjustment of the 24-hour cycle, when it occurs, is called *acclimatization*. Even a small interruption—for example, a weekend off—in which a normal day-activity, night-sleep schedule is restored allows the body to go back to the normal day-night schedule.

Because night workers go back to daytime activity on weekends and holidays, there is no real adaptation or acclimatization to night work among most shift workers. Short shift schedules of four or five days ensure that there is no adjustment to the night shift.

A great deal of scientific literature has been written on this subject, but little hard-and-fast evidence has been gathered. Two effects of shift work that have been proved are sleep loss and digestive disturbances. Most studies of shift workers have shown that sleep patterns are disrupted during the rotation to night work. Many night workers have difficulty sleeping during the day. They are disturbed by light and by street and household noises, and are more restless and nervous. Studies of the brain waves of night-shift workers show that the workers get less deep, full sleep

---

4 L. Teleky, "Problems of Night-work, Influence on Health and Efficiency," *Industrial Medicine*, vol. 12 (1943), pp. 758–79.

than people on normal schedules.[5] This means that night workers are always tired. The fatigue builds up until the next day off and forces them to sleep more to pay off the sleep they missed during the preceding week. This debt is paid by sleeping during their spare time.[6]

Appetite and digestion are disturbed in shift workers. Eating is a social activity as well as a necessary biological one. In surveys of worker's eating habits, a large majority had poor appetites while they were on night-shift rotation. A substantial minority, about 10 to 20 percent, had disturbances of other periodic intestinal functions, such as bowel movements, while on the night shift.[7] Changes in eating habits and digestive functions may cause worsening of existing intestinal disorders, or may trigger attacks of such diseases as ulcers in people who have a tendency to them.

The special mealtimes required by the night-shift workers also affect their home life. The families of shift workers frequently complain of changes in meal schedules and of the extra work of preparing separate meals at odd hours of the day.

There is as yet no scientific evidence that shift work causes serious diseases such as digestive disorders, ulcers, and nervous disorders, but no really conclusive research has been done. Workers on shift systems who become sick usually switch to day work if they are able to find it. Therefore, many of the shift workers who are left are actually healthier than day workers.[8] Surveys of currently employed shift workers generally show that they are healthier than the population in general. But so are most active working people, and self-selected shift workers are healthier than day workers.

[5] Quoted by G. S. Tune in "The Human Sleep Debt," *Science Journal*, vol. 4 (1968), pp. 67–71.

[6] G. S. Tune, "Sleep and Wakefulness in a Group of Shift Workers," *British Journal of Industrial Medicine*, vol. 26 (1969), pp. 54–8.

[7] P. J. Taylor, "Shift Work—Some Medical and Social Factors," *Transactions of the Society of Occupational Medicine*, vol. 20 (1970), 127–32.

[8] A. Aenonsen, "Shift Work and Health," Norwegian Monograph on Medical Science, Oslo, 1964.

Certain chronic diseases are aggravated by the irregular hours of shift work. Diabetics usually find that the irregularities of shift work lead to poor control of their sugar level. People with chronic bronchitis, perhaps induced by poor working conditions, may find that the colder night air aggravates their condition and that loss of sleep increases the frequency and severity of chest infections.

The emotional responses of workers to shift work depend on what irregular hours do to their social lives as well as on how much the body functions are disturbed. In a survey of male refinery workers, 48 percent thought that shift work adversely affected their health and 76 percent stated that it adversely affected their social life. The afternoon shift was the most disruptive, the night shift next. As with any survey of work habits, a few workers preferred these shifts.

The workers who were single found shift work most distasteful, because they had most difficulty meeting and dating and disliked the interruption of weekend pleasures such as sports. The irregular hours also meant that they had difficulty getting together with friends who worked days and who had forgotten what it is like to work nights.[9]

Men and women with families may find that shift work interferes with family routines. It is difficult and even cruel to discipline young children to be quiet during the day while one parent is sleeping. However, economic needs often force some working couples to look at shift work in a positive way. In a family in which both parents must work, the arrangement of work hours is planned to allow for shared child-care responsibility. While one parent works, the other baby-sits. Family life doesn't really exist for such a couple.

On the other hand, some workers find it convenient to have normal business hours available for personal business such as medical visits, shopping, or hobbies. Daytime work-

9 Taylor, "Shift Work."

ers are often forced to take nonmedical absences to carry out necessary personal business.[10]

The personal relationships that develop at the workplace are another important factor contributing to workers' attitudes toward shift work. The development of a group with close personal relations often makes a job more tolerable, or even pleasant. These groups, the informal work groups, which will be discussed later, form more easily among shift workers, who develop more team spirit while rotating to the night shift together.

Workers who have a negative physical or emotional reaction to shift work probably also suffer more stress from existing physical and chemical hazards. For some workers, this stress may contribute to the appearance of related diseases such as peptic ulcer and coronary heart disease.

Emotional and psychological pressures in industry are important causes of the stress reaction in workers. Fear of injury from dangerous chemicals or unsafe equipment can be a constant source of stress. Accidents and near-accidents are particularly frightening, even over and above the actual injury that may occur. To see a co-worker injured or killed can often be as upsetting as the loss of a member of the family. Emotionally, people respond to such loss with grief and depression. The human body responds with the stress reactions just described.

A common cause of emotional stress is job or financial insecurity. Financial worries are brought on not only by fear of loss of earnings but also by a gradually decreasing income. A person used to a certain life-style will become insecure and worry if he or she cannot maintain the same standard because of a lower wage or decreasing real income due to inflation.

It is widely known that when a business executive or a professional dies of a heart attack at age 50, a contributing

[10] P. J. Taylor, "Shift and Day Work, a Comparison of Sickness Absence, Lateness, and Other Absence Behavior at an Oil Refinery from 1962–1965," *British Journal of Industrial Medicine*, vol. 24 (1967), pp. 93–102.

cause is often stress from a life of constant tension. But
stress as a contributing cause to the heart-attack rate of
workers is not admitted at all. However, stressful competi-
tive drive, and especially feelings of personal inadequacy,
are common among working people. These feelings of inadequacy are promoted by the stereotype of the worker as
portrayed on television and in the movies—fat, opinionated,
and lazy. Such stereotypes degrade anyone who works for
a living, and imply that working for a living is a badge of
inferiority. Yet the ability to leave the working class and
enter income levels where there is more financial security is
not possible for most workers. Thus the elimination of stress
requires more than changing individual workplaces. The
attainment of dignity and self-respect will have to be
achieved by working people as a whole if individuals are to
be free from the stress caused by competition and feelings
of inadequacy.

The organization of modern industrial production has
built-in sources of emotional stress. Workers frequently feel
powerless and controlled or dominated by the complicated,
fast-paced machines with which they work. The division of
work and the organization of responsibility in modern industry are not designed to satisfy the needs of those who do
the work. A worker may be assigned to a boring, repetitive
task that has no clear relationship to the end product, and
the product itself may seem—especially to those who produce it—of little use and low quality. This can lead to a lack
of a sense of importance and a lack of personal satisfaction
and accomplishment. Not surprisingly, refusing to allow
workers to have control over the work process can actually
cause inefficiency, because it restricts workers' initiative
and interest in the job. But management, in its blind fear
of giving up some of this control, often settles for this inefficiency, at the cost of the workers' health and dignity.[11]

11 R. Blauner, *Alienation and Freedom: The Factory Worker and His Industry* (Chicago: University of Chicago Press, 1964).

Large companies are often bureaucratic and impersonal, and treat their employees as numbers or objects. This can be especially infuriating to workers. Even though management, as would be expected, attempts to keep workers isolated and unable to control their work, the workers who work together in a particular process or in a certain location will often join in a group. These social groups form in spite of the totally arbitrary assignments made by management, who place workers from widely different social backgrounds in adjoining work areas and processes. The informal work groups that form are often known by the workers as "cliques." Generally, each such group has someone who acts as the spokesman—a leader.[12]

Changes in family life and in the workplace in the past two decades have increased the importance of these informal social groupings. The comradeship of the work group is important for workers who have found that the material gains won over the past several decades are not a great source of satisfaction in their lives. The scattering of large families (which included grandparents, aunts, and cousins, as well as parents and children) has meant that an isolated person can no longer depend on her or his extended family for emotional support, understanding, or defense from the stresses of life, especially the stresses of the workplace. Also, the tasks and technology of the workplace have become more and more complex. Individual tasks on specialized machines, although often extraordinarily dull or repetitive, might still be impossible for a worker to describe to anyone, even close family members. Only a few fellow workers, those in the work group, really understand the major part of each other's lives and emotions, yet they may know little of each other's families or homes. Thus the stabilizing social force of the traditional family has been partially replaced by that of the work group.

[12] L. Sayles, *Behavior of Industrial Work Groups* (New York: John Wiley & Sons, 1958).

For these reasons, and because workers spend so much of their lives in these work groups, they may develop the feeling that "an insult to one is an injury to all." This feeling is backed up by the unspoken agreement that the group as a whole will help any member who is in need. Pressure on any group member at work or at home creates stress in all the members. Just to be a witness of another's harassment, overwork, or degrading treatment creates stress. Loss of a group member through illness, accident, or layoff is felt by all the members. The grief of loss may turn to anger and cynicism toward the authorities responsible for the loss— the employer, a foreman, or an ineffective union.

A union should recognize the existence of these informal work groups. The group need not be officially represented by having its informal leader as steward. It may even be better for the group if the steward is from outside the group, or not one of its leaders. But the steward must be selected by the group. It is important and necessary that the group can see the elected steward and interact with her or him during most of the working hours. In this way the work group can work with and even discipline the steward to represent their interests on the shop floor and in the union.

When all the informal work groups in a workplace collectively achieve control of their working conditions through collective bargaining and other processes, they can begin to change all the conditions that impose stress on working men and women. They have the strength that comes from the solidarity and unity of purpose and spirit of the work group.

Alcohol and Drugs

The use of alcohol and drugs is unfortunately one means by which people try to escape from emotional stress. Alcohol in moderate amounts may relieve the tension that builds up from on-the-job and personal stress. Alcohol exerts this effect by depressing all functions of the central nervous system.

The main danger of moderate alcohol intake is for workers who handle the chlorinated hydrocarbon solvents, such as carbon tetrachloride and tetrachloroethane. These solvents and alcohol are metabolized by the liver. The presence of alcohol and solvent together in the liver increases the damage that either alone might cause. The combination can lead to liver damage, chemical hepatitis, which, if repeated, may cause cirrhosis, or scarring of the liver.

Chronic use of excessive quantities of alcohol leads to addiction. Alcoholics commonly need alcohol early in the day. They may substitute alcohol for food. They drink because of psychological illness, and also to avoid withdrawal tremors, hallucinations, seizures, and other symptoms. Chronic alcoholism leads to cirrhosis of the liver, pancreatitis, intestinal ulcer, and neurological illness.

Heroin is a direct depressant of the nervous system. When first used, the drug causes a brief euphoria—relief of psychological pain and anxiety. With repeated use at increasingly higher doses, the euphoria becomes harder to achieve. Even without euphoria, the drug itself forces its continued use because it is addictive, and when it is stopped the user suffers severe withdrawal symptoms: fatigue, running nose, tearing, sweating, restlessness, insomnia, waves of gooseflesh, shivering and muscular twitchings, hot and cold flushes, generalized muscular aches, nausea, vomiting, and diarrhea. The symptoms come on gradually after 16 hours of withdrawal and reach their maximum intensity in 2 or 3 days. The symptoms may continue for 7 to 10 days; the insomnia and muscle pains may continue for several weeks.

Heroin use can lead to serious accidents in the workplace, since it depresses responses. Accidents may injure co-workers as well as the drug users. Heroin users often die from an overdose. They sometimes suffer bloodstream infections from bacteria and viruses. They may develop heart-valve infections by injecting infectious bacteria; infection from the injection of viruses can cause fatal hepatitis.

The use of alcohol or drugs is obviously not the answer to relief of job stress. Achievement of security and self-respect through changes in the organization and management of the workplace as well as in the workers' position in society are the more difficult yet correct ways of relieving stress.

Thus we see there are many factors that contribute to the stress of being a worker. The relief of this stress is a challenge that workers must undertake because it really involves changing the nature of the workplace. Through their unions and their collective will, workers must start to raise the questions of controlling the speed of their work and of being part of how and why certain jobs are done. In this way work can be made less impersonal, monotonous, and taxing so that workers need not be under stress from their jobs.

Of course, as long as workers have to work overtime or continue to have financial problems, these sources of stress will remain. The struggle to eliminate all stress will be a long one.

CHAPTER FOUR
NOISE AND VIBRATION

SOME TECHNICAL TERMS

audiometer: instrument used to measure hearing ability

decibel: unit of sound pressure or loudness

dB: abbreviation for *decibels*

frequency: characteristic of sound that is measured in cycles per second, or *hertz;* high-frequency sounds are shrill and low-frequency sounds are deep

hertz: unit of frequency, equivalent to 1 cycle per second

Hz: abbreviation of *hertz*

permanent noise-induced threshold shift: permanent loss of hearing from noise

PNITS: abbreviation for *permanent noise-induced threshold shift*

sound-level meter: instrument used to measure noise levels

temporary threshold shift: temporary loss of hearing from noise

TTS: abbreviation for *temporary threshold shift*

IN OUR DAILY LIVES, at home, on the job, traveling, even while sleeping, we are all exposed to noise. Noise is often described as unwanted sound. As research on the effects of noise continues, it becomes more and more obvious that noise pollution is a dangerous matter. This chapter will describe the effects of too much noise exposure on hearing, and a very important part will explain how noise affects the rest of the body in ways that few people are really aware of. The effects of vibration will also be discussed, as exposure to noise and vibration combined is commonly encountered in industry, and researchers have found that this double exposure is especially harmful to the body.

What Is Sound?

Sound travels through the air in the form of waves, very similar to the waves that are made when you throw a pebble into a pool of water. Just as the pressure of the pebble forces the water to spread out in waves, so the pressure of someone speaking or a shot going off forces the air to spread out in waves. If instead of a pebble, you were to throw a large stone into the pool of water, you would form larger waves that traveled farther. Likewise, if instead of whispering you shouted, you could create sound waves in the air that were larger and traveled farther.

One characteristic of all waves, including sound waves, is their frequency. A whistle makes a shrill, high-pitched sound that has a high frequency, while a tuba produces a deep, low-frequency sound. An example of middle-frequency sound is typical conversational tones. Frequency is measured in units of cycles per second, or *hertz,* which is the equivalent term. As people develop hearing loss, they generally lose different amounts of hearing ability at different frequencies. Thus after many years at work, a worker may still be able to hear a conversation which takes place in the middle-frequency range, but may no longer be able to hear a child's voice or the high notes in an orchestral performance.

Figure 9 is a graph of the frequencies corresponding to the waves that make up the ranges of normal speech, vibrations and ultrasonic waves whose frequencies are too high for the ear to hear. The sound pressure refers to the loudness, and the line drawn on the graph shows the amount of sound pressure, or loudness, that people with good hearing need in order to hear a sound.

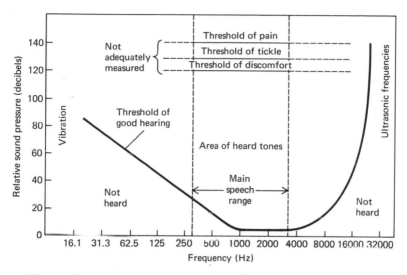

Figure 9. THE FREQUENCIES CORRESPONDING TO SOUND
THAT CAN BE HEARD

The sound pressure, or intensity, is measured in units called *decibels,* abbreviated dB. Decibels are a relative measure. That is, a sound intensity of 60 dB is actually 60 dB greater than some arbitrary reference sound, given the value 0 dB. This arbitrary zero is said to be the weakest sound that a young, sensitive human ear can hear. (Some people can actually hear better than this, and the pressure or the sound that they could hear would correspond to a −dB or negative decibel level on this relative scale.)

The decibel scale is confusing because it is not linear but logarithmic. We are used to counting things on a linear scale—20 eggs are twice as much as 10; 30 eggs are just 10 more than 20, and three times as much as 10 eggs. But if eggs were counted on a logarithmic scale, it would not work this way. A logarithmic scale is based on powers of 10. This means that each increase in 10 dB is equivalent to multiplying the intensity by 10; in other words, 20 dB is ten *times* as intense as 10 dB, and 30 dB is ten *times* as intense as 20 dB. Since 30 dB is ten times as intense as 20 dB and 20 dB is ten times as intense as 10 dB, 30 dB is one hundred times, or 10 x 10, as intense as 10 dB. An increase from 0 to 90 dB is an increase of 10 x 10 x 10 x 10 x 10 x 10 x 10 x 10 x 10, or 100 million times.

Table 10 gives the intensity of noise for industrial operations.

Table 10. ESTIMATED NOISE LEVELS AT VARIOUS MACHINES AND LOCATIONS

80–89 DECIBELS

Furnace, annealing
Grinding
Lathe, automatic
Liming machine
Machine shop
Machining, aluminum
Sand molding
Spraying, varnish, etc.
Veneer department
Welding, arc
Wood finishing, sanding, planing, jointing, etc.

90–99 DECIBELS

Boring machine
Castings, cleaning
Core room
Drill, pneumatic
Fabrication, steel, handling, cutting
Foundry operations, sand slinging, etc.
Furniture making, planers, jointers, saws, etc.
Grinders, castings, pipe, metal parts, etc.
Hammer, drop forge
Jointer, wood

Lathe, engine
Lathe, turret, other than ram type
Mill, bloomer, strip steel
Mill, strip steel
Milling machine
Polisher, metal tubes
Power plant, alternators, etc.
Ram, pneumatic, sand molds
Rivet bucking, fuselage
Router, aluminum stock
Sand muller
Sander, wood
Sawing, logs, etc.
Scarfing, acetylene welding equipment
Screw machine, automatic
Shaper, small steel parts
Shear, steel plate
Shot blast room
Steel pouring
Welder, butt electric
Welder, gas on steel
Welding machine, tube
Wire drawing

100–109 DECIBELS

Chipping, castings, etc.
Conveyor, steel strip
Forging manipulator

Table 10. ESTIMATED NOISE LEVELS AT VARIOUS
MACHINES AND LOCATIONS (CONTINUED)

Furnaces, oil, gas, electric
Grinder, pedestal on small tools
Hammer, drop, automatic
Hammer, forging
Hammer, pneumatic, peening
Hammering machine, rotary, on
  steel tubes
Hoop machine, steel wire
Jolt squeeze machine, sand molding
Lathe, automatic, wood
Lathe, turret, ram type
Mill, roughing, steel plates
Molding, push-up machines, etc.
Planer, wood
Pointing machine, steel parts
Press, pneumatic
Press, punch, automatic
Rivet, bucking, wings
Riveting gun, pneumatic, wing as-
  sembly
Riveting hammer, fuselage assem-
  bly
Sand slinger
Saw, circular, cutting metal
Saw, cutoff, circular, wood
Saw, friction, steel
Shakeout, castings
Shot blast, small castings
Tumbler, small castings
Vibrator, pneumatic, sand molds
Wrench, pneumatic

110–119 DECIBELS

Air hoist, pneumatic
Chipper, pneumatic, castings
Core blower, sand cores
Corrugating machine, sheet steel
Cutting machine, hardened tools
Decoiler, steel coils
Hammer, bumping, on thin metal
Hammer, drop, automatic
Internal combustion engine test
Nail machine
Riveting jig, wing assembly
Sandblast machine, on hand tools

120–129 DECIBELS

Chipper, pneumatic, tank
Engine, airplane propeller
Riveting gun, pneumatic, subassem-
  bly

130 DECIBELS OR OVER

Engine, jet
Riveting, hammer. pneumatic on
  steel tank

How the Ear Interprets Sound

The ear picks up sound waves and, through its intricate mechanism, translates them into nervous impulses, which it sends to the brain through the auditory nerve. The manner in which the ear handles sound is actually quite similar in many ways to mechanical things we are familiar with. Figure 10 is a simple illustration of the ear. As you can see, the ear is divided into three parts: the external ear, the middle ear, and the inner ear.

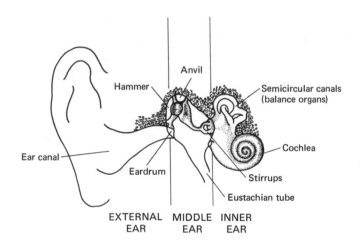

Figure 10. THE EAR

The *external ear* is the part on the outside of the head, and also includes the canal that leads into the head and up to the eardrum. At the opening of the canal there is hair and wax, which are one of the body's ways of keeping out unwanted things. The eardrum is made up of parchment-like material and is the first step in the translation of sound waves into nervous impulses. The eardrum is so called because it operates just like the snare drum in a brass band.

When a drummer hits a drum, it vibrates back and forth, and the amount of vibration depends on how hard it is hit. Likewise, when a sound wave hits the eardrum it vibrates back and forth, the amount depending on how loud the sound is. The eardrum thus translates the high-frequency vibrations of sound into the moving vibrations of a membrane.

The movement of the eardrum then goes to the *middle ear,* which consists of three delicate little bones, appropriately called *hammer, anvil,* and *stirrups* because they resemble these objects in shape. The three bones are movable, and the vibration of the eardrum causes them to hit against one another. When they hit each other like this, they conduct the vibrations from the eardrum to another membrane, finer than the eardrum, called the *oval window.* Thus the sound that was originally captured by the external ear is translated into movement by the eardrum, and now the movement is further translated and reduced in intensity by the middle ear.

This reduction of intensity along each step of the way is necessary, because the farther you penetrate into the ear, the more delicate its various parts become. The pressure of undiminished sound waves would rupture or distort the delicate inner mechanisms of the ear.

The middle ear also has another canal, the *Eustachian tube,* which connects it to the back of the throat. This second opening that the ear has to the outside world allows it to equalize pressure on both sides of the eardrum so that it can vibrate properly. For instance, when you undergo a sudden change in pressure while flying in an airplane or climbing a mountain, you can feel your ears "pop" because of the unequal pressures. You can help overcome this by swallowing, which changes the pressure in the Eustachian tube.

Finally, the oval window at the end of the middle ear connects to the *inner ear,* which is an intricate series of cavities.

The three *semicircular canals* are concerned with the sense
of balance and feed impulses to the brain. They will not be
discussed here in any detail. The other part of the inner ear
is the *cochlea,* which is shaped like a snail and is lined with
very fine, hairlike nerve fibers and fluid. As the oval window
moves, it causes the fluid in the cochlea to flow, brushing
the hairlike nerve fibers in proportion to the intensity of the
vibration of the window. These nerve fibers stimulate the
auditory nerve, which goes directly to the brain. The brain
then proceeds to handle the message.

In summary, when a sound wave hits the ear, it is transformed and translated to a nervous impulse. The parts of
the ear are all delicate; but like the other organs of the body,
unless the ear is really abused by noise or severely infected,
it can be expected to perform well for a lifetime, with only
the gradual normal loss of hearing that comes with growing
old in our society.

Effects of a Noisy Environment on Hearing

Working in the modern industrial environment, however,
is often more abuse than the ear can cope with. Let us consider what happens when the ear is exposed to loud noise.
If the noise is extremely loud and sudden, the impact of the
sound pressure on the eardrum can be so great that the eardrum breaks. A small hole in the drum may mean a minor
loss of about 5 dB in hearing ability; a larger hole may
cause a significant loss of 20 dB or more.

If the noise is not loud enough to rupture the eardrum,
but is louder than about 80 dB and continuous, then the
sound transmitted through the ear puts so much strain on
the nerves of the inner ear that they become fatigued, just as
any other part of your body gets tired. When these nerve
fibers are fatigued (some experts even think they become
traumatized or shocked), they cannot send messages to the
brain as well and hearing ability is decreased. This decrease
of hearing ability is called a *temporary threshold shift*

(TTS), because the level, or threshold, at which a sound can be heard is raised. If you experience a 25 dB TTS, this means that a sound has to be 25 dB louder than before the TTS in order for you to hear it. It is interesting that these effects seem to begin at about 80 dB, a noise level well below that of most industrial workplaces (see Table 10).

The amount of TTS that a person suffers is related to the type of noise experienced, its intensity, and the length of exposure, as can be seen from Figure 11. The greatest amount of hearing loss occurs during the first hour of exposure and seems to level off after that. It also appears that interrupted noise, or intermittent noise, as it is called, causes about half the TTS that continuous noise does.

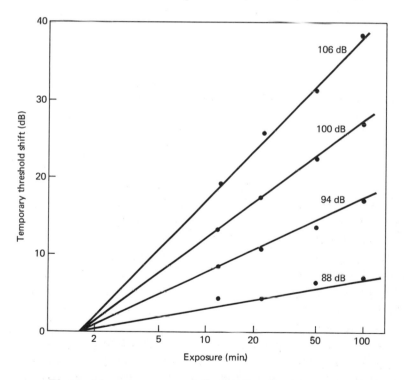

Figure 11. HOW THE TEMPORARY THRESHOLD SHIFT GROWS WITH TIME

The reason that the hearing loss first experienced after being exposed to noise is called a temporary threshold shift is that the ear has the ability to recover from a loss if allowed to rest. The length of time needed for recovery depends on how great the initial loss was: the greater the initial loss, the longer the time needed to recuperate. But the body's ability to repair itself depends on whether the ear is given sufficient time to recover. A worker who works continuously under conditions noisy enough to cause a TTS can probably expect that the temporary hearing loss that he or she suffers will become permanent after about ten years. Of course, as with anything concerning health, this depends on individual susceptibility, and is not a hard-and-fast rule.

Permanent hearing loss due to noise is called a *permanent noise-induced threshold shift* (PNITS). The relationship between the TTS and the PNITS is still a matter of controversy among scientific workers in the field. What is no matter of controversy, however, is that workers in noisy industries are losing their hearing abilities.

The probable connection between the temporary threshold shift and permanent loss can be illustrated by an example. Suppose you get on a noisy subway car or enter a noisy room. When you get off that car or leave the room and return to a more quiet area, you will notice that you can't hear as well as before. What has happened is that you've experienced a TTS. But let's suppose that you are the motorman on the train, or a worker in that room, and that you have worked under these conditions eight hours a day, five days a week, for ten years. What you can probably expect is that the temporary loss you experienced the first time you were exposed to noise will become permanent loss after those ten years at work. If your work involves entering and leaving the noisy environment, the hearing loss will probably take about twice as long to occur. This means that rest periods and job rotation can at least prolong the time it takes to develop a hearing loss.

It is important to differentiate intermittent noise from impact and impulse noise. Intermittent noise is non-continuous noise with the sound-pressure level changing slowly, while impact and impulse noises are characterized by very rapid changes in intensity. A drop forge creates impact noise, for it changes the sound-pressure level extremely fast, whereas a steady hammer riveting creates impulse noise with sound-pressure levels changing at a moderate rate. It is not clear at present what the long-term effect of fluctuating noise is in comparison with continuous noise. Workers exposed to impact and impulse noise do show a hearing loss, but whether it is equal to, greater than, or less than the loss that results from steady noise is not known. The time-averaged noise-exposure standards refer to *steady noise only*. The maximum level for impact noise is 140 dB, an almost unbearably high level, at which workers will almost certainly develop hearing loss and probably other illness. This means that workers exposed to impact and impulse noise are virtually unprotected by the law.

The frequency of the noise is not thought to affect the hearing loss, but here, as in so many areas of occupational health, a significant amount of research has not been carried out. High-frequency noise can be especially irritating, and though it may finally be shown that it is no more harmful to the ears than other noise, exposed workers who find it more annoying or disturbing still have a valid health and safety complaint. No one should be made to pay an emotional price for working, and emotional stress can lead to physical disease as well.

Other Effects of Noise

Besides its effects on hearing, noise affects other bodily functions. There is a great deal of evidence that when the auditory nerve, which transmits sound impulses, is stimulated by excessively loud sounds, it in turn stimulates other parts of the nervous system. This means that noise acts very much like other stresses upon the body. These other effects of noise are a highly controversial matter, and the present American standards are based only on the effects of noise on hearing. The government denies non-auditory effects to such an extent that in 1970 two Public Health Service scientists wrote an article in the *American Industrial Health Association Journal* questioning the lack of American research on these effects, since foreign research seems to indicate that they *do exist*.[1]

As was pointed out in Chapter Three, when the body is subjected to noise, as well as to other stresses, many biological changes take place as the body seeks to defend itself against the source of the stress. The blood vessels constrict in all organs except the muscles and brain. This constriction is especially noticeable in the intestines. There is an increased output of adrenaline, which in turn increases the pulse rate, blood pressure, and rate of breathing. The adrenaline also releases into the bloodstream other chemicals that supply the muscles with energy. The blood-clotting ability of the body is also increased. Extra fats may be released into the bloodstream, and the combination of these events may predispose a person to a heart attack. The rate of movement of the intestines changes, and generally the body's response to stress requires much more energy than normal bodily functions.

Because of the lack of controlled conditions, long-range studies of the sickness rate of workers in noisy industries do not indicate how much illness can be attributed to the effects of noise alone. In almost all cases studied, there were other contributing factors—anxiety over danger from accidents, inadequate ventilation, hot working conditions, poor lighting, and so on. However, in one study that compared 1,005 German workers in a less noisy factory with a similar number in a more intensely noisy factory, significant dif-

---

[1] J. R. Anticaglia, and A. Cohen, *American Industrial Health Association Journal*, vol. 1, no. 3 (1970), p. 277.

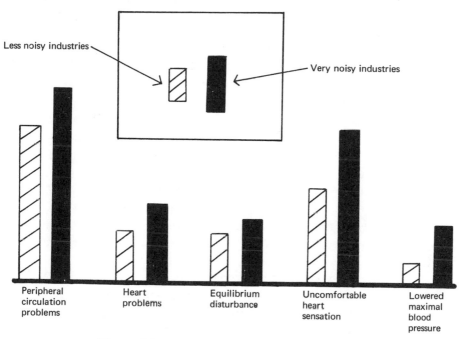

Less noisy industries

Very noisy industries

Peripheral circulation problems

Heart problems

Equilibrium disturbance

Uncomfortable heart sensation

Lowered maximal blood pressure

Figure 12. COMPARISON OF EFFECTS OF
DIFFERENT LEVELS OF NOISE

ferences appeared. Workers in the noisier factory showed
a higher percentage of circulatory, heart, and equilibrium-
disturbance problems. Other data, collected in Russia, show
that workers in noisy heavy-industry environments suffer
from a higher than normal incidence of digestive and circulatory problems. They also have psychiatric difficulties,
as well as disturbances of the nervous system and metabolism.[2] An Italian study of 73 workers exposed to noise and
vibration showed deterioration of the digestive tract after
1 to 2 years of exposure. Signs of this deterioration appeared

[2] A. I. Andrukovich, *Gig. Tr. Prof. Zabol,* vol. 9 (1965), p. 39. N. N. Shatalov,
A. O. Saitanov, and K. V. Glotova, Report T-411-R, N-65-15577
(Toronto, Canada: Defense Research Board, 1962). A. B. Strakhov, Report N-67-11646 (Washington, D.C.: Joint Publications Research Service,
1966).

in X-rays in 65 percent of the workers, and 60 percent com-
plained of symptoms.[3] Changes in the blood pressure and
pulse rate of people working in noisy conditions have been
documented.

Although none of these studies is proof positive that noise
alone caused the observed higher incidence of illness, it
seems reasonable to conclude that noise and the concurrent
stress contributed significantly. In addition, much of the
controversy may be without value, because in fact one can
almost always expect anxiety and physical stress in a noisy
shop. And it is most important to remember that the object
of industrial hygiene research should not be to supply absolute proofs, but to uncover *potential* dangers and see to
it that they are eliminated. Absolute proof requires that
workers become sick or die so that scientists can then determine what it was that injured or killed them.

Noise also interferes with communication between workers. This can create safety hazards when one worker cannot
warn another of danger. In addition, not being able to
speak to each other deprives workers of the normal personal
relationships that all people need. It also limits their ability
to discuss mutual work problems, and may even be considered a barrier to concerted activities, which are a right of
all workers. Finally, since some speech communication is
necessary even in noisy shops, many workers in noisy industries develop throat disorders and voice problems from
having to shout to one another to be heard.

How Vibration Affects the Body

Vibrations are transmitted to the body vertically through
the feet or the seat when a person is sitting or standing on
a vibrating surface. When vibrations are less than 3 cycles
per second, they cause the entire body to move. A common
response of the body to this type of vibration is motion sick-

---

3 A. Tarantola, A. Grignari, M. Lalli, and Santarelli, *Lavoro umano* (Naples),
vol. 20, no. 6 (1960), p. 245.

ness. Vibrations in the range of 4 to 12 cycles per second can cause the hips, shoulders, and abdominal parts to move, depending on the direction of the vibration and the position of the affected person. Vibrations of this frequency can also cause specific damage to the skeletal and muscular systems. The vibrations are transmitted to the bones and joints, where they cause alternating movement of the bone ends against the *cartilage*, the smooth material of the joint surfaces. The bone impulses squeeze the cartilage that lines the joints and cause some of the living cartilage cells to be damaged or destroyed. Damage to these cells causes the area to become inflamed, which adds to the joint damage. After repeated exposures, permanent joint damage known as *arthritis* results. Workers whose whole body is exposed to vibration while standing or sitting can suffer damage to the joints of the spine. For example, a tractor driver may develop low-back pain, which may be the result of damage to the joint cartilages of the spine. A slipped disc is a complication that can come about from damage to the spinal cartilage. In this disorder, the cartilage pads that serve as cushions for the bones of the spine (*vertebrae*) slip out of place. A slipped disc may frequently cause pain by pressing on the nerves that leave the spinal cord through special holes in the spinal bones. These joints of the backbone seem to be most seriously affected by vibrations of 4 to 5 cycles per second. The organs of the abdomen also vibrate at these frequencies. Workers exposed to these vibrations complain of more stomach and digestive problems than workers not exposed to vibration. However, no specific diseases of the stomach due to vibration have been described by scientists because no one has carried out properly designed studies of the subject.

As the frequency of vibration increases and reaches 20 to 30 cycles per second, the skull vibrates. This causes loss of visual ability. If the vibrations at these frequencies are intense enough, the whole body will again begin to vibrate and permanent physical injury can result.

Workers who handle vibrating tools have been found to have four special problems.

Injury to the bones. X-rays of the hands may show small holes due to loss of calcium in the small bones of the wrists. These changes may also occur from other forms of heavy manual work. They are thought to cause no symptoms and not to lead to increased breaking of the bones.

Injury to the soft tissue. The soft tissues of the hands— the muscles, nerves, and connective tissue—may be injured by vibrations. This may result in loss of the use of the hand muscles, or cause the tendons of the hand to contract and thicken. The hand may become weak and contracted and have limited movement. This form of vibration injury is said to be rare.

Injury to the joints. The joints of the hands, wrists, and elbows may develop *osteo-arthritis*. This type of arthritis results from the wearing out or degeneration of the cartilage that normally provides a smooth surface for joint action. Osteo-arthritis is a common disability of normal aging, but it comes on earlier as a result of added wear and tear on the joint surface from vibrating tools.

Injury to the circulation. This form of vibration injury is quite disabling. The vibration apparently causes injury to the mechanisms that control circulation in the fingers. The damage becomes progressively worse. At first the worker notices that one or another of his or her fingers has become white. This lasts for about 15 minutes to an hour. While the finger is white, it is numb except for a tingling feeling. When the worker tries to use his fingers, his movements are clumsy and may be painful. The injury may spread to the other fingers, and the whole hand may become involved. If the circulation to one area is severely impaired, the tissue in the finger may turn blue from lack of oxygen. In rare cases, a finger may die from lack of circulation (*gangrene*). This impairment of circulation is aggravated by cold, because cold causes the blood vessels to constrict. It also occurs more frequently in workers who grip vibrating tools tightly. Decreased circulation in the fingers may begin or progress after the exposure to vibration has ended. Many types of pneumatic tools have caused this type of injury: air hammers, compressed-air chisels, compressed-air drills, jack hammers, riveting guns, pounding-up machines, and stone-cutting hammers, among others.

As with so many industrial exposures, there is also a non-specific, or general, response of the body to vibration. People exposed to vibrations become uneasy, fatigued, and irritable and experience vague discomfort. They bring these feelings home with them, which means both they and their families have to pay an emotional price for the exposure. Since vibrations are frequently accompanied by noise, all the body responses to noise are occurring at the same time.

Ultrasonic vibrations are those above 20,000 cycles per second. Their use in industry can be expected to increase rapidly. Very little is known about the effects of ultrasonic vibrations on humans. Thorough studies of their effects should be made when any ultrasonic equipment is used.

Control of Noise and Vibration

As with all kinds of pollution, the best way to control noise and vibration (a common cause of noise) is at the source. If machines were designed in the first place to minimize noise, much of the problem would be eliminated.

Despite the existence of federal and state standards, the noisiness of a machine has almost never come into consideration in industrial design. Probably because of the non-enforcement of these standards by the authorities, manufacturers have never been compelled to produce quiet machines. It is interesting that when the communities surrounding airports took public action to diminish airport noise, jets such as the 747 were developed to incorporate noise-reduction technology. If community groups who are able to exert only limited pressure can bring about such changes, then certainly workers can bring about changes in the design of industrial machinery, since without their labor none of the machines would work at all.

Redesign of equipment for noise control often does not require a great deal of inventiveness. For instance, if the noise is due to one piece of metal hitting against another, a thin film of polypropylene or a felt gasket placed between the parts may solve the problem. Where feasible, plastic may be substituted for metal. If a machine or a fan is not set on level ground or if it vibrates against the floor, it can be mounted on springs or other vibration isolators to reduce the noise from this source. Such vibration-isolating techniques are also available for punch presses and similar equipment. Sound-absorbing floor material may also help.

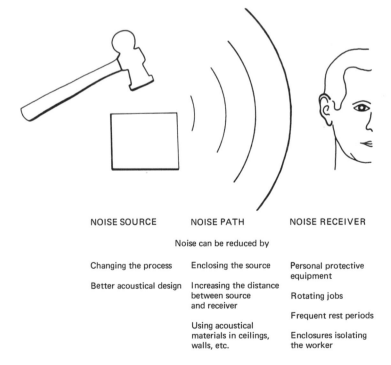

NOISE SOURCE	NOISE PATH	NOISE RECEIVER
	Noise can be reduced by	
Changing the process	Enclosing the source	Personal protective equipment
Better acoustical design	Increasing the distance between source and receiver	Rotating jobs
		Frequent rest periods
	Using acoustical materials in ceilings, walls, etc.	Enclosures isolating the worker

Figure 13. NOISE CONTROL

One source of noise is air rushing from jet streams on air-pressure lines. The noise from these lines is often high-pitched, damaging, and extemely irritating, and eliminating it can make a great deal of difference in the plant environment. To quiet this source of noise, inexpensive mufflers are available, both as air-inlet filters and as air-discharge mufflers. Mufflers can also be installed on all motor vehicles, such as fork-lift trucks, and on other motored equipment.

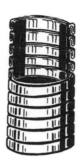

The noise-reducing effect of mounting equipment on vibration isolators can be lost if the equipment is attached to rigid pipes, electrical conduits, or shafting. This can be corrected by using rubber or plastic, where possible, or a flexible metallic hose, as illustrated.

Design for noise control should not be added on as an afterthought, but should be included in the original design of equipment. Where this has not been done and where it is impossible to redesign equipment, the next best step is to isolate noisy equipment or processes from the people who work with or around them. Noise travels many paths within a building, through openings or thin walls, and off non-absorbent surfaces. To avoid exposing workers to noise, the paths it travels must be rerouted or the sound must be absorbed. Noise-absorbing partitions or walls can surround equipment, or else equipment can be operated by remote control with the operator in a soundproof room or booth. If only one piece of equipment or process on the shop floor is noisy, it should be acoustically isolated from the other workers and operations. Distance is an effective sound reducer, since the intensity diminishes with the square of the distance. That is, doubling the distance cuts the sound down by one-quarter; quadrupling the distance reduces the sound 16 times.

Sometimes equipment is noisy because it is not maintained properly. As with all other aspects of industrial health and safety, preventive maintenance is essential for keeping a plant a safe place to work in. Even the best-designed machinery will become noisy and dangerous if it is not maintained. Unfortunately, shutting down machines and operations for preventive care is not always monetarily profitable to companies, so machines are allowed to wear down and become hazardous as long as they can still keep up production. Workers must insist on proper preventive maintenance as a regular part of the production schedule.

If all other measures fail to bring the noise in the plant down to an acceptable level, protective devices can be used as a last resort.

Table 11. EFFECTIVENESS OF VARIOUS
EAR-PROTECTIVE DEVICES

DEVICE	AMOUNT OF NOISE REDUCTION
Ear plugs	
cotton wool	8 dB
waxed cotton wool or glass-fiber wool	20 dB
individually molded acrylic	18 dB
individually molded silicone rubber	15–30 dB
mass-produced rubber plugs	18–25 dB
semi-insert silicone rubber	14 dB
Ear muffs	
heavy	40 dB
medium	35 dB
light	25 dB

Not all protective devices are equally effective. A measure such as putting cotton in the ears should not be used at all, because it is practically useless. How much the sound is diminished by the various protective devices is shown in Table 11.

The best protective device is the earmuff. Earplugs are also effective to some extent. It is important to get molded earplugs that are fitted to an individual's ear because, as was mentioned in the discussion on how the ear works, the canal that leads into the head from the outer ear is curved and irregular. It changes shape each time you chew, talk, or move your jaw in any way. Unless the earplugs are individually molded, they will slip out of place when the canal changes shape, and noise will leak in.

Unfortunately, protective devices are not really a solution to the problem but only a compromise. All protective devices, whether for the ears, mouth, or whatever, are uncomfortable to wear, especially if the plant is hot. Some workers develop fungus infections in their ears from constantly keeping them covered. A study has shown that workers find it harder to tell which direction a sound is coming from when they are wearing earmuffs.[5] If this is generally true, it could increase the number of industrial accidents. If this equipment must be worn, it will be necessary to rotate jobs and provide adequate rest periods—with, of course, no financial loss to the worker.

In summary, the best way to eliminate the hazards of noise is to incorporate noise-control technology into the industrial environment. In a society as technically advanced as the United States, this is not an unreasonable demand to make. There are always advertisements for consumers, telling of quiet air conditioners or quiet cars or quiet dishwashers. If the technology is available for making quiet products, then certainly it is available for quieting the

[5] G. R. C. Atherley and W. G. Noble, *British Journal of Industrial Medicine*, vol. 27, no. 3 (1970), p. 260.

production of these products. Workers must insist on these changes.

Prevention of health hazards arising from vibrations depends primarily on proper equipment design. An operator using a mechanized process rather than a vibrating tool is much better off. Equipment design should allow for padding and insulation to prevent the transmission of vibrations. Hand tools should be held in a normally comfortable position with a loose grip. Although gloves do little to diminish the transmission of vibration, they do help to keep the hands warm and prevent additional injury from cold. Rest periods, short work days, and job rotations decrease a worker's exposure and delay the onset of vibration effects.

Evaluating the Noise Problem—Surveying the Plant

There are two important aspects of evaluating an industrial noise problem. One is finding out how noisy the plant is. The other is measuring the extent of hearing loss in exposed workers, as well as trying to determine if other diseases are being caused by noise exposure.

Figure 14. A SOUND-LEVEL METER

In order to evelute the extent of noise pollution in a plant, it is necessary to use a sound-level meter to measure the sound. Figure 14 is a photograph of a typical sound-level meter that is approved by the government. It is a relatively simple instrument to use and calibrate, but it is quite costly—about $700 for the set. For measuring impact noise, where the changes in the sound-pressure levels are too rapid for the meter, special attachments are available. There are also inexpensive meters available from Radio Shack and RCA, among others, which can be used for approximate measurements of noise.

All that is necessary for taking a noise reading is to hold the microphone on the meter at a right angle to the path of the noise. This avoids having the sound reflected. It is also important for the person who is taking the measurements not to stand in the path of the sound. The decibel level, or sound-pressure level, is then read from the meter dial. With a little training, practice, and common sense, anybody can become adept at taking sound measurements.

The readings are usually taken on the slow-response "A" scale of the meter in order to conform with government standards. The "A" scale is weighted to filter out low-frequency noise and is thought to give a better indication of loudness, annoyance, and speech interference as felt by people exposed to the noise. All decibel standards set by the government refer to readings taken on the "A" scale and are often written as dB(A).

Any plant with a noise problem should be surveyed with a sound-level meter, and the decibel levels for various processes should be posted. A common trend throughout industry is for the company to survey a plant, find hazardous noise conditions, and then require the use of earmuffs, in order to meet federal noise-exposure standards. This is not the intent or purpose of the Occupational Safety and Health Act of 1970, which clearly states that protective devices are to be used *only* if engineering or process changes are *not possible.*

In addition, in any unionized shop, no changes in working conditions can be made without prior consultation with the union. The wearing of earmuffs should be considered a change in working conditions, and before the union agrees to their use, it should insist that attempts be made to eliminate or reduce the noise. The union should insist that an independent engineering consultants' firm be brought in, or perhaps the workers themselves may have specific suggestions for eliminating the noise. Before agreeing to the use of protective devices, the workers and their union should see that all other possibilities have been exhausted.

It should also be realized that even if all the workplaces in the country were to meet the legal standard, American workers would still be suffering a hearing loss. The 90 dB(A) American standard is 10 dB higher than that of most other industrial countries, which represents an increase of *10 times* the allowable noise. It is based on the attitude that although many people develop a 25 dB hearing loss at frequencies of less than 2,000 Hz, and unlimited loss at higher frequencies, this is not to be regarded as "impairment of hearing." Impairment is apparently defined as not being able to communicate. Therefore, even though a worker with a 25 dB hearing loss can only hear sounds that are more than 100 times as loud as before that loss developed, he or she does not qualify as having impaired hearing.

Figure 15 shows the percentage of a population of 6,835 workers that were found to suffer a 25 dB hearing loss as a result of age and exposure to noise in industry, as opposed to people not exposed to noise. It is important to note that the highest noise level listed is 92 db(A), very close to the legal limit. One must realize that many plants in this country do not even approach the legal standard. It is a sad commentary on any government that it should set a health standard that guarantees disability to its people, and that the people should have to struggle to have even that unsatisfactory standard instituted.

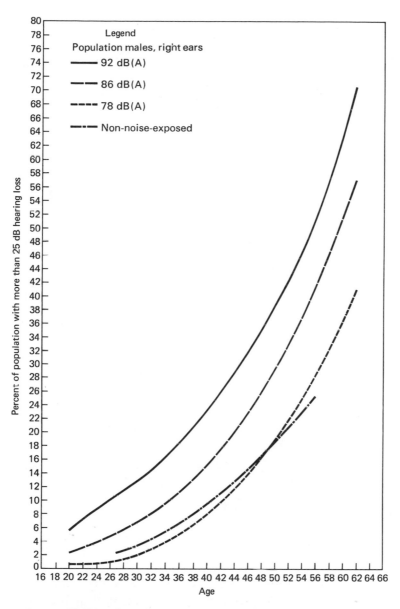

Figure 15. THE PERCENTAGE OF INDUSTRIAL
WORKERS SHOWING 25 DB HEARING LOSS.
6,835 WORKERS FROM INDUSTRIES WITH VARIOUS
NOISE LEVELS WERE EXAMINED AND COMPARED.

A Medical Program

In addition to surveying the plant and attempting to correct unsafe and unhealthy conditions, it is also important to determine whether the workers have developed hearing loss from excessive exposure to noise. This is done by audiometric testing. Figure 16 is a photograph of an audiometer, which is simply a device that produces tones of a specific frequency. The loudness of the tones can be varied. By screening a patient at each frequency and determining how loud the tone has to be before the listener can hear it, the hearing ability of the listener can be determined. The particular instrument pictured in Figure 16 can also be used for doing tests other than the one just described, which is called an "air conduction test." These other tests will not be discussed here. Administration of an audiometric test is quite simple; in fact, in many plants it is the company nurse that gives hearing tests.

Figure 16. AN AUDIOMETER

Because these tests can be administered by any intelligent person who has been given the proper training, it becomes practical to suggest that trained workers themselves, through their unions, begin to give hearing tests and keep records of the results. In fact, the Oil, Chemical and Atomic Workers International Union (OCAW), which has been pioneering many programs in health and safety, is carrying out such a program now. It is their aim to have each district in the union purchase an audiometer, sound-level meter, and calibrator, and designate workers who will receive intensive training in administering audiometric tests. Each person in the union will then have his or her hearing tested.

Preliminary experiments with this project have worked well, and initial surveys have revealed similar hearing losses in people in the same craft or job. It is also the aim of OCAW to obtain a hearing test of every person who becomes a member of the union, so that baseline statistics will be available on all OCAW members.

Such a survey can have several important results. First, when sufficient data are obtained on large numbers of people, it becomes possible to prove that particular working conditions are leading to specific diseases. This is fully discussed in Chapter Ten. If the OCAW survey finds that a large percentage of pipefitters develop a similar hearing loss, the people involved can then begin to file and win compensation cases on a company-wide or even nation-wide basis. They can go to the bargaining table with scientific evidence as justification for explicit demands for changes in their working conditions. They can use this evidence to increase concern among some of their fellow workers who may not be as anxious as they to fight for change. This combination of compensation cases, negotiation demands, and the determination of the people will begin to force companies to take noise abatement in the plants seriously.

Another important aspect of this survey is that while people are being tested for hearing loss they can also be surveyed for incidence of heart attacks, gastric disorders, circulatory problems, and other effects of noise unrecognized at present by American official standards. Such a survey should be able to correlate hearing loss with the sickness rate and to establish that people exposed to noise do suffer a higher incidence of the diseases previously discussed.

All the data taken across the country are fed into a central data bank so that results can be compiled on a nation-wide basis. A complete description of the OCAW program is available by writing to OCAW, 1636 Champa Street, P.O. Box 2812, Denver, Colorado 80201. There is no reason why other unions across the country cannot carry out similar programs. Workers should insist that their unions contact the appropriate professional advisors or the OCAW and set up similar programs for them.

In summary, the effects of noise are serious and far-reaching. Workers will suffer hearing loss and a higher incidence of many other illnesses from noise exposure, especially in combination with the other environmental hazards present in the workplace. There are many ways to reduce or eliminate noise exposure. By using techniques and programs such as those described in this chapter, workers can pressure their employers to improve their working conditions so that they have a truly safe and healthful place in which to work.

CHAPTER FIVE

THE EFFECTS OF HEAT AND COLD

A VERY COMMON COMPLAINT from workers is that their workplace is either too hot or too cold. These complaints are verified by the fact that each summer, workers in many plants across the country walk off the job because they simply can't stand the heat. In 1959, the United States Appeals Court upheld a National Labor Relations Board ruling that gave workers the right to refuse to work in conditions that were so cold as to be uncomfortable. But in spite of these job actions and court decisions, workplaces with climate control are very rare. Workers are still being subjected to temperature conditions that take a physical and emotional toll of them.

The body is comfortable in a very narrow range of temperatures. The average comfort temperature is 73° F. at 45 percent humidity. Since the body generates different amounts of heat for different tasks, comfortable temperatures for varying levels of activity have been estimated. Because body activity generates heat, lower temperatures are preferred for more active work. Humidity, or the amount of moisture in the air, and air movement also play a role in body comfort, because they affect the rate at which the body loses heat.

Table 12 gives some estimated comfortable temperature

ranges for different levels of work activities. Outside of these temperature ranges, a person notices that he or she is too hot or too cold. At temperature levels far outside of those in the table, there are definite effects on the ability to work. Industrial management has been interested in these effects. At uncomfortable temperatures, especially in heat, there are more accidents and more mistakes. This means lower work quality as well as a higher absence and lateness rate. Workers may even be so discontented that they hold work actions and strikes. All these things decrease productivity, which greatly disturbs management.

Table 12. ESTIMATED IDEAL TEMPERATURES
FOR VARIOUS TYPES OF WORK

Relative humidity: 30–60%
Air velocity: 21 ft./min.

WORK	TEMPERATURE
Clerical personnel	65–72° F.
Light but active industrial work	60–67° F.
Heavier industrial work	55–65° F.

Monotonous or repetitive work in each category is best done at temperatures a few degrees cooler than those listed.

For workers, however, the relationship between temperature and productivity is not the most important thing. An accident that results from mental fatigue, irritability, and decreased manual dexterity due to uncomfortable temperatures means far more than "bad statistics." It may have serious health consequences. In addition, if work makes a worker feel tired and irritable, it also affects his or her ability to enjoy spare time with family and friends. Productivity is not the criterion that people should use in judging working conditions. The standard should be the workers' mental and physical well-being.

Heat Stress

The hottest work is encountered in mining, steel and glass manufacture, agriculture, and roadbuilding. There is often severe environmental heat stress in these occupations. However, these are not the only industries where heat stress can occur. The temperature in factories in temperate climates during the summer months, and in subtropical climates all year round, may also subject workers to environmental heat stress.

The human body produces its own heat from chemical energy, and either loses heat to the environment or gains heat from the environment. It operates under the same physical laws that govern heat exchange from any warm object. Heat is lost from the body by three principal means: convection, radiation, and evaporation.

Convection, the least important for the body, means simply the flow of heat away from hot areas toward areas of lower temperature. Convection currents transfer heat through air movement. Air is warmed by a hot object. The warm air expands, becomes lighter and rises, and is replaced by cooler air. The ability of the body to lose heat through the mechanism of convection is increased by air movement, ventilation, or wind.

Radiated heat is transferred through invisible waves. These waves are part of the spectrum of waves that make up light, sound, and X-rays. The sun gives radiant heat to the earth. As a rule, a surface or substance that reflects light also reflects radiant heat. A dull black surface, which reflects no light, absorbs radiant heat. Human skin has the characteristics of a dull black surface. When it is heated, it emits a large amount of radiant energy, but this radiant heat is not emitted as visible (colored) light. Therefore, the pigmentation of the skin, white, black, or brown, makes no difference in the skin's ability to radiate heat.

Evaporation is the body's most efficient method of heat loss. When water is heated, it uses the heat as energy and is transformed into water vapor. As the water in sweat turns to water vapor, it takes the heat away from the skin, leaving the skin cooler.

In spite of warmer temperatures in the air around it, the body must maintain its normal internal temperature, 98.6° F. During the first few days of exposure to heat, its temperature-regulating mechanisms are very inefficient. Its first reaction to heat is a rise in its internal temperature, or fever. This leads to an increase in the heart and pulse rate, and may cause a strain on the heart in older people. Complete failure of the body to regulate its temperature may lead to heat stroke. A person with heat stress is uncomfortable and distressed. A worker under these conditions is less alert, more careless, and less efficient. Consequently, he or she will make more mistakes and have more accidents.

If the exposure to heat continues, the body makes certain adjustments that enable it to maintain its normal internal temperature—a process called *acclimatization*. This process takes four to six days, depending specifically on how much heat exposure there is and what activity is being done. Acclimatization occurs best in young people. Older people never acclimatize to heat completely. When the exposure is discontinued, acclimatization is lost. If a worker leaves the environment for one week, he or she may lose between one-quarter and two-thirds of the acclimatization. A three-week absence from exposure usually results in the loss of all acclimatization.

The body regulates its internal temperature through two mechanisms: circulatory changes and changes in the volume and composition of the sweat. In acclimatization, the body tries to make these mechanisms more efficient. Its efforts to get rid of heat begin with dilation of the small blood vessels

of the skin. This allows the heat of the blood to be transferred through the skin back to the environment. Dilation of the blood vessels increases the work load for the
heart, which must pump blood through a larger circulation
area. This increased work load is partially compensated for
by constriction of the blood vessels to the liver, stomach,
and intestines, which allows less blood to flow. Sometimes
the liver is damaged when active work in a hot environment
causes such intense blood-vessel constriction that the organ
does not receive enough blood and oxygen. When the body
is acclimatized, these changes in circulation are well balanced and the work load on the heart remains near normal.

The body loses heat mostly by evaporation of sweat. A
rise in skin temperature stimulates sweating even before
there is a rise in the over-all body temperature. An unacclimatized person has inefficient sweat production. Not all
of the 2.5 million sweat glands are "turned on," and some of
them secrete no sweat. This means there is not enough sweat
to maintain a normal body temperature. Those sweat glands
that are functioning produce sweat with a high salt concentration. Loss of excess salt from the body can lead to heat
exhaustion. The sweat glands may even produce too much
sweat in certain areas of the body, saturating the skin and
clothes to such an extent that no evaporation takes place
and no body heat is lost. A person acclimatized to specific
temperature conditions and activity levels sweats at a rate
that is just balanced with the rate of evaporation. The sweat
has a low salt concentration. An acclimatized person also
has a greater maximum sweat production than one who is
not acclimatized.

Acute Reactions to Heat Stress

There are several different heat reactions, depending on
the conditions of exposure, the degree of activity, and the
way the individual's own body responds.

Heat stroke is characterized by a sharp rise in body tem-

perature, confusion, angry behavior, delirium, and even convulsions. There is no sweating and the skin is warm and dry.
Unacclimatized people who are moderately active in a hot
environment can get heat stroke. It is more likely to happen
to people who are older, who have pre-existing heart or
circulatory disease, who are overweight, or who begin with
slight dehydration—as does a person who has had a lot of
alcohol to drink. A person who gets prickly heat in hot
weather has poorly functioning sweat glands and is also
likely to develop heat stroke. It can be fatal. The victim
should be sponged down or given cold baths as rapidly as
possible to lower the body temperature. Heat stroke should
be treated in a hospital.

Heat exhaustion, heat fainting, or "heat syncope," is the
mildest form of heat exhaustion. It can happen without any
elevation of the body temperature and with only moderate
heat exposure. Heat syncope is the type of reaction that
occurs when the New York City subways stop underground
during rush hours in a midsummer electrical blackout. People who are not used to heat, and who may be physically
unfit as well, develop heat fainting. The victim feels tired,
giddy, and nauseated and may complain of feeling chilly.
He or she sighs and yawns and may have rapid, shallow
breathing, a weak, slow pulse, and moist, clammy, cold, pale,
or even bluish skin. The blood pressure is low. In this condition, the blood vessels *all over the body* are dilated, and
there is not enough blood to circulate through these tremendously enlarged blood vessels. A person will recover rapidly
when removed from the heat and allowed to rest with the
head lowered.

Heat shock is a common form of heat exhaustion that
occurs in healthy persons working in hot climates to which
they are partially or totally unaccustomed, or unacclimatized. This reaction takes place when the body loses an
excessive amount of fluid or salt because of inefficient sweating and inadequate fluid replacement. As a result of this

loss, there is not enough fluid in the body to maintain circulation to all the organs. This is the medical condition known as shock. An unacclimatized person working in a hot environment will usually lose more salt than water in the sweat. Failure to take salt pills or drink salty beverages will result in heat cramps, weakness, nausea, headache, fatigue, and dizziness. He or she will become irritable and suffer from considerable muscular weakness. This condition is treated by removing the victim to a cool environment and replacing the fluid and salt losses.

Heat fatigue is less well defined by scientists than heat exhaustion or heat stroke. It is probably also the most common of the three disorders. It is the emotional reaction to a hot environment. The lassitude, irritability, and easy fatigability that occur on hot days are familiar sensations. A person with heat fatigue does not work as well, produces less, makes more mistakes, and has more accidents. The feelings get worse if the fatigue is not relieved by rest. Heat fatigue may affect a person's personal relations on the job and at home as well. Thus in addition to paying a physical price for working in a hot environment, a worker will pay an emotional price.

Heat and Chronic Illness

No careful scientific studies have been made of the long-term effects of heat exposure or the aftereffects of heat disorders. A survey in 1920 of workers in the English shoe and boot industry found symptoms of nervous depression, lethargy, muscular inefficiency, disturbance of the digestive functions, heart strain, and susceptibility to infectious disease. The authors blamed these disorders on inadequate ventilation and a warm environment. However, other factors may have been involved as well. Scientists have raised the question whether workers breathing hot dry air develop lung disease. This is under investigation, and no conclusions have been reached. It is probable that all stresses to which the body is subject add up. Heat stress, noise, and chemical and other physical insults, as well as emotional stress, each use up part of the total energy reserves of the body.

Health and Safety in Hot Environments

Certain types of industrial operations as now carried out require exposure to extremes of heat. These extremely hot environments often have open flames and other sources of radiant heat. The temperatures may be so high that even a brief exposure without protection can be immediately fatal. Workers exposed to these environments must wear protective clothing that is properly designed for the particular type of activity and the particular conditions of heat and humidity encountered. The design must allow for the type of movement needed for the job and for the amount of heat the body generates in doing it. In addition to the special protection of clothing, dangerously hot operations should have all the usual safety precautions, such as adequate guard rails and non-slippery floors. General safety and health measures for hot workplaces should also include adequate training and acclimatization periods before beginning work involving heat stress, and the provision of rest periods during and between periods of hot work. Cool rooms with cool beverages and salt should be available for these rest periods.

Cold

Workers don't need to be told what a cold climate is—the sensation of cold is an accurate indication of coldness and of danger from the cold. The body can tolerate very little exposure to cold without the protection of appropriate clothing. The body has two mechanisms for maintaining its internal temperature in the cold. First, the blood vessels supplying the skin, hands, and feet constrict so that less blood flows to the surface of the body and less heat is lost. The hands and feet become numb and cold, and may even be injured if they become too cold. In addition, someone with cold, stiff, numb, or painful hands cannot perform manual tasks with any dexterity or skill. The second protective mechanism is shivering. The rapid muscle contractions of shivering generate heat and help to maintain the inner temperature of the body. Shivering also makes it difficult to do any work requiring skill.

The body as a whole does not acclimatize to cold as it does to heat. This means that it does not become more efficient at withstanding cold temperatures. However, workers who are continuously exposed to cold do develop local acclimatization in their hands. The blood flow to the hands increases, so that the hands stay warmer and can therefore function better and have more dexterity.

In addition, people in cold climates or workplaces get used to the cold, a process called *habituation*. This means they are less uncomfortable in the climate to which they are accustomed and therefore can work somewhat more effectively.

Acute Reactions to Cold

Frostbite is the result of local tissue damage, usually to the skin and muscles of the hands and feet. The hands and feet are affected because the body keeps them cold in order to save heat for the rest of the body. They actually freeze, with crystals of ice forming in the tissues and damaging them. The small blood vessels are especially likely to be damaged because they are blocked with tissue debris, which causes further injury to the tissues they are now unable to supply with blood. Frostbite is often irreversible, and amputation is sometimes the only remedy. If the foot or hand is not too severely injured it may heal, but not without chronic symptoms that are present in any climate. The limb may sweat excessively, or be painful, numb, and have abnormal color. There may be joint pains for years after the injury. All these symptoms are worse at cold temperatures.

Immersion foot is a condition that results when the foot is exposed to moisture for long periods of time but not frozen. It has usually occurred in shipwreck survivors or soldiers. The injury is primarily due to spasm of the blood vessels, which stops circulation to the foot. In contrast to frostbite, the skin and blood vessels are not injured. The spasm injures the muscles and nerves. This injury, though not so severe as the injury in frostbite, may cause extensive damage to the muscles of the foot, and gangrene may develop. When the foot heals, it is often more sensitive to cold exposure and pain, especially when walking or standing. This sensitivity can last for many years.

General body reactions to cold. When a person is exposed to cold air for long periods, or is briefly exposed to cold water, the body and brain temperature may be lowered. This can lead first to strange behavior, then to loss of consciousness, and then to coma. A body temperature of 64° F. stops the heartbeat.

There are very few studies of the health effects of exposure to cold over many years. Some research has shown that exposure to temperatures below freezing leads to chronic lung disease and sinus irritation. Other possible effects of cold work may be arthritis and an increase in virus infections. But as yet there is no scientific proof of these effects because adequate research has not been done.

Protection from Cold

There are two main ways of protecting workers from cold. One is by work design, and the other is by protective clothing. Work should be designed to minimize cold exposure. There should be convenient areas of warmth, or at least protection from the cold and weather. Where possible, work should be done elsewhere and then moved to the cold location. For instance, parts should be prefabricated and maintenance jobs should be performed in warm areas. There should be rest periods in warm places, with warm beverages supplied.

Cold-weather clothing is designed to insulate the body and provide a barrier to the cold. Because the body also creates its own heat, especially when muscular work is being done, cold-weather clothing must also be designed to allow body heat to escape. There should be a balance between heat production and cold protection. Removing layers of clothing or opening vents in the clothing are simple ways of obtaining balance. The accumulation of sweat in the clothing leads to overheating. The sweat later evaporates, chilling the worker once he or she has stopped working.

It should be obvious that consideration of the temperature of the workplace is very important. Because of the many physical and mental effects of hot and cold working environments, climate control should be the subject of collective-bargaining sessions. Workers should not be asked to give up their well-being simply because industry finds it "too expensive" to provide decent working conditions. The question that should be asked in this case, as well as in all the other situations discussed in this book, is, What is worth more expense than the health, well-being, and dignity of the workers of America?

CHAPTER SIX

LIGHT, X-RAYS, AND OTHER RADIATION

SOME TECHNICAL TERMS

curie: unit for the rate at which radioactive materials throw off their radioactivity

electromagnetic radiation: term that includes all kinds of ionizing and non-ionizing radiation

footcandles: unit used to measure illumination

half-life: amount of time needed for a radioactive particle to lose half its radioactivity

ion: an electrically charged particle

ionizing radiation: high-energy radiation, such as X-rays, that can disrupt and destroy living matter

IR: abbreviation for *infrared radiation*

non-ionizing radiation: radiation, such as light, that does not possess enough energy to disrupt living matter and form ions

photosensitization: process of making the skin more susceptible to the harmful effects of sunlight

rad: standard unit for measuring the dose of radioactivity an exposed person receives

RF: abbreviation for *radiofrequency*

rem: measure of the effective dose of a particular radioactive material

roentgen: unit that measures the dose of radioactivity an exposed person receives; this unit has been superseded by the *rad*

UV: abbreviation for *ultraviolet radiation*

IN OTHER CHAPTERS of this book we discuss various chemical hazards, dusts, gases, and the effects of heat and cold. Most of these hazards are fairly readily detected—you know whether you are hot or cold, whether you are having trouble breathing or are working with dusts. In this chapter we treat a different type of hazard—radiation—which has very serious effects yet cannot readily be sensed.

Radiation is generally divided into two categories: ionizing and non-ionizing radiation. Ionizing radiation is the region of radiation (see Figure 17) that corresponds to X-rays, alpha, beta, and gamma rays, and neutrons. The energy of these rays is so great that when they travel through space and interact with the atoms that make up all matter, they actually break the atoms up, remove electrons from them, and cause them to develop an electric charge. A charged particle is called an *ion,* and hence this type of radiation is called *ionizing radiation,* because it turns atoms into ions. When ionizing radiation, such as X-rays or the radiation from radioactive material, interacts with the body, it ionizes the atoms of the body and damages them, sometimes seriously or even fatally. Thus ionizing radiation is a very serious hazard.

Ionizing radiation is widely used in industry, but it presents a problem to citizens generally as well. In 1970 there were 50 nuclear power reactors either built or in the planning stage. This number will increase tremendously. Each of these plants is a potential ionizing-radiation polluter, especially if not properly maintained. In addition, more than half the population is given medical or dental X-rays at least once a year. More than three-quarters of a million people are given radioactive material for medical reasons. The hazards of ionizing radiation are therefore of interest to everyone.

The other types of radiation in Figure 17, those waves with less energy than X-rays, do not possess enough energy to ionize atoms, and are therefore classified as *non-ionizing*

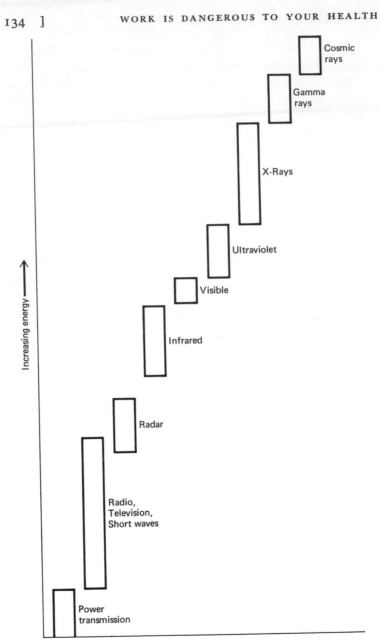

Figure 17. THE ENERGY OF DIFFERENT TYPES
OF RADIATION

radiation. Non-ionizing radiation can cause burns, however, and damage the eyes, skin, and other organs of the body. It is also called *electromagnetic radiation,* because it generates both an electric and a magnetic field in space. That is, if you had a magnet or an electric-field detector, you would be able to detect electromagnetic radiation by the presence of the electric and magnetic fields that it generates.

Before the effects of radiation can be discussed, it is important to understand just what radiation is. Almost everybody has seen sunlight separate into the different colors of the rainbow. These colors—red, orange, yellow, green, blue, indigo, and violet—make up *visible light,* or light that can be seen. They are all forms of radiation. In addition to visible light, there are kinds of radiation that the human eye cannot see. If you were to move a thermometer slowly along the rainbow of colors from violet toward red, the thermometer would continuously rise in temperature. It would even continue to rise once you passed the red end of the rainbow. This region beyond the red is the *infrared* region. Although infrared (IR) cannot be seen, it exists and has energy that is given off as heat and is "seen" by the thermometer's rise in temperature.

Similarly, if you went beyond the violet end of the rainbow, you would come to the *ultraviolet* region of radiation, also invisible but still a form of energy. Other forms of radiation are radio waves, microwaves, and so on. All these are types of radiation that travel through space, but they differ in the amount of energy they possess, as is shown in Figure 17.

Non-ionizing Radiation

The parts of the eye are shown in Figure 18. The eye is the most susceptible organ of the body to non-ionizing radiation. One probable reason for this susceptibility is that the lens of the eye acts like a camera, focusing and intensifying the effects of the rays. All workers exposed to non-ionizing radiation should have frequent eye examinations, and the union should keep the results of these examinations for the entire exposed group. In addition, infrared (IR) and radio-frequency (RF) waves, as well as microwaves, have thermal effects on the body: that is, they heat the body up. The IR and RF waves heat the skin, while microwaves go beyond the skin and penetrate deep into the body. Finally, ultra-violet light (UV) and some visible light burn and redden the skin. Table 13 lists some common industrial uses of these sources of energy and some of the major effects they have on the body. Each type will be discussed separately.

Ultraviolet radiation (UV) is the portion of sunlight that causes sunburn. The "black light" lamps used in night-clubs are sources of intense UV. UV causes reddening of the skin, which experts have found a "convenient" measure of the amount of exposure—convenient for the scientists, but hardly so for the workers involved.

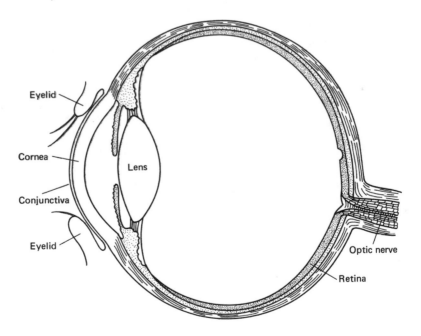

Figure 18. THE EYE

Table 13. Sources and Effects of Non-Ionizing Radiation

RADIATION	SOME INDUSTRIAL EXPOSURES	MAJOR EFFECTS
Ultraviolet (UV)	Sunlight, electric-arc welding, germicidal lamps, "black light" used in blueprinting, laundry-mark identification, dial illumination.	Irritates and damages eye tissue; can cause painful sunburn and possibly skin cancer.
Visible—lasers	Used in construction industry as reference lines, in medicine for surgery, in communications, in holography; may be used in drilling or wherever a high-energy beam is useful.	Extremely hazardous to the eye because it concentrates light intensely on the retina.
Infrared (IR)	IR is given off by all heated bodies. Welders, steelworkers, glassblowers, etc., are exposed. Also used for drying and baking paint, varnishes, and enamels.	Can cause damage to parts of the eye. Workers may develop a condition called "heat cataract."
Microwaves	Military, radio navigation, radar communications, food ovens, certain drying processes, medical diathermy.	Eyes and testicles most susceptible; genetic effects and effects of long-term low level exposure unknown. Microwave generators may also give off X-rays.
Radiofrequency waves (RF)	RF heating equipment used for hardening metals, soldering, and brazing. RF can be used in woodworking for bonding, laminating, and gluing. Also used for sterilizing containers, thermo sealing, and curing plastics.	Improper operation can lead to electrical shock and burns. If operator has wet skin, he or she can be electrocuted.

The symptoms of overexposure to ultraviolet radiation are the same as those of severe sunburn: reddening of the skin, blistering, and pain. Workers who work outdoors, such as construction workers, roofers, farmers, and pipeline workers, all suffer from overexposure to UV from too much sunlight. Roofers, who are also exposed to chemicals such as coal tars or cresols, have an additional hazard to contend with: these chemicals are *photosensitizing* agents, which means that they make the skin especially sensitive to the effects of UV. Such workers have a high rate of skin cancer. It is also likely that long-term exposure to UV alone can increase a person's chance of developing skin cancer.

Ultraviolet radiation has serious effects on the eyes. A person who is being exposed will not realize it is affecting his or her eyes. Then, six to eight hours after exposure has ended, the victim's eyes will feel as if there were sand in them. The outer membranes, the *conjunctiva* and *cornea,* will show small areas of damage, or *lesions,* and the eye will be irritated. This condition is very painful. Welders who do not wear the appropriate shade lens in their eye-protective equipment, or who do not wear any eye protection, damage their eyes in this way because UV is given off by many welding processes. Long-term exposure to UV can lead to partial loss of vision.

There is no legal limit for exposure to ultraviolet radiation, although several standards have been "suggested." These standards are based on the reddening of the skin in response to UV. A good rule of thumb is that if there is any reddening of the skin, there is too much exposure and the process that is giving off the radiation should be shielded or changed. Ordinary window glass will stop most UV, but where intense radiation is generated, as in welding, shaded glass may be needed. If there is *no* way to shield the operation, proper eye protection and covering for the body must be supplied, and the operation should be isolated from other parts of the plant as much as possible.

An additional hazard that arises from ultraviolet radiation is the reaction it has with chlorinated hydrocarbons, such as perchloroethylene and trichloroethylene and other chemicals commonly used as degreasers. When these chemicals are exposed to UV, they form phosgene, a highly toxic gas used as a nerve gas during World War I. This is another important reason for shielding or isolating operations that produce UV, such as welding.

Infrared radiation (IR) is the heat waves given off by all bodies that radiate heat. Whenever you feel heat coming at you from some source, you are actually feeling infrared rays. Infrared rays are mostly absorbed by the skin, and they may burn the skin in a manner very much like sunburn.

Like UV, infrared radiation has seriously damaging effects on the eyes. If the IR source is really intense, you can sense it burning the eyes. However, low doses of IR over the years may not be felt at all and yet may cause serious permanent damage to the parts of the eye. IR can burn the lens and cause it to become opaque. This condition is called *cataract,* and if it comes from IR or heat waves, it is called *heat cataract.* Many persons who work near industrial ovens in steel operations, or as glassblowers, suffer from heat cataracts.

There are many ways of protecting industrial workers against exposure to infrared radiation. Ovens and other sources can be shielded with shiny materials such as polished aluminum, which will reflect the heat back to its source. Engineers in Poland have developed an ingenious water screen for iron and steel mills; IR is absorbed by water, and therefore water acts as a barrier against it. These screens are both mobile for emergency situations and stationary for permanent processes. Such protective techniques should be developed in this country as well.

Regular clothing protects the skin against infrared radiation. There are also goggles to protect the eyes against IR. Because of the high temperatures, workers may justifiably be reluctant to wear protective clothing and devices. Obviously, engineering methods of shielding the IR source and of air-conditioning as well are the preferred ways of handling an IR problem.

Microwaves. There is a great deal of exposure to microwaves among people in the armed forces because they are the form of radiation used for radar. With the advent of microwave ovens, more and more people are now being exposed to this kind of radiation at home as well as at work. Microwaves are also used as a medical treatment in diathermy machines.

The property of microwaves that allows them to be used for medical treatment is their ability to heat the body. This heating ability is a hazard to overexposed and unprotected workers. Microwaves penetrate deeply into the body and cause its temperature to rise. If the intensity of the microwaves is great enough, this can lead to permanent damage to the affected areas. If the body temperature rises high enough, a person can go into a coma and die, just as would happen from running an extremely high fever with an infection.

As discussed in Chapter Five, even a moderate rise in body temperature places stress on the body. It causes the blood vessels to dilate in order to try to dissipate the heat. This means that the heart must work harder to pump blood through the dilated vessels. The rate of breathing increases because the body needs a great oxygen supply, and the body in general is under a great deal of stress. Long-term exposure to stress can lead to many physical and emotional disorders. People who are predisposed to heart problems can suffer a heart attack from such stress.

In addition to its effects on the whole body, microwave radiation particularly affects the eyes and the testicles. The lens of the eye, which serves as its camera, has no blood supply of its own and hence has no mechanism for dissipating heat. It is therefore particularly seriously affected by microwaves. There are many reported cases of formation of cataracts due to exposure to microwaves. The testicles are susceptible to damage because they must remain at a temperature that is lower than the rest of the body in order to function properly. This is why they are external to the body. Microwave radiation will increase the temperature of the testicles, causing the cell lining to degenerate. It is thought that these effects are reversible, but that is not certain.

It is feared that microwaves may cause genetic damage, that is, that children fathered by men exposed to this hazard may have abnormalities. One study has shown an increased incident of mongoloid children among fathers exposed to microwaves, though the implications of this study are not certain. In view of this, the utmost caution should be exercised in the use of microwaves, as it is possible that this hazard may have terrible effects, not only on the men who are forced to work under these conditions but also on their innocent children.

Foreign studies also indicate that exposed workers have developed symptoms of heart, thyroid, and nervous system disease, as well as eye damage. This could follow from the generalized stress reaction discussed previously. However, American studies do not come up with the same findings. It would be interesting to find out why the discrepancy in research findings exists. Until a satisfactory answer to this question is found, the only logical conclusion is that microwaves must be treated with a great deal of respect and caution until the genetic effects and the effects of long-term, low-level exposure are understood.

Microwave sources can be effectively shielded by thin metal screens such as copper mesh or thin steel plates. Any source should be periodically checked with a microwave detector. Since the doors on microwave ovens tend to become loose, ovens must be checked for leakage. Microwave generators also produce X-rays, which are a serious hazard that will be discussed in the second part of this chapter.

The heating effects of microwaves present a further safety hazard if flammable chemicals are present. The microwaves can heat such chemicals past their flashpoint and cause a fire or explosion.

The legal limit for exposure to microwaves had been set at 10 milliwatts per square centimeter. The United States Army and Air Force and the C-95 Committee of the American Standards Institute changed this to an average exposure over six months and permitted exposure of 100 milliwatts per square centimeter for 36 seconds, which gives the same average. The standard was adjusted to meet the increased power output of newer microwave systems. It is not at all clear that this is a safe limit, and it may represent another instance where the safeguarding of workers has been sacrificed to the necessities of the military, profit, and "progress."

Radiofrequency waves are used as heating sources in such processes as hardening metals, sterilizing containers, and molding plastics. Exposure to these waves does not seem to have any medical effects. However, there are serious safety hazards involved in radiofrequency power, which will burn a person who comes into contact with it. Electrical burns are extremely painful and difficult to heal. Safety precautions, proper grounding, and discharging of capacitors are important. There must be adequate training of all workers using the generators and frequent maintenance and safety checks on the equipment itself. The details of the maintenance and safety precautions should be posted and thoroughly explained to all workers involved with the equipment.

Lasers are a recent invention that are finding more and more uses in industry and medicine. The word *laser* stands for *light amplification by stimulated emission of radiation*. This means that laser light is regular visible light that has been made into a very coherent, concentrated beam, all of one color. At the beginning of the chapter we discussed how light is made up of a rainbow of colors, each with different energy. Laser light, however, is all one color, hence of the same energy and more concentrated than any other source. If you look directly at the beam, or at its reflections off a shiny surface, serious, permanent eye damage will result.

Laser beams are used in the construction industry for reference lines, since they travel in straight lines. They are used in welding, cutting, and dulling. Surgeons have used lasers for delicate operations. There are many other uses for lasers, and undoubtedly more and more applications will be developed.

If the laser beam is aimed at someone's eye, the energy of the beam is focused by the lens of the eye onto the retina. The rear portion of the eye is essential for sight. The lens thus further intensifies an already intense beam, and tissues of the retina are destroyed. This can lead to blindness. As even low-powered lasers can cause this damage, all necessary precautions should be taken. The beam should never be aligned by eye, and it should not be focused into a mirror or on any highly reflective surface. When the beam is aligned, it should be focused onto a dull, non-reflecting object. Eye goggles should be worn that are designed for protection from the particular kind of laser being used, and the laser ought to be permanently mounted so that it cannot accidentally be swung around. Enclosure of the beam and remote-control operation are probably also wise precautions.

Visible light has no directly damaging effects on the body, but physical and emotional problems can arise when there is either too much or too little light. Some sources of intense light are the sun, arc welding, and highly glowing bodies. Laser beams are of course an extremely hazardous form of intense visible light. There is obviously too little light when illumination indoors is inadequate. The key index of whether the lighting in a workplace is proper or not is whether the workers have visual comfort working in the area.

Inadequate lighting can be an important cause of indus-

Table 14. RECOMMENDED LIGHTING LEVELS

AREA OR OPERATION	FOOT-CANDLES *	AREA OR OPERATION	FOOT-CANDLES *
Assembly		Materials handling	
Rough easy seeing	30	Wrapping, packing, labeling	50
Rough difficult seeing	50	Picking stock, classifying	30
Medium	100	Loading, trucking	20
Fine	500†	Inside truck bodies and	
Extra fine	1000†	freightcars	10
Building (Construction)		Offices	
General construction	10	Cartography, designing, detailed drafting	200
Excavation work	2	Accounting, auditing, tabulating, bookkeeping, business machine operation, reading poor reproductions, rough layout drafting	150
Building exteriors			
Entrances			
Active (pedestrian and/or conveyance)	5		
Inactive (normally locked, infrequently used)	1	Regular office work, reading good reproductions, reading or transcribing handwriting in hard pencil or on poor paper, active filing, index references, mail sorting	100
Vital locations or structures	5		
Building surrounds	1		
Garages—automobile and truck			
Service garages			
Repairs	100		
Active traffic areas	20		
Parking garages		Reading or transcribing handwriting in ink or medium pencil on good quality paper, intermittent filing	70
Entrance	50		
Traffic lanes	10		
Storage	5		
Inspection		Reading high-contrast or well-printed material, tasks and areas not involving critical or prolonged seeing such as conferring, interviewing, inactive files, washrooms	30
Ordinary	50		
Difficult	100		
Highly difficult	200		
Very difficult	500†		
Most difficult	1000†	Corridors, elevators, escalators, stairways	20
Loading and unloading platforms	20	Storage rooms or warehouses	
Freight car interiors	10	Inactive	5
Locker rooms	20	Active	
Machine shops		Rough bulky	10
Rough bench and machine work	50	Medium	20
		Fine	50
Medium bench and machine work, ordinary automatic machines, rough grinding, medium buffing and polishing	100		
Fine bench and machine work, fine automatic machines, medium grinding, fine buffing and polishing	500†		
Extra-fine bench and machine work, grinding, fine work	1000†		

trial accidents. Glare, either from the light source or reflected from the work, dark shadows, and eyestrain can seriously increase the accident rate. In addition, if workers have to go from bright areas into dark ones they may have accidents because their eyes don't have time to adjust properly.

It is important to realize that working with inadequate light is a strain on the eyes and is emotionally taxing as well. Workers should not have to work in poorly lighted, depressing surroundings, because this takes an emotional toll from them. Of course, if a worker cannot see well enough to do his or her work properly, that worker will be less productive. This is a secondary consideration, however, to safety and emotional well-being.

There are legal standards for the amount of light required for various operations. Illumination is measured in footcandles, and any light meter, such as one purchased in a camera shop, will measure the amount of footcandles. The recommended standards are given in Table 14.

Ionizing Radiation

X-rays, alpha, beta, and **gamma rays,** and **neutrons** are all forms of ionizing radiation. Radioactive materials emit this radiation, but different types of materials emit radiation of different energy. Ionizing radiation is a very serious hazard because it injures the cells of the body and can lead to many illnesses, including cancer and genetic damage. It can even be immediately fatal in large doses. Moreover, peo-

Notes for Table 14

* Minimum on the task at any time.
† Can be obtained with a combination of general lighting plus specialized supplementary lighting. Care should be taken to keep within the recommended brightness ratios. These seeing tasks generally involve the discrimination of fine detail for long periods of time and under conditions of poor contrast. The design and installation of the combination system must not only provide a sufficient amount of light, but also the proper direction of light, diffusion, color and eye protection. As far as possible it should eliminate direct and reflected glare as well as objectionable shadows.

ple cannot "feel" radiation, so that they may be unaware of being exposed. Because of the serious effects of ionizing radiation, the utmost care and maintenance of equipment must be practiced. Special training and education also are essential for personnel who work in industries using X-rays or radioactive material.

Radioactive materials and X-rays are used quite extensively in industry and medicine. They are used as tracers in pipes and tanks to check the flow and level of liquids. They are used in quality control to test the thickness of fluids and the texture of mixes. Pipes are X-rayed to check on welds and for thin spots. Uranium is a source of atomic power, and many workers are exposed to radioactivity in the mining, processing, shipping, and use of uranium and other radioactive materials for atomic energy. Radioactive materials are widely used in medicine, so that workers involved in the manufacture of these medicines and in the administering of them to patients are also exposed to radiation. Finally, many industrial processes such as welding, microwave, and radiofrequency generators also produce X-rays.

In addition to industrial exposure, everybody is exposed to X-radiation at one time or another for medical purposes. At one time X-ray pictures were taken quite freely with little protection for the patient or technician. Children used to have their feet X-rayed for shoes, and pregnant women routinely underwent X-ray examinations that were not really essential. Now with the growing realization that every X-ray has a physical price, good doctors require them only when necessary. It is important to be sure that the X-ray equipment being used is reliable and has been checked to prevent excessive X-radiation. Mobile van units, company X-ray machines, and machines in private physicians' offices are often not maintained and serviced frequently. If you negotiate X-ray exams for your group, or if they are required by the company, your union should insist that the equipment used be checked by a reliable health physicist.

The various types of radiation have different energies and penetrating powers. Alpha radiation has the least amount of energy and can be stopped by a thin sheet of paper. Beta rays have more penetrating ability and need at least a quarter of an inch of aluminum to stop them. Gamma rays, like X-rays, have a deep penetrating ability and need heavy shielding, such as several inches of lead. Everyone is familiar with the fact that X-ray technicians, while taking pictures, stand behind a thick lead shield to avoid exposure to radiation. The penetrating ability of X-rays is the reason for this. Finally, neutrons are so penetrating that special, very absorbent materials are necessary for shielding.

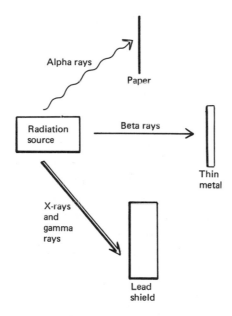

Figure 19. SHIELDING NEEDED FOR IONIZING RADIATION

Not all forms of alpha, beta, and gamma radiation are the same. For instance, gamma radiation emitted by radioactive uranium is slightly toxic, while gamma radiation from cobalt is classified as moderately toxic. The penetrability of the rays is related to their toxicity.

Because alpha particles are not very penetrating, they are a hazard only when taken *internally,* that is, eaten or breathed in. Some weak beta rays are also only internal hazards. Other beta rays, gamma rays, X-rays, and neutrons are so penetrating that they are called *external* hazards. Just standing near them allows them to travel through space and penetrate the body, damaging its cells. Once inside the body, external and internal radioactive materials have the same damaging effects. (Some authorities believe that alpha radiation is actually more harmful inside the body.) The only difference is the precaution and shielding that must be used as protection.

For example, uranium workers are exposed to radioactive dusts from the mining and machining of uranium. Radioactive uranium decays to form **radon gas,** which is also radioactive. As a result of inhaling these gases and dusts, uranium workers have a high incidence of lung cancer. In fact, the Public Health Service has recently estimated that between 600 and 1,100 of the 6,000 uranium miners will die of lung cancer during the next 20 years.

Different radioactive substances may be selectively absorbed or may selectively damage specific parts of the body. One such substance is **strontium,** which is deposited in the bones, leading to an increase in bone cancer and leukemia. Radioactive strontium that is eaten in milk products replaces the calcium in the bones and stays in the body, emitting radiation to the bones and surrounding tissues. During the years when the world's nuclear powers were testing atomic bombs in the atmosphere, there was a great deal of public concern about the spread of strontium throughout the world. Although atomic tests have stopped, workers exposed to strontium are still affected by it, but the concern has diminished.

When a radioactive particle passes through a cell, it interacts with the atoms and molecules that make up the cell and breaks them up. It may destroy the cell permanently, or else cause it to stop functioning properly. How radiation causes this damage is not well understood, but the results of exposure are widely documented. One effect of exposure to radiation is an increased incidence of cancer. The government has set a legal limit of 5 rems per year for radiation exposure, but a worker who is exposed to this much radiation has a 50 percent greater chance of developing cancer than an unexposed person. According to recent estimates, there is a 1 percent increase in the incidence of all types of cancer for each rem of exposure to radiation.

It is a sad commentary on our standards-making system that the legal limit, in essence, guarantees a higher incidence of cancer for exposed workers. Independent researchers have never been able to find a level of radiation that *did not* produce some biological damage. In other words, any exposure to radiation has an effect. Of course, when we take X-rays or get treatment with radioactive materials, we are voluntarily exposed to radiation, but for what are in most cases beneficial reasons. Also, the amount of radiation involved in such a dose is very low. However, the only "benefits" derived by workers who are exposed to radiation are their jobs and livelihood, which are really their right in the first place. It is unacceptable for any legal exposure standards to exist in which there is the possibility of disease or disability developing. The Occupational Safety and Health Act of 1970 guarantees every worker the right to a safe and healthful workplace, yet standards exist that do not meet this guarantee. The reason for these dangerous standards in radiation exposures may very well be that the Atomic Energy Commission is charged *both* with setting the standards and with promoting the expansion and use of atomic energy.

In addition to a higher incidence of cancer, workers who are exposed to radiation have a decreased life expectancy. In some way that is not understood, ionizing radiation makes the body age more rapidly, and radiation workers suffer medical changes that come with aging, in the blood, muscles, and other parts of the body, at a younger age than unexposed people. It is estimated that each rem of exposure decreases life expectancy by two to five days. The genetic material of the cells, which determines the inherited characteristics of a person's children, is also damaged by ionizing radiation. There is a marked increase in birth defects and prenatal deaths among radiation workers. People who have had a dose of between 10 and 100 rems of radiation in their lifetime are thought to have a 4 percent rate of defective children as well as a 4 percent prenatal death rate. This is *double* the incidence for the population as a whole.

Protection and Safeguards Against Radiation

It is obvious that a great many safeguards against exposure are necessary whenever radioactive materials are being used. There should be no unnecessary exposure, such as occurs if radioactive wastes are left around or if a system using radioactive material is not maintained properly. The two key words to proper industrial hygiene are *isolation* and *shielding*. Whenever possible, processes using radioactive material should be isolated. As with many other things in nature, such as sound and light, the strength of the radiation wave varies with inverse square $(1/D^2)$ distance from the source. What this means is that if the radiation level is 1.0 roentgen per hour at a spot 1 foot away from the source, doubling the distance from the source to 2 feet reduces the radiation four times to 0.25 roentgen per hour (2 squared is 4 and the inverse of 4 is $\frac{1}{4}$). Tripling the distance to 3 feet reduces the radiation to $\frac{1}{9}$ of its original strength, or 0.11 roentgen per hour (3 squared is 9 and the inverse of 9 is $\frac{1}{9}$). Thus, isolating the process is an effective means of protection in some cases.

Isolation alone is not sufficient protection, however, because workers going near the source will still be exposed at full strength. It is also necessary to shield the process with the proper kind and thickness of material. Tables of average radiation-shielding materials can be found in radiation handbooks. If the radioactive source is a dust or gas, it is essential that there be proper ventilation to keep the work area dust-free and gas-free. It is also important to make sure the ventilation system keeps all the radioactive material inside it so that it can be disposed of properly. It makes no sense whatsoever to remove radioactive material from one work area only to dump it into another.

It is always important to measure the radiation in the air continuously to see that the levels are well below the "safe" limits. Measuring devices must be accurately calibrated in order to be useful. Alpha rays need special meters with enough windows so they do not stop the alpha particle from entering the meter. *Geiger counters* and *scintillation counters* can be used for beta, gamma, and X-rays. If the radioactive source is a dust or gas, an air sample at the breathing level of the exposed worker can be taken, using the equipment and techniques described in Chapter Ten. This sample can then be evaluated to see how much radioactivity it contains.

All people who work near radioactive material must wear radiation-film badges that give a measure of the absorbed dose of radiation and an idea of the kind of radiation the workers are exposed to. These badges cannot measure alpha radiation, which is stopped by the paper that covers the emulsion on the film. Because of this difficulty in monitoring for alpha-radiation exposure, all air samples should be taken when working with alpha-ray sources. Other devices for monitoring the absorbed dose of exposure are mobile personal monitors and dosimeters (which can be placed in a worker's pocket). The dosimeter measures the accumulated dose of X-rays and gamma rays, while the monitors can be used to measure the radioactivity on the hands and clothing of exposed workers.

When there is a dust or gas source of radioactivity, workers should have special clothing that is worn only in the radiation area and that does not leave the area or come in contact with clothes and shoes that are worn home or in other parts of the plant. Washing facilities must be made available.

For emergency use, there should be protective clothing and respirators that are specifically designed for the kinds of radioactive materials used. Workers should insist on seeing the specifications on this equipment to make sure it will safeguard them if an emergency arises. Also, since every exposure to radiation is really a health compromise, workers should demand the results of the film-badge readings or whatever other readings are taken. The worker and the union can then keep accurate records of the exposure, and also accurate medical records on all exposed personnel to make sure that similar sickness trends are not developing among exposed workers.

Workers who use radiation should try to bring the level of exposure down to a minimum. Even if the workplace meets the government standards shown in Table 15, the level of exposure should be decreased, for all the reasons we have discussed. The government limits set a maximum value, not a minimum. Therefore, a good tactic to use during collective bargaining is to attempt to set new standards for exposure in the work contract. This will, of course, mean more investment for the company in process control and shielding and other precautions, and the negotiation will be difficult. But the price that workers will pay for exposure to radiation is worth the cost of the struggle across the bargaining table. Workers should ask themselves what good a few cents extra per hour will be, or what the extra benefits of the pension really mean if they never live to collect the payments.

Table 15. LEGAL STANDARDS FOR EXPOSURE TO RADIATION

	ALLOWABLE DOSAGE *
Whole body exposure	5 rem/year
Gonads and red bone marrow	5 rem/year
Skin, thyroid, bone	30 rem/year
Hands and forearms; feet and ankles	75 rem/year
All other organs	15 rem/year

* Recommended by the International Commission on Radiological Protection (adopted September 17, 1965).

CHAPTER SEVEN

CHEMICAL HAZARDS

SOME TECHNICAL TERMS

aliphatic compounds: all organic compounds that are not in the benzene family

aromatic compounds: all organic compounds that are derived from benzene

carboxyhemoglobinemia: a change in the blood caused by carbon monoxide, which blocks the oxygen-carrying function of hemoglobin. The victim has a cherry-red color. Cigarette smokers have up to 8 percent carboxyhemoglobin in their blood.

carcinogens: substances that cause cancer

flammable: easily set on fire

halogens: chemical family that consists of chlorine, bromine, fluorine, and iodine

inorganic compounds: all chemical substances that do not contain carbon

metal fume fever: illness with symptoms similar to flu that can be contracted from contact with some metal fumes

methemoglobinemia: change in the blood from chemicals such as aniline, in which the hemoglobin is altered so that it cannot carry oxygen. This causes the victim to turn blue.

organic chemicals: all chemical substances that contain carbon

photosensitization: process of making the skin more susceptible to harmful effects of sunlight

pneumoconiosis: lung disease that results from breathing in dusts

ppm: concentration of a chemical in *parts per million*

pulmonary edema: condition where the lungs fill with fluid and swell

pulmonary fibrosis: thickening and scarring of the lung tissue

reversible: capable of being healed or repaired
sensitization: development of an allergy
threshold limiting value: legal limit of exposure to toxic chemicals
TLV: abbreviation for *threshold limiting value*
toxic: poisonous or harmful
volatile: tending to vaporize rapidly

THE WORKPLACE is filled with thousands of chemical fumes, dusts, mists, and vapors, some of which are very harmful. As most workers are unaware of how these chemicals affect the body, this chapter discusses many of the substances and tries to explain their effects. It does not attempt to give an exhaustive list of all the chemical hazards that workers encounter, but rather to break them down into categories or classes and to discuss the general toxic effects of each class. Many individual chemicals are briefly described, or covered in tables throughout the chapter. All the chemicals mentioned are listed in the Index of Substances. The chemical hazards associated with different types of work are listed in the appendix on pages 368–419, under occupational headings. It is suggested that you first read the introduction to each section dealing with a class of chemicals, and then read about those individual chemicals that are of interest to you. Skipping parts of this chapter will not affect your understanding of the rest of the book.

The effects of the vast majority of substances in common use are not really known, and little is being done to expand our knowledge. Each year about 3,000 new chemicals are introduced into industry, yet their effects on the workers who will use them are never tested. Standards for allowable exposure exist for less than 500 of the toxic chemicals currently in use. These standards are supposed to be the maximum concentrations to which workers can be exposed for eight hours a day, every day, without developing disease. They are based on average exposures, which means it is not illegal to be exposed to higher levels at certain times so long as the average exposure is below the legal limit. This limit is known as the *threshold limiting value,* or TLV. The chemicals for which standards have been established are listed, with their TLVs, on pages 261–74.

Many drawbacks are associated with these TLVs, the most obvious being that there are simply not enough inspectors to see that the standards are enforced. Workers, in essence, have to depend on voluntary compliance with the law by their employers, and unfortunately, such voluntary compliance is not widely practiced. In fact, without an established system of monitoring chemicals in the air, setting up TLVs is almost meaningless. Another major drawback is that the TLV list covers only a fraction of the approximately 15,000 chemicals in industrial use. Finally, the TLVs have been formulated to be applied to the healthy working person and not to those disabled in any way. Although workers are not for the most part disabled, they are usually exposed to more than one toxic chemical at a time. The effects of this multiple exposure are not known, and the TLVs do not take into account the possibility of interaction or cumulative effects.

In addition, since the amount of toxicological research done in the United States is relatively small—especially as compared with research in other areas—many TLVs are based simply on the best conclusions that formulators can draw from experience and not on experiments. Sometimes the formulators are forced to revaluate—just recently, the mercury standard was made considerably more stringent. The philosophy seems to be that workers and unions must prove that a standard is dangerous, rather than the makers of standards having to prove that the legal limit of exposure is safe. When sufficient data do not exist, a standard is set anyway, and it is hoped that the level is not too high. A more logical approach would be to commission an immediate study to find out the possible effects and, for the time being, to set the most stringent standards so as to eliminate unnecessary risk.

Another interesting fact is that a chemical such as carbon monoxide or sulfur dioxide is given one acceptable exposure level in the TLVs and another, much lower level in community standards for air-pollution control. The argument used is that the young, the old, and the disabled are more susceptible than working people, though the validity of this argument has not been substantiated. Thus the worker is exposed to higher concentrations of toxic substances at work and continues to be exposed to lower concentrations of the same substances at home.

One word of caution: as you read through the descriptions of chemical hazards, you should notice such phrases as "no reports of illness from industrial exposure." This does not mean that illness is not possible. What it means is that no scientist or doctor has ever written up a case of illness due to the substance. Even though workers may know that they or their co-workers are sick from exposure to the substance, officially it may still be considered harmless.

GASES

Industrial gases can be health hazards either because they have direct toxic effects on the body or because they take up space in the atmosphere and keep the body from getting the oxygen it needs to survive. For example, when a gas such as nitrogen, which is not in itself harmful or irritating, is used to purge a tank, it can become so abundant that it reduces the oxygen in the air that is necessary for breathing. A worker in this atmosphere can rapidly suffocate and die. Some asphyxiating gases, as they are called, are listed on page 158. In high concentrations, some of them may also have an anesthetic effect.

acetylene *	hydrogen *
argon	methane *
butane *	neon
carbon dioxide	nitrogen
ethane *	nitrous oxide
ethylene *	propane *
helium	* flammable

Some gases are irritating or damaging to the lungs. How soluble an irritant gas is in water determines what part of the lungs it will injure. The more soluble the substance, the more quickly it dissolves in the watery passages of the breathing tubes, and therefore the higher up the respiratory tract the damage occurs. Cadmium oxide fumes, which are relatively insoluble, are breathed in all the way past the air tubes, which do not react with them, and reach the air sacs, where they cause a severe reaction. Nitrogen dioxide and phosgene, which are slightly soluble, dissolve slowly in the water vapor of the air tubes and air sacs and cause severe burns. Finally, highly soluble gases like fluorine, chlorine, and sulfur dioxide mainly affect the upper airways, which go into spasm and prevent the gas from entering the lungs, thus protecting the air sacs from injury. Certain other gases are absorbed from the lungs into the bloodstream, and damage other organs of the body.

Irritating Gases

Some gases produce both acute irritation by a single, high-level exposure and chronic lung disease by repeated exposures at low levels. Since cigarette smoking and urban air pollutants cause similar changes in the lungs, a physician will often not recognize the occupational cause of this irritation in a patient. It is difficult to pinpoint such effects even in large-scale, long-term studies of worker groups with similar exposures. Low-level exposure to all irritating gases, including those in cigarette smoke and urban air, probably produces a continuous minor irritation of the medium-sized air tubes, causing overproduction of mucus by the bronchial lining and resulting in chronic cough with phlegm. This makes a person more susceptible to infection of the air tubes, or *acute bronchitis.* After repeated infections, he or she may develop *chronic bronchitis,* which means that the flow of air out of the lungs may be chronically slowed down. The back-pressure of air trapped in the lungs causes the air sacs to rupture, leading to *emphysema.* The dilated air sacs do not efficiently transmit oxygen into the blood, and the victim of chronic bronchitis and emphysema is always short of breath. These two conditions usually occur together, and are known as *chronic obstructive pulmonary disease.*

Nitrogen oxides are used in the nitration of aromatic compounds and the pickling and etching of metals, and are a by-product in the manufacture of explosives, dyes, lacquers, and celluloid. The high temperatures used in processes such as welding and detonation oxidize the nitrogen in the air into the various nitrogen oxides. The ordinary type of soda-lime activated-carbon masks are *not* satisfactory protection against these substances.

The nitrogen oxides are extremely irritating to the eyes and upper respiratory tract. Exposure to concentrations above the threshold limit causes cough and chest pain immediately. Lower concentrations may produce only mild immediate symptoms, which the worker may ignore, so that exposure is continued until a large dose is received. All the nitrogen oxides gradually change into **nitrogen dioxide.** Five to 12 hours after exposure, the nitrogen dioxide has had time to react with the watery atmosphere of the respiratory tract and air sacs and produce **nitric acid,** which burns the lungs and air sacs. As in response to any burn, the injured tissue literally pours out fluid from the bloodstream. Fluid accumulation in the lungs interferes with oxygen exchange, and the victim may suffocate. The

symptoms of this severe reaction, called *pulmonary edema*
(*edema* means swelling caused by fluid), are weakness, cold
sweat, nausea, cough with frothy, yellow-brown sputum,
severe shortness of breath, air hunger, and anxiety. Death
is common. Even if the victim recovers, there may be permanent damage mainly to the small air tubes and air sacs,
resulting in frequent lung infection, chronic cough, and
shortness of breath.

Two other reactions to nitrogen dioxide are possible. One
is acute but reversible, with shortness of breath, chest pain,
nausea and vomiting, dizziness, sleepiness, and then loss of
consciousness. There is no pulmonary edema, and the victim recovers when exposure is ended. The second reaction
results from sudden exposure to a high concentration of gas,
and is rapid and fatal. The victim suffocates, goes into convulsions, and stops breathing.

The effects of repeated low-level exposures have not been
studied, but such exposure probably causes chronic bronchitis.

Sulfur dioxide has a distinctive and unpleasant odor. It
is produced in the smelting of sulfide ores and the processing of sulfur-containing fuels. Magnesium workers may be
exposed to large amounts of this gas, since it is used to prevent oxidation of the metal. It contributes very largely to
atmospheric pollution in large cities and in areas surrounding smelters and oil refineries. The average individual can
detect 0.3 to 1 part per million (ppm) mainly by taste, 3
ppm by odor, and 6 ppm by immediate sharp irritation of
the nose and throat. When the concentration reaches 20
ppm, there can be inflammation of the eyes. Some scientists
think sulfur dioxide is the part of smog that irritates the
eyes.

As sulfur dioxide is very soluble, its main action is in the
upper respiratory tract—nose, throat, windpipe, and bronchi.
These tissues may swell and block the passage of air, but
occasionally the air sacs also are injured and pulmonary

edema results, which may be fatal. The long-term effects of low concentrations are not known, though chronic irritation of the nose and throat *(nasopharyngitis)*, changes in the senses of taste and smell, and increased fatigue have been documented. Chronic irritation of the air tubes may cause chronic bronchitis and emphysema.

Hydrogen sulfide has an odor of rotten eggs. It is encountered in industries where sulfide ores are used, and in the manufacture of chemicals, dyes, and pigments and the refining of petroleum. It irritates the eyes and upper respiratory tract and can cause pulmonary edema, but these effects are overshadowed by its effects on the brain. It is absorbed rapidly into the bloodstream, but as long as the exposure level is under 700 ppm, the blood can convert the gas into a harmless compound. At concentrations above 700 ppm, the body cannot cope with the excess gas, which reaches the brain and causes breathing to stop. Suffocation occurs in a matter of minutes unless the victim is removed and given artificial respiration. Long-term exposure to low levels may cause chronic lung disease.

Hydrogen selenide belongs to the same chemical group as hydrogen sulfide and is believed to cause similar symptoms. However, it is toxic at much lower concentrations.

The term **halogens** refers to a particular set of chemical elements: **fluorine, chlorine, bromine,** and **iodine.** These cause very similar reactions, but the rate and severity of reaction are quite different.

Fluorine is highly irritating to the skin, eyes, and lining tissues of the mouth, throat, and respiratory passages, and may burn any of these. Burns of the lungs result in pulmonary edema. The chemical burns caused by the reaction of fluorine with the skin may be accompanied by flash flames that cause further burning of the skin *(thermal burns)*. The chronic effects of repeated low-level exposures are not known.

Chlorine, at concentrations of 3 to 6 ppm, is markedly irritating and damaging to the skin, eyes, nose, nasal sinuses, throat, larynx, and larger air tubes. It may cause chronic inflammation of the sinuses and continual hoarseness. If greater concentrations reach the lungs, pulmonary edema may result. However, the gas is so irritating that workers usually avoid such high concentrations. Repeated low-dose irritation probably causes chronic bronchitis and emphysema, although this is not certain. Chlorine may cause ulcers to form in the center section of the nose and on the skin, and may produce acne and skin infections.

Chloride of lime, which releases chlorine gas, has toxic properties similar to chlorine.

Bromine causes coughing, nosebleed, a feeling of oppression, dizziness, and headache. These symptoms may be followed several hours later by stomach pains, diarrhea, and a blistering skin rash that looks like measles. Skin contact causes burns and painful ulcers. Continuous exposure to levels near the legal limit will also bring on many of these symptoms. After several years of exposure, a worker may develop thyroid abnormalities, chronic sore throat, heart disease, and low blood pressure. The effects of much lower concentrations are not known.

Iodine is a solid, but it vaporizes so easily that the solid form is always accompanied by vapor. Iodine vapor irritates the lungs and causes them to fill rapidly with fluid (pulmonary edema). It irritates the eyes and causes tearing and swelling of the eyelids. Inflammation of the upper respiratory tract may lead to runny nose and sores in the mouth and throat. A skin condition similar to acne is thought to be an allergic response to iodine.

The chronic effects of long-term, low-dose exposure have not been studied, though it is known that iodine is absorbed into the body through the lungs and that prolonged exposure may produce nervousness and weight loss. This is due to the action of iodine on the thyroid gland. The thyroid, which is in the neck, regulates the rate at which the body utilizes oxygen to produce energy, and iodine damage increases this rate.

Phosgene was used during World War I as a nerve gas. Although extremely useful in many chemical reactions, it is often an unwanted side product. It is produced whenever a compound containing chlorine comes in contact with a flame or hot metal, as occurs in some types of welding, and is very hazardous in a closed space. Many deaths among firemen are thought to result from the use of carbon tetrachloride fire extinguishers. The carbon tetrachloride is converted to phosgene by the flames and the firemen are poisoned by the gas.[1]

Low concentrations (2–10 ppm) are mildly irritating to the eyes, nose, throat, and respiratory passages, and skin contact causes dermatitis. Symptoms of slight gassing may occur: dry burning throat, numbness, vomiting, chest pain, and cough with phlegm. There may be shortness of breath. Phosgene is slowly soluble in the watery mucus of the air passages, and the reaction to a moderately high exposure, though initially irritating, is delayed for several hours. When the gas dissolves in the fluid of the lung tissues, it causes severe burns. Large quantities of tissue fluid then flood the lungs and pulmonary edema results. Large doses can produce immediate acute responses, be corrosive to the lungs, and cause sudden death. Chronic effects have not been studied, but phosgene probably causes chronic lung disease.

Ammonia has an odor familiar to almost everyone because it is present in so many household cleaners. Ammonia gas and solution are intensely irritating to the eyes, with symptoms ranging from watering to closing and swelling of the lids and damage to the cornea, the delicate covering of the lens. Ulceration of the cornea can lead to blindness. Moist skin in contact with ammonia is severely burned. Inhalation of the gas is strongly irritating to the air tubes and lungs, usually forcing people to leave the vicinity. If this is

[1] These fire extinguishers are no longer widely used.

not possible or if the dose is large, there is marked burning of the small air tubes and air sacs, and the lungs fill with fluid, resulting in pulmonary edema. This is frequently fatal.

Effects of long-term exposure to ammonia in low doses have not been studied. Constant irritation of the air tubes probably results in chronic bronchitis and emphysema.

Ozone is a sweet-smelling gas that may be produced from oxygen in the air by lightning or by electric-arc welding. It is irritating to the eyes and the lining tissues of the throat, air passages, and lungs. Prolonged inhalation above 0.05 ppm can cause burning of the lungs. The level at which ozone is irritating is about one-quarter of the legal threshold value. Higher doses can cause pulmonary edema. Animals exposed repeatedly to 1 ppm over a long period show scarring and thickening of the smaller air passages, which partially blocks the air flow out of the lungs—a condition called *chronic obstructive lung disease* in humans.

Most important, ozone affects the genetic structure of the cells in a manner very similar to X-rays. No exposure is really safe, because any dose will affect the cells in some way. There is a good possibility that ozone causes cancer.

Non-irritating Gases

The following gases do not irritate the upper respiratory tract or the eyes, but still have definite, serious toxic effects.

Carbon monoxide almost always accompanies any combustion unless it is carried out at really high temperatures. The distillation of coal or wood and operations near furnaces, ovens, stoves, forges, and kilns (especially when first starting up) all produce this substance.

Carbon monoxide is colorless and odorless and has no locally irritating effects. Its toxic effects come from its being absorbed from the lungs into the bloodstream, where it replaces oxygen by attaching chemically to hemoglobin, the red chemical that carries oxygen in the blood. If the body tissues do not receive a constant supply of oxygen, they stop functioning and die. The brain is the tissue most dependent on this supply, and most of the initial symptoms of carbon monoxide poisoning are due to brain misfunction caused by lack of oxygen. The first such symptom is headache. Further exposure causes throbbing headache, reddening of the skin, weakness, dizziness, dimness of vision, nausea, vomiting, and at even higher concentrations, coma, suffocation, and death.

Effects of long-term low-dose exposures have been studied. Swedish workers exposed to concentrations that caused carboxyhemoglobin levels of 10 to 30 percent showed such symptoms as headache, dizziness, decreased hearing, visual disturbances, personality changes, seizures, psychosis, palpitation of the heart associated with abnormal rhythms, loss of appetite, nausea, and vomiting.

Medical tests for poisoning are quite simple, involving measurement of the carboxyhemoglobin level—that is, the amount of hemoglobin that contains carbon monoxide. It should be noted that people who smoke inhale as much as 700 to 800 ppm of carbon monoxide at each puff.

Hydrogen cyanide is a side product from blast furnaces, gas works, and coke ovens, and is used as a fumigant. It is absorbed into the body very rapidly through the lungs. The liquid and possibly the vapor are absorbed through the skin. When there is a possibility of absorption, the legal threshold limit is much too high. Once absorbed, hydrogen cyanide is the most rapid-acting of all known poisons. It exerts its poisonous effect by interfering with the body's use of oxygen for energy production. This causes all other chemical body processes to stop, and the victim dies rapidly. **Cyanogen, sodium cyanide, potassium cyanide, copper cyanide,** and **calcium cyanide** affect the body in the same way.

Symptoms of poisoning start at low doses with headache, weakness, mental confusion, and occasionally nausea and vomiting. Higher doses cause the victim rapidly to become unconscious and stop breathing. The cyanide content of blood and tissues can be measured, and characteristically the blood in the veins is very red because of the high content of unused oxygen.

Emergency treatment of mild poisoning consists of artificial respiration and simultaneous inhalation of amyl nitrate vapor from ampoules crushed in a handkerchief and held to the victim's nose; several ampoules may be used in the first half hour. The victim should be washed with large quantities of water to remove any residual dissolved cyanide. In more severe poisoning, nitrite and thiosulfate solutions must be injected into the veins. These medications should be part of the standard safety kit whenever there is a chance of exposure to hydrogen cyanide or its derivatives.

The halogenated cyanides, such as **cyanogen chloride** or **cyanogen bromide,** are also highly toxic to the body's chemical processes. In addition, they are highly irritating at very low concentrations, causing both rapid and delayed inflammation of the air passages and lungs that may result in pulmonary edema. The **nitriles,** such as **acetonitrile (methyl cyanide), acrylonitrile, methacrylonitrile, isobutyronitrile,** and **propionitrile,** can cause the same general symptoms as hydrogen cyanide, but the onset is apt to be slower. They are also more likely to irritate the skin and eyes. They can be absorbed rapidly through intact normal skin, and workers exposed to these compounds from the decomposition of polystyrene have suffered liver and blood damage as well as going into shock. A number of other chemically related materials, such as **cyanamide, calcium cyanamide, cyanates, isocyanates, isonitriles, thiocyanates, ferri-** and **ferrocyanates,** and **cyanoacetates,** do not have the typical toxic properties of cyanide and nitriles and harm the body by other mechanisms.

DUSTS

Dusts usually cause disease only in the lungs. Lung disease that results from dusts is called *pneumoconiosis*, derived from the Greek *pneumo-*, meaning "lung," *coni-*, meaning "dust," and *-osis*, meaning "reaction." The development of pneumoconiosis depends on the type of dust and the susceptibility of the individual worker, and is also affected by duration of exposure and by exposure to other chemicals or to cigarette smoke at the same time. Particle size affects the toxicity of dusts: the smaller the particle, the farther down into the lungs it can go, and also the greater the amount of dust that will be retained by the body. A general rule is: The more dust retained, the more severe the resulting illness. Dust particles may be so small they are visible only by microscopic analysis. Particles this size behave essentially like air, passing practically unimpeded through the protective barriers of hair and mucus in the upper airways. The air sacs and small air tubes therefore receive a very large dose, and the intensity of lung reaction is greater than if the same quantity of larger particles had been inhaled.

Inorganic Mineral and Chemical Dusts

Silica. Inhaling the free, finely divided dust of silicon dioxide causes a form of pneumoconiosis called *silicosis*. The dust may be in crystalline form, as in **quartz, cristobalite,** and **tridymite,** or in non-crystalline form, as in **opal.** The smaller and more compact crystalline forms (quartz, tridymite) cause the greater lung scarring since they have the most dangerous particle size: 0.05 microns to 5 microns (1 inch contains 25,400 microns).

Silicosis may develop either rapidly or slowly. The rapid-developing variety occurs most commonly among workers in the manufacture or packing of abrasive soap powders, among sand blasters, especially those working in enclosed tanks), and among tunnel workers using high-power drills on rock. These workers are exposed to tremendous amounts of very finely divided crystalline silica. The chemical interaction between silica dust and lung tissue may be heightened by the presence of alkali or acid.

In rapid-developing or acute silicosis, symptoms may appear as soon as 8 to 18 months after the first exposure. The main symptom is shortness of breath, at first occurring only during physical activity, but soon appearing after less and less exertion, until eventually the victim is short of breath even while at rest. This is caused by the many small round lung scars that develop from irritation by silica dust. These hard, inelastic scars—just like those on the skin that result from an operation—make the lungs stiff, so that it takes more work to inflate them with air. The scars also thicken the walls of the air sacs, blocking the transfer of oxygen into the blood; low blood oxygen is a characteristic finding in silicosis. The area surrounding each scar becomes stretched and distorted, breaking down the normally tiny, delicate air sacs so that they form larger, thicker-walled sacs—a form of localized emphysema. Further reaction to the silica may cause the scars to join into larger scars; some may occupy the entire lung. This process, *progressive massive fibrosis,* is frequently accompanied by increased susceptibility to tuberculosis and other infections. Finally the heart, which must pump blood through these stiff, inelastic lungs, becomes weakened and enlarged and fails to pump effectively.

The course of chronic or slowly progressive silicosis is very similar to that of acute silicosis, except that rather than developing quickly, it arises after many years of exposure and may take many more years to become worse.

The legal standard for silica exposure depends on the form of the dust. Protective apparatus is useful only for emergencies, and would be difficult to work in for a full working day.

The prevalence of silicosis in workers exposed to silica dust is probably very high. A survey showed that in the period 1950–1958 there were 12,000 cases in 26 states. This may be a gross underestimate of the true numbers, as the study was based mainly on workers over 50 and already disabled.

Another variation of rapidly progressive silicosis results from exposure to **diatomite**, a non-crystalline silica used in filters, insulating materials, absorbents, and polishes. It produces a very fine dust capable of causing severe illness. *Shaver's disease* is a similar, rapidly progressive disease caused by inhalation of **bauxite** fumes. These fumes are emitted by electric furnaces in the production of corundum or refining of aluminum ores, and consist of extremely small particles (0.02–0.05 microns).

Coal dusts. Exposure to soft-coal dust causes *simple pneumoconiosis,* or *black lung.* Soft (bituminous) coal contains little silica. Large amounts of the black dust are deposited in the lungs, especially at the end of the small air tubes just before they enter the air sacs. This peculiar location and the properties of the dust itself give the disease the characterestics both of bronchitis and emphysema and of silicosis. Like emphysema, it shows up late or not at all on chest X-rays and in most simple lung-function tests. Like silicosis, it leads to lung scarring, which makes the lungs stiff and may lead to heart strain.

"Complicated pneumoconiosis" results when the small scars of simple pneumoconiosis join together to form large masses of scar tissue. These masses, which are stony hard, may occupy more than one-half of the lung. As the scars pull together, they rip the air-sac walls behind them and cause severe emphysema. Respiratory infections, and especially tuberculosis, may be a fatal complication of black lung. Among workers exposed to coal dust, smoking cigarettes appears to make black lung more frequent and more severe.

Anthracite, or hard coal, which contains more silica,

causes a lung disease more like silicosis. It shows up on X-rays fairly early and leads to symptoms of respiratory disease only years later. Large scars taking up almost the whole lung are a particularly common effect of hard-coal dust.

For many years, doctors thought the amount of lung scarring or disease that resulted from coal dust depended on the exact silica content of the dust. But ample information shows that silica-free coal can produce both the very small scars and the very large ones. Some doctors, many of whom work for the compensation insurance companies or the coal companies, still do not believe that bituminous coal causes any illness. This shows that doctors are not immune from bias in the interpretation of scientific facts, especially when the facts have economic consequences for major industries.

The standard for coal dust, under the Coal Mine Safety Act of 1969, is 3 million particles per cubic foot of air. However, the Bureau of Mines has not conducted enough inspections to ensure that this limit is maintained, not even the weekly spot inspections of "especially hazardous mines" required by law. It has either not given fines, or has failed to collect on fines it has given. With this sort of enforcement, it is not surprising that coal miners are dying from accidents at a rate of approximately one man every other working day, a rate higher than that of 1969, before the Coal Mine Safety Act went into effect. Young miners exposed to coal dust in mines where inspectors never come will develop black lung that will show up at least 10 years from now.

Carbon black is produced by burning creosote oil, tar oils, or petroleum distillates in a shallow iron pan and then collecting the resulting soot. Slight variations in the process produced **channel black, furnace black,** and **acetylene black.** In England the word "carbon black" is used for the same products as the word "soot." (Most of the research on the health effects of this dust comes from England.) In this country, "carbon black" refers to the specific product of the incomplete burning of natural gas or oil: essentially pure carbon, containing less than 1 percent mineral impurities and *no* silica. Exposure to large quantities, as in the manufacture of rubber or inks or the production of carbon black itself, causes pneumoconiosis identical to that produced by soft coal, with emphysema, abnormalities of the small air tubes, and lung scarring that may lead to progressive massive fibrosis.

Graphite, also called plumbago, has numerous industrial uses as a lubricant, a conductor of heat and electricity, and a carbon carrier in steels. Natural graphite can contain from 2 to 10 percent silica after it is freed of the rock from which it is mined. Synthetic graphite contains less than 1 percent silica and often none at all. Pneumoconiosis can result from both natural and synthetic graphite. The illness and the medical findings are similar to those found in black lung. Synthetic graphite and graphite with low silica content produce lung disease like that caused by soft coal. Graphite with higher silica content produces a disease more like silicosis.

Asbestos is a fibrous mineral that, like cotton, can be made into thread and cloth. Unlike cotton fibers, its fibers are as strong as piano strings. It is virtually indestructible—heatproof, fireproof, and resistant to most chemicals. Because of these properties, it has found its way into more than 3,000 products, from potholders and children's toys to welding rods and industrial and household insulation. No factory or home is without asbestos, and almost no group of workers is without potential exposure.

Asbestos fibers are extremely fine. Fibers that can be seen with the naked eye or under an ordinary light microscope actually consist of thousands of smaller fibers called fibrils. It takes a specially powerful microscope, called an electron microscope, to magnify enough to make individual asbestos fibrils visible, as they are so small that there are 1 million of them to an inch. (For comparison, there are approximately 630 human hairs, side by side, to an inch.) When such very fine fibers get into the air, they float like water vapor and never settle. They are not trapped by the mucus or hairs of the nose or air passages, and can enter the air sacs of the lungs without any hindrance. Based on the usual method of dust counts, which measure only fibers larger than 5 microns and leave uncounted the majority of smaller and more dangerous fibers, it is estimated that in an 8-hour day, at the legal limit of 5 fibers per cubic centimeter of air, a person breathes 15 million fibers. Many of these remain trapped in the lungs. Once they are inside the body they remain as indestructible as they were outside it. An asbestos fiber you breathe when you are 18 years old may stay in your lungs until you die.

Although the body's mechanisms cannot destroy the asbestos fiber, they try very hard to isolate it. Over a period of 10 to 20 years individual fibers or groups of fibers may be surrounded with a thick wall of scar tissue. When these scars form in the air-sac walls, the lungs become inelastic, hard and contracted, smaller than normal. This means they hold less air and the body must work harder to fill them. The heart must work harder to pump blood through them. Since the scars are situated in the delicate air-sac walls, they block the transfer of oxygen to the bloodstream. Low blood oxygen and decreased lung volume are what doctors find in a patient with asbestosis. The main symptom is shortness of breath. Frequent respiratory infections are an additional problem, as the slightest infection can cause severe shortness of breath and even death in a person with asbestosis. Cigarette smoking appears to increase the frequency and severity of asbestosis.

Extensive studies have been made of asbestosis in asbestos-insulation workers, most of whom spend only about half their working time handling asbestos. This is a low exposure compared with workers who handle asbestos all the time. In spite of this "low" exposure, 1 out of every 10 asbestos-insulation workers dies of asbestosis. Occupational groups that may have generally higher exposures to asbestos dusts include shipyard workers, asbestos-manufacturing workers, and asbestos-textile workers. The British Threshold Standard for asbestos is 2 fibers per cubic centimeter. The American standard is 5 fibers per cubic centimeter, to be reduced to 2 in 1976. There is already scientific evidence that these standards are inadequate to protect workers from asbestosis.

Asbestos dust has another, even more serious effect: it causes body cells to turn cancerous. Scientists do not know exactly why this happens. The reason may be the long residence of the dust in the lungs, or the ability of asbestos to adsorb other harmful chemicals onto its surface. The surface of asbestos fibers may hold carcinogenic (cancer-producing) chemicals in contact with the body's cells, which become malignant many years later. Workers develop cancers after at least 20 to 30 years have elapsed.

Asbestos can cause cancer in several parts of the body. *Mesothelioma* is a cancer of the membrane lining of the chest or abdomen. Normally this membrane is microscopically thin and acts to lubricate the walls of the chest and abdomen so that the lungs and intestines can move without rubbing or friction. In mesothelioma the lining becomes thickened and the chest or abdomen fills with fluid. Eventually the cancer overgrows the body, the way a weed overgrows a garden, and kills the victim. There is no known cure, and the cancer usually cannot be removed surgically. Mesothelioma is believed to occur only after exposure to asbestos. Prior to the introduction of asbestos into our society at the turn of the century, mesothelioma was so rare that it was unheard of by most doctors. Then the only cases appeared to result from exposure to natural deposits of asbestos, or to asbestos-rich soil in farming country. Mesothelioma can result from small doses of asbestos, well below the current threshold limit. There are reports of mesotheliomas resulting from only one day of work in a shipyard, or developing in asbestos workers' wives who were exposed when washing their husbands' work clothes. Insulation workers get mesothelioma whether they smoke or not. One in 10 asbestos-insulation workers dies of mesothelioma.

Asbestos also causes cancer of the air tubes in the lungs. Lung cancer occurs mainly in asbestos workers who also smoke cigarettes. An asbestos worker who smokes has 92 times the chance of dying of lung cancer that a person has who neither smokes nor works with asbestos. One out of every 5 asbestos workers dies of lung cancer, and about 1 out of every 30 smokers dies of it. Cancers of the intestinal

Table 16. CAUSES OF DEATH AMONG
ASBESTOS-INSULATION WORKERS

CAUSE	PERCENTAGE OF TOTAL CAUSES OF DEATH
Cancer, all types	40.4
Lung cancer	17.3
Mesothelioma	3.3
Cancer of the intestines	11.1
Other cancers	7.7
Asbestosis	5.5
All other causes	54.1

tract are also common in asbestos workers. They have about 3 times as many cancers of the stomach, large intestine (colon), and rectum as the general population.

Because asbestos is so common and so hazardous, special methods of preventing exposure must be used. The important and necessary goal of these methods is that *no* asbestos

dust should be released into the work or general environment, and none should get into anyone's lungs.

One way to prevent a harmful substance from causing
disease is not to use it. Numerous substitutes for asbestos
have been tried. Each has its own hazards. Some are touted
as harmless, but they have been in use for too short a time
for long-term effects to be seen. A general rule seems to be
that any substance with the properties of asbestos, any substance as strong and indestructible, does not belong in your
lungs and will do harm if it gets in. Substitutes now in use
are fibrous glass, rock wool, sodium silicate, and polyurethane. The rules for handling fibrous glass are identical to
the rules for handling asbestos. The British have experi-
mented with asbestos-free calcium silicate block and pipe
covering; this substitute is not generally available in the
United States.

Modifications of asbestos may prevent some dust exposures. Various coatings have been designed to keep installed
asbestos—or fibrous glass—from flaking. These should be
used on all asbestos blocks, and especially on the asbestos linings of air-conditioning ducts. (Contamination of the air in
urban office buildings from asbestos-lined ducts is being
studied in several cities.) Dust suppressants are also used in
asbestos cloth such as that in insulation blankets. (Fibrous
glass has been substituted for asbestos as the filler in the
blanket.) These blankets are not only cleaner to install but
also cleaner to rip out.

Work practices specifically designed to release little or
no dust into the air have been introduced for processes requiring asbestos. Their use for all operations involving asbestos (or other hazardous substances) will allow workers to
control the safety of their own workplaces, and free them
from dependence on all-too-infrequent inspections. If the
correct, safe work practices cannot be followed, the job
should not be done.

Isolation of a hazardous process is an important way of

limiting the number of people exposed. Isolation of asbestos processes has an added avantage: it puts most of the work in shops equipped with proper ventilation and with tools designed for the special problems of asbestos. Shops can be equipped with strong general ventilation and hoods that even the best-ventilated field equipment cannot rival. Prefabrication can be done in such shops, or in the factory from which the asbestos supplies come. (The factory, of course, is really a giant shop and must also have properly designed and functioning equipment and ventilation.) Prefabricated blocks can be shipped, wet, in plastic bags labeled with the exact area where they will be used. Insulation blocks can be bought prescored, or scored in the shop with a properly ventilated gang saw. Pads and blankets for valves and machinery can also be made up in the shop. The fibrous glass or asbestos-cloth blanket is unrolled, wetted, and rerolled. Thoroughly dampened by this procedure, the blanket is then cut to shape in a well-functioning lateral-draft hood.

Repair work can also be done in the shop. Convenient-sized sections of pipe assembly can be brought to the shop, the insulation stripped, the pipes or insulation fixed, and new or repaired lagging put on. Only small repair jobs should be done at the job site.

Unfortunately, all work cannot be brought to the shop. Demolition and some removal operations must be performed on the job site. These operations too must be isolated, both from other workers and from the general community. Other workers must be kept off the job site during the operations. Trade crafts not directly involved in the rip-out should not be scheduled to do any work while the area is contaminated. The surrounding community can be protected by plastic tenting that completely encloses the work area and is equipped with a portable industrial vacuum cleaner to create a negative-pressure exhaust ventilation within it. All of the dust liberated within the tent stays there, because of the inward pull of air created by the exhaust system.

The spraying of insulation cement must also be isolated on the job site. Many cities are now abolishing the use of asbestos-containing cement for spraying operations. However, work rules for the use of substitutes such as fibrous glass are the same. Only materials thoroughly mixed with water should be used. Mixing must be done in an enclosed system, such as a plastic bag with a valve for a hose. Cement shot dry from the nozzle of the sprayer and then passed through water onto the steel is dangerous because it sprays the entire workplace. It should not be used. The spray area should be sealed off from the outside and from the rest of the building with large plastic or plastic-coated tarpaulins. The tarpaulins must reach through window areas into the building to prevent any loss of hazardous dust at the bottom of the enclosure, and must seal off elevator shafts and stairwells. No other workers should be in the area while it is contaminated. During spraying operations, air levels of asbestos may be higher than 100 fibers per cubic centimeter, and the sprayer must wear a powered air-purifying or air-supplied respirator. Wet cement that falls on the floor should be immediately shoveled or vacuumed up and disposed of.

Large power tools, such as band saws and gang saws, can be equipped with good local ventilation. The usual system for a fixed installation shop is a high-volume exhaust system that draws the dust into a baghouse. Special tables for cutting asbestos have perforations, and a down-draft pulls the dust down through the holes into a disposable container. Larger pieces of asbestos debris go into a plastic bag at the side of the table.

Some of this equipment has been successfully modified for field use. Smaller power tools, such as saber saws or rotary saws for cutting cement board, can be fitted with an industrial vacuum cleaner that creates a high-airflow, low-volume draft leading to a bag. The air stream can be moved past the saw blade at a rate three times faster than the saw blade moves.

One word of caution: these ventilated power hand saws are not as safe as the larger ventilated shop tools, and should be used only for those jobs that cannot be brought to the shop. They should never be substituted for band saws, which can be attached to dust collectors of much greater efficiency. The perforated down-draft worktable can also be operated on the job site with an industrial vacuum cleaner creating the draft.

Whether such dust collectors are installed on power equipment or are part of hoods or of a general ventilation system, they must be tested with dust counts to see if they are safe. If they are found to be operating properly, a pressure gauge can be installed on the exhaust duct to indicate whether adequate ventilation is being supplied for each job. All the workman has to do then is see if the pressure as shown on the meter is at a safe value. If it is, the job is being done in a safe environment. If not, the job should wait until the exhaust equipment is working properly.

Asbestos cement can be prepackaged in convenient-sized plastic bags with a valve for the nozzle of a hose. Water is introduced and the cement is kneaded right in the bag. (If the water is introduced the night prior to use, no kneading will be necessary.) Very little dust is liberated by this procedure.

Housekeeping is another important part of preventing exposure. Good housekeeping begins before the job starts, with the laying of plastic dropcloths. At the end of the day, all debris can be wrapped up and thrown away. In between, the smaller dust should be vacuumed up with industrial vacuum cleaners into strong, single-use inner bags that can be removed and sealed. Asbestos or other harmful dusts should never be swept up by hand. Cleanup can be performed either by the workers doing the job or by special cleanup crews.

Disposable work clothes, made of material such as resin-impregnated paper, are recommended. No one else has to handle them, and they can be placed in plastic bags at the end of the day. They help prevent the spread of contamination outside the work area. The clothes should be removed and thrown out before leaving the work area, or in a special "dirty" locker room, so that street clothes do not come in contact with them. In field operations, the work van can have a "clean" locker room for street clothes and a "dirty" one for work clothes. Asbestos brought home on work clothes has been responsible for disease in wives and children of workers.

All these special procedures are to be used in addition to the general procedures of industrial hygiene discussed in Chapter 9. Safe use of a substance as harmful as asbestos shows that both simple and sophisticated techniques can be developed and applied to protect workers.

Mineral wool, rock wool, and **fibrous glass** are synthetic materials (silicates) with properties similar to asbestos, which makes them both useful and harmful. There are over 70 varieties of fibrous glass. Since diseases resulting from these substances probably take a long time to show up, and the substances have only been in widespread use for 10 to 15 years, it is not surprising that little disease due to them has actually been reported. Studies on animals, however, show that they can cause lung scarring and also some cancers. This should be warning enough. Fibrous glass should be handled as carefully as asbestos and the same protective measures applied to its use. It also causes skin irritation, and may cause acute bronchitis and asthma if inhaled.

Talc is very similar to asbestos. Its particles are mostly flat, like little plates, but may also be fibrous. It is usually found in deposits mixed with asbestos, and it is not clear if the health effects of talc are from the asbestos or from the talc itself. (American cosmetic talc tends to have less mixed asbestos than French.)

The effects of talc are similar to those of asbestos, although they have not been as carefully studied. Talc is also a very fine dust, with some particles so small that they cannot be seen with the usual optical microscope. These smaller particles, the most dangerous ones to the body, are not counted by routine dust counts. Once in the lungs, talc stays there and cannot be removed by body mechanisms. The effects show up 20 or 30 years after exposure.

Talc causes lung scarring similar to that in asbestosis. Mesotheliomas have also been reported, and cancer of the lungs and the intestinal tract may also occur, though there have been too few studies for definite proof. Talc has found its way into some foods. Scientists, concerned that talc may cause intestinal cancer, have requested that talc additives in foods be banned by the Food and Drug Administration.

Beryllium causes lung, skin, liver, and kidney disease; however, the effects on the lungs are the most prominent and disabling. *Acute berylliosis* is a severe, pneumonia-like lung inflammation that can occur after brief, intensive exposure, or prolonged exposure to low concentrations. It may result from extremely low exposure in some workers, depending on individual susceptibility. The typical symptoms are cough with phlegm that can be bloody, severe shortness of breath, and weight loss over the first few weeks. The chest X-rays may be normal initially, even though the worker is seriously ill. The lung inflammation may be severe enough to block oxygen transfer into the blood. This disease is frequently fatal, and those who recover may have sustained permanent damage to the lungs.

Chronic berylliosis may develop after 1 year or even up to 20 years of exposure. Chest X-rays may be abnormal for many years before symptoms begin. At first the symptoms may be mild, slowly progressing to a more severe and disabling illness in which chronic inflammation makes the lungs stiff and unable to transfer oxygen to the bloodstream. The heart is affected by years of straining to pump blood through the inelastic lungs. The final stage of the disease is similar to asbestosis.

Iron, barium, tin. *Siderosis* (iron), *baritosis* (barium), and *stanosis* (tin) are caused by compounds such as **iron oxide, tin oxide,** and **barium sulfate,** which give rise to dusts that are insoluble in lung tissue but not irritating. Because these dust deposits are visible on chest X-rays, they are included under pneumoconiosis, but they do not cause symptoms and apparently do not produce disability.

Organic Dusts

Organic dusts may contain plant fibers or fungi, bacteria, and their products. They act upon the lungs in a different manner from mineral dusts. The diseases they cause are frequently the result of an allergic reaction to the inhaled dust.

For example, the disease called *byssinosis* or *brown lung* is caused by an allergy-like reaction to cotton dusts (and also to the dusts of flax and hemp). The severity of reaction depends on the susceptibility of the individual worker to the allergy-producing qualities of the dust.

Although a susceptible cotton worker may develop this reaction at any time in her or his working life, it usually begins after only a short exposure. The first reaction generally occurs when the worker returns to the mill after a few days off—a weekend or a vacation. Because of this pattern of reaction, the initial stages have been called "Monday fever." The first reaction is like an asthma attack, with shortness of breath, severe air hunger, chest tightness, and a dry cough. These symptoms are caused by the narrowing of the medium and small air tubes and by the accumulation of excessive amounts of thick mucus in the tubes. The air has difficulty moving *out* of these narrowed tubes, and makes a wheezing sound. A single such attack is extremely frightening and uncomfortable, but causes no permanent lung damage. Repeated attacks, however, may damage the lungs.

In the initial stages, these attacks occur only on the first day back at work. But as exposure continues over months and years, the attacks last longer, until at last the sensitized worker has lung trouble all week long. Repeated exposure has led to chronic inflammation of the air tubes, or *chronic bronchitis.* The obstruction of the small air tubes causes back-pressure on the air sacs and rupture of their delicate walls, leading to *emphysema.* Chest X-rays, however, are often completely normal. In appearance, this chronic bronchitis and emphysema is the same as that resulting from cigarette smoking or long-term exposure to industrial substances like nitrogen dioxide and chlorine.

It is estimated that there are 17,000 cotton workers with byssinosis in the United States. In many states there is no compensation for this disease, whose existence is not recognized by cotton companies or many of their doctors. The voluntary limit for cotton dust in this country is 1 milligram per cubic meter of air. This is not a legal limit. There is still much dispute between industry-oriented doctors and other scientists about what the limit for cotton dust should be and whether it should apply to total dust, some of which never enters the lungs, or just to the very fine dust that actually causes damage. This fine dust is both harder to measure and harder and more expensive to control. As this dispute is waged, workers are unprotected by the law and are being exposed to great quantities of cotton dust.

Farmer's lung is caused by dust from moldy hay and other vegetable products. *Bagassosis* is caused by the fibers of sugar cane after the juice has been extracted. Both are acute illnesses like pneumonia and usually start abruptly, a few hours after exposure. The worker becomes short of breath and has a dry cough, chills, sweating, and fever, with weight loss later in the illness. Recovery usually follows a first episode, but repeated episodes may lead to lung scarring, especially around the air sacs. X-rays may or may not show the scarring, but tests of lung capacity show a decrease in the volume of air the lungs can hold. The scarred air sacs may be unable to transfer oxygen into the blood.

Many other natural substances cause reactions similar to farmer's lung or bagassosis: redwood sawdust, mushroom spores, detergent enzymes, and molds or fungi that contaminate air-conditioning systems of office buildings.

ORGANIC CHEMICALS

Organic chemicals are any chemical substances that contain carbon. As their name implies, they are the basic chemicals that make up living or organic matter. Literally thousands of them are used throughout industry as solvents, intermediates, and starting materials. Not all of them are harmful, of course; the food we eat is made up of organic chemicals. Many, however, do have serious toxic effects. For example, those which are used as solvents can cause severe skin irritation. The very fact that they are solvents means they can dissolve fats and oils, including those in the skin, so that exposure may lead to a dry, scaly, painful dermatitis. In addition, solvents or other chemicals that vaporize easily can be breathed in, enter the bloodstream, and damage other parts of the body. Carbon tetrachloride and perchloroethylene, for example, have extremely toxic effects on the liver after their vapors have been inhaled, and there are many other chemicals that affect the liver. Workers who are exposed to these chemicals and then drink alcoholic beverages, especially after work, greatly increase their risk of permanent liver damage. The solution here is not to abstain from drinking, but to demand working conditions in which there is no exposure to potentially liver-damaging chemicals.

Some chemicals, such as phenol, have the ability to penetrate right through the skin's protective barrier and enter the bloodstream by direct absorption. Still others, when inhaled, have directly irritating effects on the lungs. Long-term, low-level exposure can lead to chronic bronchitis and other diseases.

This section is arranged so that related organic chemicals are treated together. The best way to use this section is to skim through it rapidly, determine which chemicals may affect you, and then read those parts.

Alcohols

In general, as the size of the alcohol molecule increases, the alcohol becomes less volatile and the worker is therefore less likely to inhale its vapors.

Methanol (methyl alcohol, wood alcohol, wood spirit). During Prohibition many deaths were attributed to drinking bootleg whiskey contaminated with wood alcohol. If swallowed, methanol affects the nervous system, causing nausea, headache, blindness, delirium, and death. Industrial exposure to the vapor is apparently much less serious, but may cause irritation of the upper respiratory tract that, if severe enough, can lead to bronchitis or other lung disease. Poisoned workers show signs of drunkenness, trembling, nausea, and blurred vision. Serious intoxication can lead to liver and kidney damage. However, studies of workers exposed to vapor concentrations at about one-quarter of the TLV indicated no apparent toxic effects. Therefore methanol must be used in well-ventilated areas to keep the vapor concentration low. The odor is distinctive only at high concentrations and hence cannot serve as warning of exposure.

Ethanol (ethyl alcohol) is the alcohol that is part of alcoholic beverages, where it is produced by fermentation. Industrial ethanol is obtained by chemical processing. The intoxicating effects of ethanol are well known, and excessive drinking can lead to liver and other bodily damage. Industrial exposure to the vapor, unless intense, simply irritates the eyes and mucous membranes. Intoxication occurs only at very high levels. Prolonged skin contact can dry out the skin and produce a scaly dermatitis.

Butanols (n-butanol, isobutanol, sec-butanol, tert-butanol) can cause skin irritation and, in poorly ventilated conditions, serious inflammation of the eyes. Since they are not as volatile as the lower-molecular-weight alcohols, inhalation of their vapor is not as common. If vapor concentration exceeds the TLV, it can be very irritating to the respiratory tract.

Amyl alcohols (Pentasol, fuel oils) apparently act on the nervous system, causing dizziness, nausea, vomiting, and double vision. They can also affect the digestive tract, leading to diarrhea and other signs of illness. Excessive exposure can be fatal; deaths of workers inside tanks of lacquer containing amyl alcohol have been reported. The vapors are quite irritating to the eyes and upper respiratory tract.

Benzyl alcohol (phenyl carbinol) is irritating to the skin. Exposure to its vapors seem to have caused violent headaches, stomach pains, vomiting, and loss of weight, among other symptoms.

Cyclohexanol is irritating to the eyes, nose, and throat. Excessive exposure may cause vomiting and tremors.

Allyl alcohol (vinyl carbinol) has a very pungent odor and causes the eyes to water. It is extremely irritating to the upper respiratory tract and eyes, and can cause blurred vision and damage to the cornea. If exposure is terminated, cornea damage can sometimes be reversed. Skin contact can cause painful irritation and slow-healing burns. There may be muscle spasms at the site of skin contact.

Beta-chloroethyl alcohol (ethylene chlorohydrin) is extremely dangerous; several deaths have been attributed to it. It enters the body either by inhalation of the vapor or by absorption through the skin, which it readily penetrates. It can also penetrate rubber, so that rubber gloves are no protection against skin contact. Fatal cases show extensive liver and kidney damage, and symptoms before death include nausea, vomiting, loss of coordination, and weak pulse. Non-fatal cases show the same signs, with cough and rash. The chronic non-fatal effects are not known.

Ethers

The ethers are quite flammable and present a potential fire hazard. When inhaled, the smaller ether molecules have a pronounced anesthetic effect; the larger molecules are more irritating and can cause serious lung disease. Some of the ethers most commonly encountered are discussed below, starting with those having the smallest molecules.

Methyl ether volatilizes very easily. When inhaled, it has an anesthetic effect and depresses the nervous system.

Ethyl ether is highly flammable and even explosive. It is commonly used as an anesthetic gas, and occupational exposure will also anesthetize a person. It can dry out the skin and cause dermatitis. Repeated exposure above the TLV causes nose and throat irritation, and also loss of appetite, dizziness, excitation, and then drowsiness.

Isopropyl ether forms flammable and explosive mixtures with the air. It has an anesthetic effect and is irritating to the nose and throat. Repeated exposure of the skin can lead to dermatitis.

Vinyl ether is highly volatile and flammable. If used as an anesthetic, it can cause the patient to go into a deep coma and suffer paralysis of the respiratory system. However, no one has observed this effect in industrial workers.

Chloromethyl ether is highly irritating to both skin and eyes. It forms formaldehyde in the body, which may explain its irritating effects. Inhaling the vapor can cause the lungs to fill with fluid, leading to pulmonary edema and pneumonia. The odor should serve as a warning to workers: if they are irritated by the smell, they should refuse to work with this chemcial.

Sym-dichloroethyl ether is explosive when heated above room temperature; therefore storage procedures must be carefully followed. Although not irritating to the skin, it may penetrate it and cause severe, even fatal poisoning. It is especially irritating to the eyes and lungs. Several hours after exposure, a worker may develop pulmonary edema.

Dichloroisopropyl ether is explosive when heated. It can be absorbed directly through the skin and damage the liver and kidneys. Inhaling the vapors can also lead to kidney and liver damage.

Guaiacol ignites only it heated to high temperatures. It is extremely irritating to the skin, and inhaling the vapors can lead to severe inflammation of the breathing passages.

Dioxane (diethylene ether) is irritating to the eyes and respiratory tract. It does not irritate the skin, but can pass right through it without leaving any sign. Excessive exposure can damage the liver and kidneys and may even be fatal. The chronic effects of exposure have not been studied.

Pyridine and Derivatives

Pyridine is a foul-smelling chemical obtained from the distillation of coal tar and is widely used as a solvent. It is absorbed into the body through the skin and by inhalation of vapors. Workers exposed to high concentrations in the air will have symptoms of digestive disturbance such as diarrhea, stomach pain, nausea, and weakness, and also headache, irritability, and inability to sleep. Even if the exposure is not high enough to produce these signs, there will still be damage to the liver. Pyridine also irritates the skin, the mucous membranes of the breathing passages, and the cornea of the eye. Pyridine compounds producing similar symptoms are **2-chloropyridine, 2-aminopyridine, picoline, vinyl pyridine, piperidine,** and **quinoline.** In addition, 2-aminopyridine causes convulsions.

Glycols

Glycol derivatives include the **cellosolves,** which have had severe effects on test animals, though their effects on humans have not been widely studied. The **glycol ethers** are broken down by the body into the glycol from which they were derived, and have the same effects as the parent chemical.

Ethylene glycol and **diethylene glycol** are flammable liquids; exposure to the vapors occurs when they are heated. Inhaling the vapors has the same effects as drinking alcohol; the person may feel dizzy and even lose consciousness. These glycols can be absorbed through the skin. They are irritating to the eyes, but not to the skin.

Ethylene glycol monomethyl ether, more commonly known as **cellosolve** or **Dowanol EM,** is flammable. If large amounts are absorbed either through the skin or by inhaling the vapor, anemia and other blood abnormalities may develop. The nervous system is also affected, resulting in reflex and behavior changes. Long-term exposure may result in neurological changes such as staggering gait, tremors, forgetfulness, fatigue, mental confusion, and headache.

Ethylene glycol monoethyl ether (Dowanol EE), another cellosolve, is less toxic than Dowanol EM, but exposure to large amounts of the vapor can lead to lung irritation, pulmonary edema, and pneumonia. If absorbed through the skin, it can affect the kidneys and also depress the nervous system.

Other **ethylene glycol ethers** are mildly irritating to the skin and can be absorbed right through it. They can cause very painful irritation of the eye membranes. Excessive absorption can affect the nervous system and also cause liver, kidney, and blood damage.

Diethylene glycol ethers are less toxic than ethylene glycol ethers, but can have the same toxic effects if the exposure is high enough. Some of the diethylene glycol ethers are listed in Table 17.

Table 17. EFFECTS OF DIETHYLENE GLYCOL ETHERS

SUBSTANCE	EFFECTS
Propylene glycol	Low toxicity to animals. Not irritating. Used as a food additive.
Other propylene glycols	Do not volatilize easily, therefore vapors are not easily inhaled. Experimental animals have not suffered from skin irritation or eye damage even after long exposures. No studies of effects on humans.
Propylene glycol ethers	Markedly irritating to eyes, nose, and throat. Irritation may be so severe that lung damage is avoided because people shun further contact.
Butanediols	Animals exposed to 1,4-butanediol have suffered kidney damage. Not found to be irritating to animal skin, but can be absorbed through skin into bloodstream. No studies of effects on humans.
Polybutylene glycols	Exposed animals have suffered kidney damage. These glycols not absorbed through skin and not irritating to animals' skin or eyes. No studies of effects on humans.

Aromatic Hydrocarbons

Aromatic hydrocarbons are compounds derived from benzene (benzol). Many of them have similar effects on the body, but they vary greatly in their toxicity. All the liquids in this group are irritants and can cause dermatitis after prolonged skin contact. The vapors are irritating, and inhalation can lead to damage to other organs. Although, as their name implies, the aromatics have distinctive odors, the nose soon loses the ability to discern them, so that their smell cannot be depended upon as a warning against excessive exposure.

Benzene is an extremely toxic chemical that is not treated with the respect it deserves. In high doses, it acts on the nervous system, causing drowsiness and loss of consciousness. Exposure to a high dose can lead to chronic benzene poisoning, symptoms of which are headache, dizziness, fatigue, loss of appetite, irritability, and nervousness.

Benzene acts on the bone marrow and destroys its ability to produce blood cells. This can lead to anemia, a shortage of white blood cells, and a shortage of platelets, the cells that help the blood to clot. Blood damage can occur without any outward symptoms. Similarly, a person may have symptoms of excessive exposure without any noticeable effects on the blood. In addition, benzene has been associated with *leukemia* in susceptible people. It has also been found to damage the chromosomes, which contain the genetic material that passes on heredity to one's children and controls the day-to-day production of material for the body. People under 18 and people who drink heavily or have any history of blood or liver disease should not work with benzene. Pregnant women should not be exposed, as there is a good possibility that benzene can damage an unborn child.

Blood examinations of exposed workers and air samples of work areas should be taken periodically. Wherever possible, less toxic materials should be substituted for benzene. If this cannot be done, proper ventilation and the safest engineering processes are essential. It should be noted that the odor level of benzene is well above the TLV.

Toluene is a powerful narcotic that causes symptoms similar to those of getting drunk. It also irritates the skin, eyes, and upper respiratory tract. Many of its chemical properties are the same as those of benzene, but it apparently does not do as much damage to the blood. It would therefore make good sense to substitute toluene for benzene wherever possible.

An indirect hazard of exposure to toluene is that a worker's judgment and reflexes are severely impaired, so that she or he is much more prone to industrial accidents. Although toluene has a distinctive odor at the threshold level, the nose soon loses the ability to discern its smell, so that the warning properties are gone.

Xylene (xylol, dimethylbenzene) is far less toxic than benzene, but it also can cause a decrease in the number of red and white blood cells. It irritates the skin and may lead to a dry, scaly, cracked dermatitis. In high doses it acts as a narcotic and is irritating to the eyes and upper respiratory tract. Individuals respond differently to xylene, but chronic exposure can lead to a decrease in resistance to its effects. Women may develop problems with menstruation, and repeated exposure may lead to heart problems. This chemical should not be blindly substituted for benzene, even though it is apparently less toxic.

Cumene is more toxic than benzene in high concentrations. The effects of long-term exposure have not been studied. Animals exposed to cumene in relatively low doses have shown no effects.

Naphthalene volatilizes easily, and has a familiar "moth ball" smell. High concentrations of the vapor irritate the skin and eyes and can cause a worker to develop a chronic allergy. It also affects the blood and causes anemia. Exposure can lead to liver and kidney damage, as well as damage to the optic nerve. Adequate ventilation is necessary to prevent such exposure.

Tetralin (tetrahydronaphthalene) has caused painters to develop headaches, fatigue, and irritation of the eyes, nose, and throat.

Trimethylbenzene (mesitylene) is irritating and can cause an attack of asthmatic bronchitis. Although no research has been done on its chronic respiratory effects, it is likely that repeated exposures and asthma attacks lead to chronic lung disease. This chemical can also cause blood-coagulation disease, which can lead to nosebleeds, easy bruising, and other hemorrhaging. An exposed person may develop anemia.

Decahydronaphthalene (Decalin) has not been widely

investigated. It is known to cause dermatitis and probably damages the kidneys as well. Urinalysis of an exposed worker may show cells and protein, signs of kidney misfunction. Other effects are similar to those of tetralin, described above.

Diphenyls

The diphenyls have many uses in industry, and their hazards are not really well known. Those which contain chlorine may cause a skin condition similar to acne, which can become infected. The diphenyls are of great environmental concern because, like DDT, they are very stable and remain in the environment for long periods of time. They may have long-lasting effects on birds, fish, and other creatures. Scientists have found that they cause cancer in some laboratory animals, and there is concern that they will have the same effect on people.

Phenols and Cresols

Phenol is a derivative of benzene and a constituent of coal tar. The other compounds in this class are derivatives of phenol, and many of them are chemically made from it. In the past, some of these chemicals were used in medicine as disinfectants, but this use has been largely discontinued because of their serious toxic effects. This has led some authorities to believe that human exposure to phenol is now rare. However, interviews with chemical workers involved in the use and manufacture of phenols show that in fact many workers are needlessly and dangerously exposed.

One danger in handling the phenols is that they are so easily absorbed through the skin. Extensive absorption can affect both the central nervous system and the circulatory system. Depression of the nervous system may be so great that the nerve signals sent out by the brain to the rest of the body become very weak. The nervous impulses that cause breathing are also weakened, and the victim often dies

Table 18. EFFECTS OF SOME DIPHENYLS

COMPOUND	EFFECTS
Diphenyl	Effects not known, but not believed to be hazardous because it does not volatilize easily.
Diphenylmethane	Same as diphenyl.
Diphenyl ether	Very irritating to skin after repeated or prolonged contact.
Dowtherm (diphenyl ether and diphenyl)	Extremely irritating to eyes and air tubes. High doses may cause irreversible kidney and liver damage. Repeated or prolonged contact may cause skin irritation. When heated, vapor may leak from gaskets or valves; strong smell should serve as warning of exposure.
Chlorodiphenyls	Irritating to skin, causing an acnelike condition that can become infected. In severe exposure, there has been nausea, appetite loss, lassitude, digestive disturbances, impotence, and bloody urine. Liver involvement may even be fatal. Stillbirths may result from exposure during pregnancy. Delayed effects of such exposures not known.

of respiratory failure. Acute poisoning is characterized by other nervous disorders, such as ringing in the ears, tremors, and convulsions. If the victim survives, he often develops pneumonia. The circulatory system is severely affected; the body temperature falls, and the victim must be given stimulants just to keep the circulation going. Phenol is also locally corrosive to the skin. Very dilute solutions can cause contact dermatitis.

Chronic phenol poisoning is different; it is characterized

by vomiting, diarrhea, and difficulty in swallowing, all indicating trouble with the digestive system. There are also nervous disorders such as fainting, nervousness, insomnia, and headache. At least some liver and kidney damage is likely, and if it is too great it usually ends up killing the victim. If the skin is exposed to phenols for a length of time, it will react with them and turn brown. Of course, skin contact also means that the chemicals have been absorbed into the body.

Workers should see to it that they learn the safest techniques possible for handling the phenols. Excellent ventilation and general industrial hygiene are also necessary. Any skin contact should be immediately flushed with water. Since these compounds can be serious pollutants, workers should make sure that phenolic wastes are not simply dumped into the air or water. It has been shown that at 4 ppm (1 ppm below the TLV), fruits and vegetables grown within a mile of a plant that was polluting the atmosphere with phenol all tasted of the chemical.

Mice exposed to phenol and 50 of its derivatives have developed cancer. Although no one has studied humans exposed to phenol for the occurrence of cancer, the only prudent action is to *assume* that it can cause cancer and to strictly avoid contact with it. Further research will tell whether or not this precaution is justified, but the initial animal-study evidence is quite incriminating, and no one should take the chance of being exposed.

Hydroquinone, resorcinol, and **pyrocatechol** have similar effects to phenol, and pyrocatechol is even more toxic.

Dinitro-ortho-cresol (DNOC) is used in dyes and as a pesticide and herbicide. It blocks many of the most important body processes and interferes with the body's use of oxygen. Agricultural workers in hot, sunny climates have been fatally poisoned from repeated exposures over several days in a room.

The first signs of poisoning are fatigue, excessive sweating, and a tremendous thirst. The weakness and fatigue become worse. In a few hours, breathing becomes heavy and the heartbeat rapid, because the body is desperately trying to get enough oxygen to live. The skin may turn yellow, as the body breaks DNOC down into a yellow-colored chemical. If exposure is ended and the organs have not been damaged from lack of oxygen, recovery can be complete. Long-term effects of small doses are not known.

Dinitrophenols are explosive. They are used as intermediates in the dyestuff industry and as wood preservatives. Like dinitro-ortho-cresol, they poison the basic energy system of the body cells. They can penetrate intact skin, leaving a bright yellow stain. First symptoms of poisoning are excessive sweating, weakness, and fatigue, followed by rapid breathing and rapid heartbeat. There may be fever. In hot workplaces, poisoning will be more severe. If the victim survives and there is no damage to oxygen-starved organs, recovery can be complete. Long-term effects of exposure are not known.

Some Other Aromatics

Chlorobenzene and its derivatives are all flammable. They irritate the skin and eyes, and direct contact can lead to dermatitis. Their toxicity is increased in that they enter the body both by absorption through the skin and by inhalation of the vapors. High-level exposure can have an anesthetic effect, and may also cause liver and kidney damage.

Ortho-chlorobenzene has effects similar to chlorobenzene. Industrial workers have suffered kidney damage from exposure. **Para-dichlorobenzene** has a strong odor like mothballs and irritates the eyes and nose at levels below the TLV. Either the solid or the hot fumes can irritate the skin. High-level exposure may damage the liver and kidneys.

Benzyl chloride produces dermatitis on exposed skin. Splashes into the eyes cause intense burning, weeping, and inflammation; the vapor may also cause these symptoms.

This compound can damage the nervous system, resulting in weakness, fatigue, headache, irritability, and loss of sleep and appetite. There may also be kidney and liver abnormalities.

Chlorinated naphthalenes become more toxic as the number of chlorines in the molecule is increased, but they also become less volatile, so that there is less exposure. The main hazards are liver damage and skin disease. The compounds dry the skin by dissolving the natural oils. Exposure to sunlight can make this condition worse. Like many other chlorinated compounds, these can cause acne, which may become infected. The acne usually occurs in the areas of direct contact, but clothes contaminated with the chemical can also expose the skin underneath and cause a reaction.

Benzanthrone is used in making dyes. Its worst effects occur in hot environments where workers sweat, and on blond people with fair skin. Direct contact irritates the skin and may cause it to itch, peel, and thicken, especially in people with allergies. This chemical also affects the nervous system, resulting in weakness, rapid pulse, loss of appetite, weight loss, and decreased sexual potency.

Aromatic Nitro and Amino Compounds

Each chemical in this group is a derivative of benzene. The parent aromatic nitro compound is **nitrobenzene** and the parent aromatic amino compound is **aniline,** a common dye. Other compounds, all with similar toxic effects, are made by adding chemicals onto these parent compounds.

The most outstanding effect of this group of compounds is on the hemoglobin in the blood. By chemically binding into the hemoglobin and converting it to *methemoglobin,* they make it unavailable for carrying oxygen. This produces *methemoglobinemia,* a disease in which the oxygen level of the blood is diminished. One symptom is a condition known as *cyanosis,* in which the eyes, fingertips, ears, and lips turn blue. The intensity of cyanosis, however, is not a good measure of the amount of methemoglobin, for often a significant amount of hemoglobin must be converted before cyanosis appears.

Methemoglobinemia is a tricky disease, because the victim may turn blue without noticing it and may even feel happy and euphoric. But as the illness increases in intensity, he or she develops a headache, a feeling of weakness and dizziness, and shortness of breath under exertion. If exposure to the toxic chemicals is ended, the methemoglobin can be converted back to hemoglobin. Many authorities believe this can happen without any long-term effects and do not give much credit to the reports of liver damage and other problems found in the German medical literature. Others believe that repeated or prolonged attacks can lead to chronic anemia.

A worker who turns blue from exposure to these compounds should be removed immediately from the vicinity. Since these chemicals are easily absorbed through the shoes, clothing, and skin, the victim must remove all clothing and wash thoroughly with soap and water, especially under the fingernails and toenails. He or she must then rest in bed until blood tests show that there is no longer any methemoglobin present.

Other serious problems are involved in the use of these chemicals, especially those commonly encountered in the production of dyestuffs and their intermediates. These problems include serious skin sensitization and, for a few of the compounds, the ability to cause cancer. For years the occurrence of cancerous bladder tumors among workers involved in dye production was attributed to aniline, and the disease was commonly called *aniline bladder tumor*. Although aniline is a very toxic chemical, it is probably not responsible for this particular illness. However, **beta-naphthylamine, benzidine, 4-aminodiphenyl,** and **2-acetyl-aminofluorene** have been strongly indicted in causing bladder tumors. Really careful studies have not been performed yet to eliminate the question of the cancer-causing ability of the aromatic amines; in the meantime, workers should make sure that they come into *no* contact with any of them.

Some chemicals in this group are irritating to the eyes and skin. Others, such as **para-phenylenediamine,** cause sensitization that can result in allergic dermatitis or asthma. The **toluidines** and **5-chlorotoluidine** directly irritate the urinary bladder, resulting in painful, burning urination and a frequent urge to urinate. There is usually blood in the urine before any symptoms are felt. This condition is called *hemorrhagic cystitis.*

Although some authorities say there are no other problems in the dyestuff industry, they also state that no one has really done much toxicological research. They cite the lack of "reported" illness in this industry, which has been in production for about 100 years. Since it has not been proved that there is no danger, workers should be extremely cautious about exposure to these chemicals. Some general hygiene standards that should be maintained are:

1. Wash-up time and showers must be provided.
2. Workers must be provided with daily clothing changes, and any clothes contaminated by spillage must be changed immediately.
3. Workers must be checked daily for visual symptoms of turning blue, since they may not be aware of the onset.
4. Air sampling must be done frequently.
5. Workers must be given blood tests regularly for red, white, and differential blood counts, and must have regular urinalyses.
6. Older workers who have been exposed more than 20 years must be checked regularly for bladder tumors. The checkup involves examination of the urine for cancer cells and blood cells. These examinations should not be stopped when the workers retire, since tumors may occur 40 or more years after initial exposure.

Table 19. HEALTH HAZARDS OF THE NITRO AND
AMINO AROMATICS

COMPOUND	IRRITATION	ALLERGY	METHEMO- GLOBINEMIA	ACUTE BLADDER IRRITATION	BLADDER TUMORS	COMMENTS
Aminophenols		dermatitis	not common	no	no	
Aniline			yes	no	no	Readily absorbed through skin. Even under nails can cause blood effects. Repeated exposure can lead to chronic anemia.
Benzidine					yes	Salts also cause cancer. Absorbed through skin and also by breathing vapor.
Chloroanilines	eyes		strong			Absorbed through skin.
5-chloro-ortho-toluidine				yes	no	
N,N-diethyl aniline	no	no	strong	no	no	Effects similar to aniline.
N,N-dimethyl aniline	no	no	yes	no	no	Effects similar to aniline.
diphenyl aniline	no	no	no	no	no	Hazardous only because it usually contains impurities that cause cancer. Impurities are concentrated by distillation.

Table 19. HEALTH HAZARDS OF THE NITRO AND AMINO AROMATICS (CONTINUED)

COMPOUND	IRRITATION	ALLERGY	METHEMO-GLOBINEMIA	ACUTE BLADDER IRRITATION	BLADDER TUMORS	COMMENTS
1-naphthylamine	no	no	no	no	possibly	Absorbed through skin and by breathing. Usually contaminated with carcinogenic agents.
2-naphthylamine	no	no	no	no	very high incidence	Absorbed through skin and by breathing vapors.
para-nitroanilines	no	no	strong	no	no	
para-nitroso-N,N-dimethyl aniline	skin	dermatitis	no	no	no	Used in hair dyes and furs, which may cause dermatitis among users.
para-phenylene-diamine		dermatitis common; respiratory allergy				
toluidines	no	no	yes	moderate, can cause bleeding	no	Absorbed through skin or by breathing vapor or dust.
xylidine	no	not known	no	not known	possibly	Animals develop liver and blood disease. Effects on humans not known.

Coal Tar and Fractions

Coal tar is broken down into many other chemicals. The most severe reactions come from the less volatile fractions. All coal tar fractions react with the skin and photosensitize it—that is, make it more susceptible to sunlight and ultraviolet radiation. Exposed parts of the body develop dermatitis, which may be either acute or chronic and is accompanied by pains, swelling, rash, and blisters. Exposure to the vapors can irritate the eyes and respiratory tract. Lung irritation may occasionally cause pulmonary edema.

Acridine (10-azoanthracene) is a powerful irritant that particularly affects the eyes; it causes watering, itching, burning, and violent sneezing. It also photosensitizes the skin and causes dermatitis and skin eruptions, which may become infected.

Anthracene irritates the skin, eyes, and respiratory tract. Because it is a powerful photosensitizer, anthracene workers should either avoid exposure to sunlight or protect themselves from it with a barrier cream of zinc oxide, glycerine, kaolin, starch, and gelatine.

People who work with anthracene can develop cancer, but there is a question whether this is caused by pure anthracene or by impurities that are found with it.

Coal tar creosotes are complex mixtures of more than 160 chemicals, many of them aromatic. They are irritating to the skin, eyes, and mucous membranes of the air tubes. They do not seem to have any ill effects on the other organs of the body unless swallowed, when they act similarly to phenol. After years of repeated exposures, a worker may develop skin cancer.

Various Petroleum Products

Asphalt (mineral pitch, bitumen) is found naturally and is also produced by distillation of crude petroleum. Hot

Table 20. CANCER-CAUSING ABILITY OF
ANTHRACENE DERIVATIVES

STRONG CARCINOGENS	MODERATE CARCINOGENS
1,2-benzanthracene and methyl or alkyl derivatives 9,10-dimethylanthracene 5,9-dimethyl-1,2-benzanthracene 9,10-dimethyl-1,2-benzanthracene 5,10-dimethyl-1,2-benzanthracene 3-methlycolanthrene	5-,9-,methyl-1,2-benzanthracene 10-methyl-1,2-benzanthracene 5,9,10-trimethyl-1,2-benzanthracene 6,9,10-trimethyl-1,2-benzanthracene

WEAK CARCINOGENS	NOT CARCINOGENIC
1,2,5,6-dibenzanthracene 1,2,7,8-dibenzanthracene 3,4,7-methyl-1,2-benzanthracene 8-methyl-2-benzanthracene	1-methyl-1,2-benzanthracene 2-methyl-1,2-benzanthracene dimethyl derivatives with substituted methyl groups in the additional benzene ring

asphalt can cause severe burns because it is sticky and hard to remove from the skin; adequate eye, face, hand, arm, and foot protection from splashes is necessary. The fumes are highly irritating to the eyes and skin. An additional danger arises from benzene and hydrogen sulfide, usually present in the fumes as impurities or solvents.

The most common troublesome reactions to asphalt result from skin irritation. The skin may be itchy and inflamed and break out in pimples, which can become infected. Exposure to sunlight may make all these reactions worse; asphalt and sunlight combined can also darken the skin. Asphalt may also cause skin cancer.

Animals exposed to the fumes develop lung and air-tube

inflammation, pneumonia, and bronchitis. Perhaps from a lack of study, this reaction is not commonly reported in asphalt workers.

Lubricating oils, including **mineral oil,** are mixtures of many different hydrocarbons. Skin contacts from greasing operations and from oil mists thrown off by moving machinery can cause dermatitis, and may cause skin cancer after many years. Oil in the lungs causes *chemical pneumonia,* an inflammation of the lungs. Repeated exposures to oil mists can lead to lung scarring and a disability similar to asbestosis. Cancer has been found in the scarred lungs of workers exposed to mineral oils. The use of oils with the cancer-causing impurities removed has been proposed as a means of preventing skin and lung cancer.

Kerosene is also a mixture of many hydrocarbons. It dries out and irritates the skin by dissolving its natural oils, so that using it to wash the hands is a dangerous practice. It is extremely irritating to the lungs, and its vapors can cause lung inflammation, or chemical pneumonia.

The volatile fractions of petroleum products such as **gasoline** have a mild anesthetic effect. If inhaled in large quantities, they can cause extreme lung irritation, which may lead to chemical pneumonia.

Halogenated Hydrocarbons

These chemicals are compounds of carbon, hydrogen, and some combination of chlorine, bromine, fluorine, or iodine (the halogens). Most of them have excellent solvent properties and are quite volatile. Their effects range from relatively non-toxic to both acutely and chronically toxic, causing severe liver or kidney injury (allyl chloride, for example) and even death. One property that many of them have in common is their narcotic effect, ranging from mild drunkenness to deep anesthesia. In fact, several halogenated hydrocarbons were used in medicine as anesthetics. Many are quite irritating. A number of them may stimulate the body's production of adrenaline, causing irregular heartbeat, dizziness, fainting, or sudden death. Chloroform and trichloroethylene have caused this reaction in industrial workers, sudden death from irregular heartbeat having occurred several hours after they left the workplace. A person who has been exposed to high concentrations of these substances *should not* be given adrenaline. People with heart disease should avoid exposure to them.

The use of alcoholic beverages by persons continually exposed to chlorinated hydrocarbons can have serious effects, since both the alcohol and the hydrocarbons have bad effects on the liver and the combination greatly increases the probability of liver damage.

Methyl chloride is often a side product in the manufacture of polystyrene and is also used as an aerosol propellant. It has no odor, and the effects of exposure may be delayed. It is a narcotic that affects the central nervous system and can cause blurred vision, tremors and staggering, nausea, and vomiting. Acute exposure can result in kidney and liver damage, while mild exposure may cause drunkenness, increasing the chance of industrial accidents. The long-term effects of continual exposure have not been studied, though if illness arises from such exposure, the recovery time may be lengthy.

Methylene chloride is extremely volatile and reaches very high concentrations in a short time, especially if the temperature in the work area is high. Both the liquid and the vapor are irritating to the eyes and upper respiratory tract, and extended skin contact can cause dry, scaly dermatitis. It has narcotic effects, with consequent loss of coordination and alertness. The body converts methylene chloride to carbon monoxide, which takes the place of oxygen in the blood and deprives the body of sufficient oxygen for proper functioning. High concentrations can cause serious lung irritation, resulting in pulmonary edema and bronchitis. Excellent ventilation is necessary when this chemical is used, because of its volatility and because the nose rapidly becomes accustomed to the smell.

Chloroform is a well-known anesthetic, but its medical use has been reduced because it can cause liver injury. Kidney damage may follow either high-level or long-term exposure. Although little research has been done on prolonged industrial exposure to chloroform, it is reasonable to believe that it is an extremely toxic substance to work with. The odor cannot be detected until the concentration is 4 to 6 times the legal limit of exposure.

Carbon tetrachloride is a recognized industrial poison that affects the nervous system. High-level exposure causes loss of consciousness; lower-level exposure results in dizziness, headache, feelings of depression and confusion, and loss of coordination. Repeated exposure to low levels can severely damage the liver and kidneys, and this damage can also follow a single high-level exposure. One cannot depend on the smell for warning, since the odor level is more than 3 times the legal level for exposure. When this chemical is used in fire extinguishers, there is the added danger of its conversion to phosgene by the flames.

Ethylidene dichloride is often found as an intermediate in chemical reactions. Little study has been made of its effects on man, but preliminary work on animals indicates that long-term exposure can damage the liver.

Ethylene dichloride is used as a solvent and as a gasoline additive. Prolonged exposure to the vapors has been found to injure the liver and kidneys and the adrenal glands. It irritates the eyes and upper respiratory tract and may produce coughing. It depresses the central nervous system, and also causes nausea and vomiting. Contact with the liquid can dissolve the oils in the skin and result in dry, scaly dermatitis.

Tetrachloroethane is probably the most toxic of the halogenated hydrocarbons. The site of greatest injury is the liver, which it causes to shrink and turn yellow. Early symptoms of exposure resemble drunkenness or anesthesia. It can be absorbed right through the skin without much irritation. The odor at low concentrations is not sufficient to be a good warning of danger.

Propylene dichloride causes depression of the central nervous system with typical symptoms of dizziness, headache, confusion, and loss of coordination. Prolonged exposure can cause degeneration of the liver, kidneys, and heart. Repeated skin contact, or contact that is kept close by clothing, can produce dry, cracked, scaly dermatitis.

Trichloroethylene (acetylene trichloride) has marked depressive effects on the central nervous system, with symptoms of dizziness, fatigue, headache, nausea, and loss of coordination. An exposed person often cannot tolerate alcoholic beverages. Liver and kidney damage have been reported, but may have been due to contaminants rather than to the trichloroethylene itself. Industrial fatalities have resulted from exposure to high vapor concentrations. This substance can cause a sudden heart irregularity or stoppage in the workplace or several hours later, resulting in death. Another danger is that this chemical releases phosgene and hydrochloric acid when heated. The serious effects on the lungs of exposure to these two gases are discussed elsewhere.

Tetrachloroethylene, or **perchloroethylene,** is believed to have effects similar to trichlorethylene. It is a strong narcotic. In a plant studied, one-half the workers exposed to this chemical suffered from hepatitis, loss of appetite, weakness, and headache. Animal studies and liver-function tests of workers complaining of narcotic symptoms indicate that liver damage results from prolonged exposure. Mild intoxication, as in exposure to improperly aired dry-cleaned goods, probably occurs frequently and goes unreported.

Allyl chloride is extremely irritating to the eyes and upper respiratory tract. Liver and kidney damage result from prolonged exposure, and eye irritation may be severe and painful. Because of its toxicity, it has been recommended that the threshold limiting value of allyl chloride be lowered and that extreme caution be used in handling it.

Methyl bromide (bromomethane), used chiefly as a fumigant, has no odor-warning properties. Skin contact can cause serious burns. Even exposures below the TLV can cause weakness or paralysis of the arms and legs, from which recovery may take many months. Repeated exposure to levels slightly above the TLV can produce nausea, vomiting, fatigue, headache, and giddiness. Higher levels can cause pneumonia, pulmonary edema, convulsions, coma, and kidney and liver damage. If the victim survives an acute exposure, there may be permanent brain damage.

Authorities have suggested that a pungent gas, **chloropicrin,** be added to methyl bromide as an odor marker. They also recommend permanent monitoring devices for all operations involving methyl bromide.

Bromoform (tribromomethane) is irritating to the mucous membranes of the throat and air passages, but little is known about its other effects. Preliminary evidence, however, indicates that it is quite toxic, causing depression of the central nervous system and liver and kidney damage.

Carbon tetrabromide is highly toxic. Severe liver damage can occur after prolonged exposure to low concentrations of vapor. Low concentrations also cause tearing of the eyes. As eye contact can cause permanent cornea damage, the eye should be flooded with water immediately upon exposure. High concentrations are extremely irritating to the respiratory tract and cause extensive liver, kidney, and lung damage.

Ethyl bromide (monobromoethane) depresses the nervous system and causes chronic disease with such symptoms as staggering gait, muscle weakness, speech disorder, and trembling. The victim may improve slowly after exposure is ended. Acute exposure leads to kidney and liver damage, which may also result from long-term exposure.

Ethylene dibromide, which can be absorbed through protective clothing and shoe leather, causes serious injury to the skin. Eye contact is extremely irritating. Breathing the vapors can cause lung congestion, pneumonia, and pulmonary edema. There is also a moderate effect on the nervous system. If the compound is swallowed, kidney and liver damage will occur, but no one has observed this effect in workers who breathe the vapors.

Ketones

Ketones are used as reactants and intermediates in chemical processes, but their widest use is as solvents for fats, dyes, oils, waxes, and natural and synthetic polymers. In general, they appear to be of low toxicity, but in concentrations above the TLV they have a narcotic effect and inhalation of the vapor can cause drowsiness and dizziness. Excessive exposure may result in coma and depression of the respiratory functions, followed in severe cases by death. The odor-warning properties of ketones are good and should not be ignored, so that excessive exposure can be avoided.

A more usual result of industrial exposure is mild narcosis with its inherent danger of accidents due to weakening of judgment and control. The ketones are also irritating to the eyes and upper respiratory tract. Because of their ability to dissolve fats, they cause dry, scaly, cracked skin. Since the body apparently rids itself of them quickly, their effects throughout the body are not severe. However, long-term exposure does lead to chronic irritation of the lungs and possibly to emphysema. There are indications that alteration of the tissues of the liver, the kidneys, and sometimes the brain may occur after excessive exposure to some of the ketones.

Methyl isopropenyl ketone is a potent irritant to the eyes and respiratory tract. The warning properties are not strong enough to prevent inhalation of harmful amounts upon repeated exposure. Experimental animals have shown severe

Table 21. Some Ketones

COMPOUND	MOLEC- ULAR WEIGHT	TLV (PPM)	EFFECTS
Acetone	58	1,000	Slight irritation at levels be- low TLV.
Methyl ethyl ketone (MEK)	72	200	
Methyl-n-propyl ketone	86	200	
Methyl-n-butyl ketone	100	100	
Methyl isobutyl ketone (MIBK)	100	100	TLV may not be low enough to prevent discomfort and irritation.
Methyl-n-amyl ketone	114	N.E.	
Ethyl butyl ketone	114	N.E.	
Diisopropyl ketone	114	N.E.	
Ethyl-sec-amyl ketone	128	N.E.	Irritating to eyes. May cause nausea if used in poorly ventilated areas.
Diisobutyl ketone	142	50	May be irritant and narcotic at higher temperatures.

N.E.: not established

damage to lungs, kidneys, and spleen. Repeated irritation by this chemical possibly leads to chronic lung disease in humans. Adequate ventilation is imperative, and respiratory equipment should be available for emergencies.

Mesityl oxide. If the warning properties of this highly irritating liquid are ignored, acutely toxic amounts may be absorbed through the skin, causing lung, kidney, and liver damage. Repeated skin contact with small amounts may have the same results. At levels above the TLV, this chemical can have narcotic effects.

3-butyn-2-one is extremely dangerous. Fatal amounts can be absorbed through the skin, resulting in lung, liver, and kidney damage, and it is severely irritating to the eyes and respiratory tract. Good ventilation, impermeable clothing, and emergency respiratory equipment must be provided.

3-pentyn-2-one is another very reactive and probably very toxic chemical. It is suggested that all exposure be avoided and impermeable clothing be made available.

Acetophenone (phenyl methyl ketone, acetylbenzene, hypnone) becomes volatile at high temperatures and can be expected to have anesthetic effects on workers unless good ventilation is provided. Skin and eye contact can be irritating.

Isophorone (trimethylcyclohexanone). It is not clear what the results of exposure to this chemical are. However, it has been noted that after repeated contacts a person becomes insensitive to the odor and the irritation, so that these warning properties may not be adequate to prevent overexposure.

Cyclohexanone may have narcotic effects at levels above the TLV, which is probably double the value that would prevent irritation and discomfort for an 8-hour day. **Methylcyclohexane** may exhibit the same effects.

Chloroacetone is used as tear gas because it is intensely irritating to the eyes, throat, and upper air passages—as irritating as chlorine.

Bromoacetone is intensely poisonous and irritating, and should be handled only in closed systems. If there is a risk of exposure to the vapors, gastight chemical safety goggles and respiratory equipment should be worn.

Aldehydes and Ketals

Many aldehydes are volatile and flammable liquids. As a group they irritate the skin, eyes, and respiratory tract, the most volatile members naturally being the worst offenders. **Chloral hydrate** and **paraldehyde** are used in medicine as sedatives and sleeping pills because of their effect on the nervous system.

The ketals are extremely reactive. Their effects are unknown, but because of their high reactivity they should be used only in enclosed, properly ventilated places.

Acetaldehyde irritates the respiratory and bronchial passages. It has a slight narcotic effect, and in high concentrations can cause headache, stupor, and lung misfunction. Death can occur from kidney or liver damage, or from injury to the heart muscle.

Formaldehyde is highly irritating to the eyes, throat, and respiratory system. It can cause dermatitis. Allergies may develop, leading to nasal congestion or asthma from contact with small amounts. Formaldehyde residues in plastics and treated fabrics may cause allergic reactions in the worker or consumer.

Furfural (furfuraldehyde) and its derivatives have strongly irritating vapors, which can cause dermatitis and allergic skin response. Prolonged exposure to the vapors may irritate the respiratory tract and cause fatigue, headache, tremors, and numbness of the tongue. These symptoms may disappear if exposure is ended.

Organic Acids

The organic acids have widely varied chemical structures, with many possible substitutions and combinations, so that they have a variety of effects. They are characterized by the presence of a chemical group called the **carboxyl** group, and are sometimes referred to as **carboxylic acids.** Some derivatives of the organic acids are the result of chemical reactions with the carboxyl group: examples are acid anhydrides, acid halides, and acid amides.

Organic acids are irritants, but the degree of damage to the skin and lining tissues depends on how strong the acid is and on its solubility in water or tissue. In general, the larger the molecule, the less soluble it is and the weaker its acidic and irritant properties. For example, salicylic acid is very soluble and can penetrate tissue, where it has serious caustic effects, while stearic acid, whose molecules are much larger, is less soluble and has practically no irritant effects. If the acids are also acrylates, aldehydes, ketones, or hydroxy or nitro compounds, the chances of local tissue damage are greatly increased, since these other chemical groups are also irritants in many cases.

Certain organic acids cause allergy by altering the proteins of the skin. The body then treats the changed protein as a foreign substance and makes antibodies to react with it, resulting in chronic skin inflammation, itching, and pain.

Some organic acids, when absorbed into the body, prevent its normal chemical reactions from occurring. One way they do this is by binding and interfering with the body's enzymes. Enzymes are proteins that aid reactions in the body just as catalysts aid chemical reactions in industry. Fluoroacetate and iodoacetate, two halogenated acids, interfere in just this way with several of the body's chemical reactions.

Other organic acids interfere with the body's metabolism by substituting themselves for chemicals normally present in the body's complex reactions. Metabolism can be described as a series of cycles in which food is chemically broken down and converted to forms useful for body processes. Each step in the cycle is actually a chemical reaction, and certain toxic chemicals replace normal body chemicals in some of these reactions. Since the body's chemistry is very exact, this substitution works only for a few steps and then the cycle cannot continue. The body can no longer metabolize properly, and illness results. Some chemicals that cause this type of abnormality are acid anhydrides, acid halides, acrylamide, dimethlyl formamide, and dimethyl acetamide.

Acid anhydrides. The word *anhydride* means "without water," and that is just what acid anhydrides are—2 molecules of organic acid with water missing from the chemical structure. For example, acetic acid (vinegar) is an organic acid, and acetic anhydride is 2 acetic acid molecules with a water molecule removed. The acid anhydrides, some of which are described in Table 22, are all severely irritating to the eyes, skin, and respiratory tract because they absorb water from them and return to their original acid structure. They can also sensitize the skin, resulting in hives or chronic dermatitis.

Phthalic anhydride is one chemical of this group whose health effects on humans have been studied. Exposed workers showed chronic irritation of the eye membranes (*conjunctivitis*), hoarseness, blood-tinged nasal discharge and sputum, and chronic cough. Some had developed chronic bronchitis and emphysema. The substance sensitizes the skin, resulting in hives and dermatitis, and may cause respiratory-tract allergies such as asthma. Phthalates are being studied to see if they cause liver damage.

Halogenated acids consist of an organic acid combined with a halogen—bromine, iodine, chlorine, or fluorine. Examples are **bromoacetic acid** (acetic acid plus bromine) and **fluoroacetic acid** (acetic acid plus fluorine). These chemicals absorb water easily and produce severe irritation of the eyes, skin, and lining tissues of the mouth, throat, and respiratory tract. Delayed effects are common: for example, deep blisters, probably from the reaction of the compounds with tissue protein. Acid halides should be handled only after proper education, protective equipment, and ventilation have been provided.

Acid amides are organic acids with an amine group, 1 nitrogen and 2 hydrogen atoms, chemically added onto the carboxyl group. Most of the simple organic acid amides are apparently not highly toxic to humans, are only slightly irritating, and only rarely cause sensitization: this group includes **formamide, propionamide, butyramide, valeramide, lauramide, palmitamide, oleamide,** and **stearamide. Acetamide,** however, has been found to damage the liver and may even cause liver cancer.

Acid amides with other chemical groups added on appear to be more toxic. **Thioacetamide,** like acetamide, damages the liver and may also cause liver cancer. **Acrylamide,** one of the most important and most toxic of this group, causes an unusual disease of the nervous system. Initial symptoms are fatigue, drowsiness, and tingling in the fingers; further exposure leads to loss of balance and difficulty in walking. The symptoms are slowly reversible on removal from exposure, but recovery may take almost a year. Exposure may be from inhalation of the dust or from skin absorption, which occurs easily. Acrylamide is also irritating to the skin.

Another chemical in this group, **methyl acrylamide**, is not reported to cause human illness. **N,N-dimethyl acrylamide** causes illness in animals similar to that caused by acrylamide in humans, but its effects on humans have not been studied. **N-isopropyl acrylamide** irritates the skin and may also be absorbed through it. It is toxic to animals, but there are no reports of toxicity in humans.

N,N-dimethyl formamide is a moderately strong skin irritant. It causes liver and kidney damage in animals, but human effects have not been documented. **N,N-diethyl formamide** and **N,N-dimethyl acetamide** are also absorbed through the skin, but toxicities are unknown. They do cause sickness in animals.

Naphthenic acids are derived from petroleum and make up a derivative group of the compound naphthalene. When lead, cobalt, and manganese are part of these acids, health hazards can arise from either the acids or the metal or both. These acids are irritating to the skin and mucous membranes. Their toxicity in humans has not been investigated.

Aromatic acids are chemically derived from benzene and belong to the carboxylic group. **Sulfonic acids** are carboxylic acids that contain sulfur. As a group and individually, neither aromatic nor sulfonic acids have been studied to any great extent, but most of them are irritating to the skin, eyes, and respiratory tract.

Table 22. ORGANIC ACIDS

ACID	SOLUBILITY	CAUSTIC EFFECT	TLV	OTHER EFFECTS
A. Aliphatic Monocarboxylic Acids				
Acetic	high	severe	10 ppm	Chronic exposure to vapor can lead to chronic bronchitis.
Butyric	high	moderately severe	N.E.*	Normal part of body fats. Apparently only an irritant.
Caproic	slight	severe	N.E.	
Caprylic	poor	very slight	N.E.	
alpha-chloropropionic	high	severe	N.E.	
Diethylacetic	soluble		N.E.	Severely damages eyes of experimental animals.
2-ethylhexanoic	soluble	moderate	N.E.	Severe eye irritant, leading to corneal injury that can apparently heal.
Formic	high	severe	N.E.	
Heptanoic	soluble	moderate	N.E.	
Isobutyric	high	moderately severe	N.E.	
Isovaleric	slight	severe, undiluted	N.E.	
Lactic	high	severe	N.E.	
Linoleic	insoluble	none	N.E.	Spontaneous combustion of materials soaked in this acid or in linseed oil, which is a source of the acid, present a fire hazard.

N.E.: not established

Table 22. ORGANIC ACIDS (CONTINUED)

ACID	SOLUBILITY	CAUSTIC EFFECT	TLV	OTHER EFFECTS
Propionic	high	moderately severe	N.E.	Can be absorbed through skin.
Salicylic	poor	severe	N.E.	
Stearic	almost insoluble	none	N.E.	
Trichloroacetic	high	very severe	N.E.	
Valeric		severe	N.E.	
B. Unsaturated Monocarboxylic Aliphatic Acids				
Acrylic	soluble	very severe	N.E.	Severe eye, skin, and respiratory irritant in concentrated solution.
Chloroacetic	soluble	very severe	N.E.	Dust or vapor locally irritating to skin and eyes. Respiratory tract irritant. Exposure should be avoided and careful medical monitoring of people working with it is necessary.
Crotonic	moderate	severe	N.E.	Solid form very irritating to skin, eyes, and respiratory tract. Less irritating when diluted.
Methacrylic	soluble	severe	N.E.	Severely irritating to eyes, skin, and respiratory tract.
C. Halogenated Acids				
Bromoacetic	soluble in alcohol	severe	N.E.	Produces very severe local reactions of eyes, skin, and lining tissues of lungs, mouth, and air tubes.

N.E.: not established

ACID	SOLUBILITY	CAUSTIC EFFECT	TLV	OTHER EFFECTS
Chloroacetic	soluble	very severe	N.E.	Same as bromoacetic acid. All persons handling this should be under medical supervision.
Fluoroacetic and Sodium fluoroacetate	salt soluble		N.E.	Absorption of dust through respiratory tract affects enzymes in many tissues. Convulsions and heart damage seen in animals. Careful training and use necessary.
Iodoacetic	soluble in alcohol	severe	N.E.	Severe dermatitis and skin sensitization. Must be handled with extreme caution.
Trichloroacetic	very soluble	very severe	N.E.	Strongly corrosive, producing very severe burns of skin, eyes, or upper respiratory tract.
Trifluoroacetic	soluble	very severe	N.E.	Caustic effects like trichloroacetic. Fuming liquid can have same effects. Can easily penetrate skin.

D. Aliphatic Dicarboxylic Acids

ACID	SOLUBILITY	CAUSTIC EFFECT	TLV	OTHER EFFECTS
Adipic	slight		N.E.	No reported health effect. Slightly toxic to test animals.
Azelaic	very slight		N.E.	See adipic acid.
Citric	soluble	none	N.E.	
Fumaric	very slight	mild	N.E.	
Itaconic		mild	N.E.	
Maleic	soluble	very severe	N.E.	Has caustic effects on eyes, skin, and lining tissue of upper respiratory tract.

Table 22. ORGANIC ACIDS (CONTINUED)

ACID	SOLUBILITY	CAUSTIC EFFECT	TLV	OTHER EFFECTS
Malic	soluble	moderately severe	N.E.	See maleic acid.
Malonic	soluble	severe	N.E.	See maleic acid.
Succinic	soluble	moderately severe	N.E.	See maleic acid.
Tartaric	soluble	severe	N.E.	See maleic acid. Also, erosion of teeth from dust can occur.
E. Acid Anhydrides				
Acetic anhydride	moderate	severe	20 mg/cu.m.	Can cause sensitization. Severe eye and skin irritant. Can burn cornea and if severe enough, lead to blindness.
Butyric anhydride	decomposes	severe	N.E.*	Contact with vapor causes persistent eye irritation.
Citraconic anhydride	decomposes	severe	N.E.	Delayed severe burns result from contact of skin with liquid or with clothes contaminated with liquid.
Crotonic anhydride	decomposes	severe	N.E.	
Maleic anhydride	insoluble	severe	N.E.	Severe eye burns possible from dust.
Phthalic anhydride	slight	severe	N.E.	Exposed workers have chronic eye inflammation, blood-tinged nasal discharge, and chronic cough. Some develop chronic bronchitis and emphysema. Also skin sensitizer and can cause allergic reactions like hives and asthma.
Propionic anhydride	decomposes	severe	N.E.	Severe eye and skin irritation possible from liquid or vapor.

ACID	SOLUBILITY	CAUSTIC EFFECT	TLV	OTHER EFFECTS
		F. Lactones		
beta-propiolactone	soluble	none	N.E.	Absorbed through skin, but no human effects studied. Toxic to animals and causes cancer in animals.
gamma-butyrolactone	soluble		N.E.	See beta-propiolactone. Apparently does not cause cancer in animals.
gamma-valerolactone	soluble	none	N.E.	Readily absorbed through skin, but toxicity unknown.

N.E.: not established

Acetates

Acetates are derivatives of acetic acid, an organic acid. One of their chief uses is as a solvent for lacquers, but their solvent properties are used in many other industrial processes. The acetate esters as a whole exhibit anesthetic effects, which increase with the molecular weight and decrease with the solubility of the acetates in water. That is, amyl and butyl acetate are less soluble and have larger molecules than methyl and ethyl acetate, and their toxic effects are greater. The irritating effects of the acetates are greater than those of their corresponding alcohols. Thus, ethyl acetate is more irritating than ethyl alcohol, and methyl acetate is more irritating than methyl alcohol.

Exposure to acetates leads to irritation of the eyes, nose, and throat; this is accompanied by the slow onset of anesthetic symptoms, and the recovery from these symptoms is slow. As with all chemicals that have narcotic effects, the

Table 23. EFFECTS OF ACETATES

COMPOUND	EFFECTS
Ethyl acetate	Irritating vapor. May lead to reversible visual impairment.
Methyl acetate	No chronic effects studied. Poisoning can lead to headache, dizziness, eye burning, and tightness in chest.
Propyl acetate and isopropyl acetate	See methyl acetate.
Amyl acetate	Irritating to eyes, nose, and mucous membranes. Chronic exposure can lead to liver damage and anemia.
Butyl acetate	Moderately irritating.
Vinyl acetate	Slightly irritating. Can cause allergic response in skin.

risk of industrial accidents is greatly increased. Prolonged inhalation may cause the lungs to fill with fluid (*pulmonary edema*). There is apparently little skin sensitization and irritation from the acetates.

Pesticides

Pesticides have become a major topic of controversy in the scientific community and among ecologically concerned persons. However, it is the workers involved in the manufacture and application of these substances who suffer the greatest exposure. There are several classifications of pesticides: insecticides, rodenticides, fungicides, herbicides, and fumigants.

One of the most famous of all insecticides is **DDT (chlorophenothane)**, which appears to cause damage to animals, birds, and fish. Once it enters the soil, it does not decompose but goes on into the streams and lakes. It may be leading to the extinction of birds, as it is believed to affect the hardness of their eggshells and prevent baby birds from hatching. DDT is a *chlorinated hydrocarbon*. It and the other chlorinated hydrocarbons in common use as insecticides are stored in human fat. They enter the body by the respiratory tract, by ingestion, or by direct absorption through the skin. The long-term effects of DDT storage by the body are not known.

Other chlorinated hydrocarbons

aldrin	**lindane (gamma-benzene**
benzene hexachloride	**hexachloride)**
chlordane	**methoxychlor**
dieldrin	**terpene polychlorinates**
endrin	**toxaphene (chlorinated camphene)**
heptachlor	

Apparently the chlorinated hydrocarbons are not acutely toxic to human beings unless exposure is excessive, as in ingestion or massive body contact, when they may cause nervous excitation, convulsion, coma, and then death. Because of their effects in nature, however, it has been feared that long-term exposure may seriously affect the human body. Under certain circumstances, DDT can cause liver cancer in animals. It is not known if it causes cancer in humans. Some have been reported to cause dermatitis. A great concern of those involved in the manufacture of these chemicals is, in many cases, the solvents used for them. These are treated in the appropriate sections.

Another group of insecticides, the **organophosphate esters,** all have similar toxic effects, but differ in degree of toxicity. They all are readily absorbed through the skin and may also be inhaled or ingested. Acute symptoms following massive exposure include vomiting, dizziness, tremors, and convulsions. Such exposure can be fatal.

Chronic exposure to the phosphate esters primarily affects the nervous system. These chemicals are called *cholinesterase inhibitors,* or *antiChe* for short, because they inhibit the action of cholinesterase enzymes. The effect of this inhibition is to change the way in which nervous impulses are transmitted. Although the transmission of nerve impulses is not well understood, one step is known to involve the joining of two chemicals in the nerve cell as a means of conveying the nerve message. After the message has been sent through the particular nerve junction, or *synapse,* these two chemicals must separate again to be ready for the next message. The cholinesterase enzymes aid in this separation. AntiChe chemicals do not allow these enzymes to work efficiently, which results in continual stimulation of some of the nerves in the body, just as if they were constantly being sent messages. Symptoms include drowsiness, ringing in the ears, tingling of the skin, dizziness, irritability, depression, and a tendency to daydream. Studies have not been made on long-term effects of depressed cholinesterase levels in the blood.

All workers in contact with antiChe compounds should

be given fresh gloves and goggles daily and should work only in well-ventilated areas. To reduce the possibilities of ingestion, workers should shower and change clothing before eating or smoking. Blood tests should be given regularly to determine the cholinesterase level, and workers who show signs of excessive exposure should be rotated to another operation and avoid re-exposure until the cholinesterase level has returned to normal, which can take several weeks. Finally, the question whether these insecticides should be used at all must be considered in view of their danger to workers and consumers.

Other insecticides include **lead arsenate** and **lead arsenite,** which combine the effects of lead and arsenic, covered in the section on metals. **Pyrethrum** is not thought to be particularly toxic, nor is **rotenone,** which is similar. Both can cause skin and lung irritation. **Nicotine,** another insecticide, is extremely toxic and acts on the nervous system by first stimulating and then depressing it. It can be inhaled, ingested, or absorbed through the skin.

Dithiocarbamates are a group of chemicals commonly used as fungicides. They are highly irritating to the skin, eyes, and respiratory tract. Chronic effects of long-term exposure are not known precisely, but allergic reactions and effects on the hormones have been observed. There is also concern that these chemicals may cause cancer or damage the genetic material in the body. Some common carbamate fungicides are **ferbam, zuram, naneb, nabam,** and **zineb.**

Chemicals Involved in the Manufacture of Plastics

The use of plastics is increasing every day, and this of course means an increase in occupational exposure to the toxic chemicals involved in their manufacture. Plastics are all *polymers*. The prefix *poly-* means "many," and polymers are simply many molecules, not necessarily all the same, chemically hooked together to form a long-chain molecule. Each of these single-molecule building blocks is called a

Table 24. TOXICITY OF COMMON PESTICIDES

COMPOUND	RELATIVE TOXICITY	TLV
Chlorthion	S. T.	N. E.
DDVP	M. T.	N. E.
Diazinon	M. T.	N. E.
Dipterex	S. T.	N. E.
EPN	H. T.	0.5 mg/cu.cm.
Ethion	H. T.	N. E.
Guthion	H. T.	N. E.
HETP	H. T.	N. E.
Malathion	S. T.	15 mg/cu.cm.
Methyl parathion	H. T.	N. E.
OMPA	H. T.	N. E.
Parathion	H. T.	0.1 mg/cu.cm.
Phosdrin	H. T.	0.1 mg/cu.cm.
Pyrazoxon	H. T.	N. E.
Systox (Demeton)	H. T.	0.1 mg/cu.cm.
TEDP	H. T.	0.2 mg/cu.cm.
TEPP	H. T.	0.05 mg/cu.cm.
Thimet	H. T.	N. E.
Trithion	H. T.	N. E.
Esters of carbamic acid		
Carboryl	H. T.	5 mg/cu.cm.
Zectran	H. T.	
Pyrolon	H. T.	
Isolan	H. T.	

Toxicity ratings: H.T.—highly toxic; M.T.—moderately toxic; S.T.—slightly toxic. These ratings are based on animal studies and acute symptoms, for the most part.
N.E.: not established

monomer (the prefix *mono-* means "one"). Thus the manufacture of plastics involves exposure to the finished product, the polymer, and also to the starting materials, the monomers. In addition, it involves the use of catalysts (often metals or peroxides) and solvents. Dyes and fillers such as asbestos, diatomite, mica, and sand are used for color and texture. Stabilizers are used, and plasticizers are added for ease of shaping. Thus there is risk of exposure to these chemicals as well, and it is usually the stabilizers and plasticizers and the monomers that are most hazardous.

There are two types of polymers: those that can be molded (*thermoplastic*) and those that cannot be reshaped because they have been hardened by heat (*thermoset*). The finished polymer product rarely causes occupational hazards unless it partially decomposes while being processed. The decomposition products can have toxic properties. Also, contact with the resins or plastics before polymerization is complete can lead to dermatitis.

Acrylics are derived from acrylate esters. The acrylates are sensitizers, and **methyl methacrylate** is extremely irritating. Breathing the vapor can result in nausea and appetite loss. Many solvents used in the manufacture of acrylics can dissolve the oils in the skin and cause it to dry out and crack.

Acrolein (acrylic aldehyde) is highly irritating to the eyes and upper air passages, and the liquid can burn the skin. If the victim survives acute exposure to the vapor, there is usually chronic damage and disease of the lungs. Repeated exposure to low doses is not known to cause disease, but it has not been studied adequately. Acrolein should be handled in an oxygen-free enclosure with explosion-proof equipment. If there is a leak, all personnel must be evacuated.

Alkyds are made from the organic acids phthalic anhydride and maleic anhydride, and are irritating to the skin.

Allyls. Manufacture of the allyl resins involves the use of phosgene, an extremely toxic chemical. **Allyl alcohol,** another starting material, is severely irritating and can be absorbed right through the skin. This is very painful and can lead to muscle spasms. Liquid allyl alcohol can burn the eyes severely, and the vapors produce tearing, blurred vision and sensitivity to light, but these responses may be delayed, leading to excessive exposure. The alcohol is also extremely irritating to the upper respiratory tract. The peroxides used as catalysts can cause contact dermatitis.

Allyl chloride and **allyl bromide** have all the toxic properties of allyl alcohol, and in addition cause liver, lung, and kidney injury. The **allyl esters,** such as the **formate** and **acetate,** are irritating, and the **allyl phthalates** are very irritating.

Amino resins such as **melamine** are made from a chemical combination of various amines with formaldehyde. Formaldehyde can cause dermatitis and inflammation of the eyes, frequently observed among textile workers who use formaldehyde reaction products as part of fabric finishes. These products can also be sensitizers and cause allergies. The amine stabilizers, such as hexamethylenetetramine, can also cause dermatitis.

Butadienes are rubbers. **Butadiene** itself is very irritating, and is such an effective anesthetic that it can cause a person to stop breathing and die. It evaporates from the skin so rapidly that it can damage it by freezing it; such exposure can lead to gangrene.

Chlorobutadiene irritates the skin, eyes, and mucous membranes. It may actually destroy the cornea of the eye. In high concentrations its anesthetic effects can cause breathing to stop. Hair loss is associated with chlorobutadiene. It has low-level chronic effects on the lungs, nervous system, liver, kidneys, and heart muscle.

Cellulosics are the familiar filmy plastics used in such products as camera film. Cellulose is the major component of the cell walls in plants and gives wood its fibrous properties. Cellulose-containing materials must be treated with alkalies and acids to be made usable. These acids and alkalies are extremely corrosive and can cause severe dermatitis and irritation of the upper respiratory tract.

To produce the filmy polymer **cellulose acetate,** cellulose is treated with acetic acid, which affects the upper respiratory tract and attacks the teeth. Plasticizers are added for ease of shaping. These are often esters, such as diethyl phthalate and various phosphate esters; their effects are discussed in Table 26. The common solvents for cellulosics are acetone, acetates, and alcohols.

Epoxides. Many health hazards are involved in the manufacture of epoxides. There is exposure to solvents such as ketones and acetates, and to the plasticizers discussed in Table 26. Liquid epoxy resins are sensitizers and irritants. The glass fiber used in epoxy laminates is extremely irritating to the skin, often causing rashes and itching. Although the solid epoxy resins are inert, they can decompose when being shaped or ground, and the decomposition products can be very irritating. The curing agents are aliphatic amines, which have serious irritating and sensitizing properties.

Not much research has been done on the toxic effects of epoxy chemicals, but it is generally conceded that they are extreme irritants and may also have anesthetic and sensitizing effects. There is some argument as to whether they are cancer-causing agents that change the basic biology of cells, in the way ionizing radiation changes cellular function. There are also arguments as to whether they interfere with the production of white blood cells by affecting the bone marrow. Since these questions have not been scientifically settled, these chemicals should be treated very cautiously, and workers should not allow themselves to become part of the experimental data that supply the answers. Table 25 summarizes the properties of some epoxy compounds.

Fluorocarbons contain fluorine, carbon, and hydrogen. **Polytetrafluorethylene (Teflon, PTFE),** one of the popular fluorocarbons, is used for coatings and bearings because of its strength and inertness, which keep it from reacting with many other chemicals. It does not react with the body either, which makes it useful in medicine for internal appliances. When PTFE is shaped or ground, however, the high temperatures cause it to decompose partially. When inhaled,

Table 25. EFFECTS OF SOME EPOXY COMPOUNDS

COMPOUND	TLV	EFFECTS
Allyl glycidyl ether	N.E	Very irritating and a sensitizing agent. Excessive exposure can produce asthma-like response and pulmonary edema, with coughing, irritability, and shortness of breath.
Butadiene oxide	N.E.	Severely irritating. Can cause delayed irritation of eyes and swelling of lids. Repeated skin exposure causes cancer in animals, as well as severe blood damage.
Butylene oxides	N.E.	It is believed that below odor level, little damage will occur. At or above this level, chronic lung irritation may occur with complications.
n-butyl glycidyl ether	N.E.	Skin irritant and potent sensitizer. At high concentrations, may affect central nervous system.
Dicyclopentadiene dioxide (EP-207)	N.E.	Nothing is really known about this chemical, except that its use in industry is likely to increase because of its commercially useful properties.
Diglycidyl ether	10 ppm	This is the uncured chemical that is epoxy resin. Non-volatile, but, when heated, extremely irritating to upper respiratory tract, skin, and eyes. Can burn skin. If lung irritation is severe enough, may lead to pneumonia and pulmonary edema.

COMPOUND	TLV	EFFECTS
Epichlorohydrin	N.E.	Curing agent in production of many epoxy resins; some of the intermediates will be considered below. Intensely irritating and causes death in animals by toxic effects on kidneys. Appears to have effects on nervous system. Can cause severe skin burns, and even small amounts can cause sensitization. Irritating to upper respiratory tract; chronic exposure can probably be expected to lead to lung illness.
Epichlorohydrin + glycerin (EPON 562) (diglycidyl ether of substituted glycerine)	N.E.	Very irritating. Exposed animals have developed cancer and workers have exhibited blood damage. May affect central nervous system. Amine curing agents also very toxic.
Propylene oxide	100 ppm	Irritating to skin if held in prolonged contact, such as within a torn glove. Eye contact can burn the cornea.
Epichlorohydrin + bisphenol (diglycidyl ethers of bisphenol)	N.E.	Often used with glass fiber for laminates, so there is danger of extreme irritation from the glass as well as the amine curing agents. Cause severe dermatitis as well as sensitization. Work areas should be kept clean and well ventilated, and skin should be immediately and thoroughly washed if skin contact occurs.
Ethylene oxide	50 ppm	Extremely irritating. Skin and eye contact can cause severe burns leading to death of tissues involved. Eye contact should be scrupulously avoided and goggles should be worn. Excessive vapor can cause severe lung irritation, which can lead to lung infections. Animal studies have shown central nervous system depression, in some cases leading to death. Danger of fire and explosion when working with this chemical.

N.E.: not established

the decomposition products cause *polymer fume fever,* similar to metal fume fever, with symptoms like those of the flu—chills, fever, and muscular aches and pains. The principal decomposition product is **hydrogen fluoride,** which is very irritating to the respiratory tract. It causes the lungs to fill with fluid, which leads to coughing, severe air hunger, shortness of breath, and irritability. Exposure to the halogenated starting products can also be highly irritating and cause lung damage.

Phenolic resins. The toxic effects of **phenol** and **cresol** have already been described. Other products involved in the manufacture of these resins are **formaldehyde** and **hexamethylenetetramine,** discussed under "amino resins" above. The molding and extruding of these polymers results in the liberation of formaldehyde vapor.

Polyamides are the compounds commonly called **nylons.** The starting material is **hexamethylenediamine,** which is severely irritating to the eyes and upper respiratory tract. Polyamides are also skin irritants and can cause dermatitis. Animals exposed to high concentrations suffer serious damage to many organs. **Epsilon-caprolactam,** used to make Perla-L and Enkalon, apparently affects the nervous system; exposed workers complain of nervous irritability and feelings of apprehension. It also is severely irritating and may cause nosebleeds, sore throat, and burning lips.

Polyesters. The monomers for these resins are **phthalic anhydride** and **maleic anhydride,** which irritate the skin. The plasticizers in Table 26 are used in their manufacture, as is **styrene** (see "Polystyrenes" below) and **methyl methacrylate,** an extremely irritating substance. When polyesters are laminated, they are coated with glass fibers, which can have severe skin effects.

Polystyrenes (ethyl benzenes), as the name implies, are made from styrene, which is extremely irritating to the skin. It can also cause headache, nausea, appetite loss, and even coma. Low concentrations can damage the liver, cause blood disorders, and affect the nervous system. The solvents used in manufacturing polystyrenes are potentially dangerous, and the organic peroxides used as catalysts can produce dermatitis. **Methyl chloride,** which may be a side product in the manufacture of polystyrene foam, acts on the nervous system. Mild exposure causes symptoms of drunkenness, increasing the risk of industrial accidents. Greater exposure can cause dizziness, staggering, and even death. Polystyrene production workers have also been found to suffer from sinusitis, weakness, and irregularities in menstruation and ovulation. When polystyrenes decompose, they form **formaldehyde, benzaldehyde, methanol, hydrogen cyanide, ammonia, acrylonitrile, carbon monoxide, nitrogen dioxide, dusts,** and **styrene,** all toxic.

Polyurethane is made from **diisocyanates,** which have serious toxic effects. Besides being extremely irritating to the skin and eyes, diisocyanates such as **tolylene diisocyanate (TDI)** have severe effects on the lungs. The **isocyanates** can combine with and alter the proteins in the lungs. The body responds to the altered proteins by producing antibodies, which attack the "foreign" substance, causing tissue damage. This is called *sensitization,* or *allergy.* If the allergy is in the lungs, the sensitized person will have an asthma-like reaction. Sensitization can be immediate in susceptible people, or build up after repeated exposures. People with chronic respiratory diseases or allergies especially should not take the chance of exposure. Once sensitized, a person cannot stand any exposure to TDI and some workers become permanently disabled. As sensitization occurs at very small concentrations of TDI in the air, engineering precautions to avoid any exposure must be taken before workers are asked to work with this chemical. Self-breathing respiratory equipment must be available for emergencies.

In making urethane foam, the TDI is mixed with **freon,** which acts as a foaming agent. Freon consists of a group of fluorinated hydrocarbons that are mildly irritating to the eyes and upper air passages. The volatile fluorocarbons have a mild anesthetic effect, and breathing their vapors may cause dizziness. The **tetrafluoroalkanes,** such as **tetrafluoroethylene** and **chlorotrifluoroethylene,** cause liver and kidney damage in animals after prolonged exposure. Another danger comes when these compounds decompose after contact with an open flame or hot metal. They form **hydrogen chloride, hydrogen fluoride, phosgene, sulfur dioxide,** and **chlorine,** all severely irritating gases. Curing agents are **aliphatic amines,** which are irritating to the skin, eyes, and respiratory tract. Like TDI, they can cause sensitization and asthma.

The finished product, polyurethane, has not yet been proved harmful to workers, though in combination with certain other chemicals it causes cancer in small laboratory animals. Workers who insulate cold pipes with polyurethane spray are exposed to the hazards of the starting materials.

Vinyls. The vinyl plastics use many of the plasticizers discussed in Table 26 and in the discussion of cellulosics. **Polyvinyl acetate, polyvinyl chloride,** and **polyvinyl alcohol** are all made from monomers that are irritants.

Vinyl chloride, or some other chemical used in the synthesis of polyvinyl chloride, occasionally causes a peculiar reaction on the hands of workers who handle these chemicals. It affects the blood vessels of the fingers, causing decreased circulation, especially when the fingers are cold. They may become blue, numb, and painful. This is similar to the reaction of the blood vessels in vibration injury. Because of the loss of circulation the skin over the fingertips may become thickened and hard. Changes in the bones and arthritis may develop. X-rays of the bones will show holes, but these may heal if exposure ceases. The blood-vessel changes may heal or remain, but they will not get worse.

Table 26. Toxic Effects of Plasticizers

PHTHALATES

Dimethyl phthalate	Little toxic effect noted.
Diethyl phthalate	At room temperatures, no irritant effects noted. At higher temperatures, irritating to upper respiratory tract.
di-2-ethyhexyl phthalate	Extensive animal studies show no significant health hazards.
Other phthalates	Apparently also inert.

PHOSPHATES

Four major modes of action if toxic:
1. Central nervous system damage leading to paralysis, from which slow recovery may be possible.
2. Nervous system stimulation leading to convulsion, or acts as an anesthetic.
3. Anticholinesterase compound (see the section on pesticides).
4. Upper respiratory tract and skin irritant.

MODE OF ACTION

tri-ortho-cresyl phosphate (TOCP)	1	Extensively studied. Slowly progressing paralysis after ingestion. Occupationally, can show up as neuritis in various parts of body, as well as typical paralysis.
tri-para-cresyl phosphate (TPP)		Much less toxic than TOCP, with above symptoms not frequently encountered.
tri-butyl phosphate and tri-isobutyl phosphate	2;4	Definite stimulation of central nervous system as well as irritation of upper respiratory tract.

Nitrogen-Containing Compounds

Aliphatic nitro compounds. Three of these compounds, **nitroglycerine, amyl nitrate,** and **ethylene glycol dinitrate,**

are commonly used in the explosives industry and are obviously dangerous to handle. They are also used as drugs to treat coronary heart disease. Absorbed through the skin or by breathing, they have a direct effect on the blood vessels, especially the small blood vessels of the heart and brain. These vessels become dilated, and initially an exposed worker will have a pounding headache and feel weak. There may be psychological effects as well.

After continuous exposure to these compounds, the body tries to overcome the effects of the chemical. The blood vessels attempt to stop dilating and develop increased muscle tone—strength. After a period of 6 to 10 years this muscle tone will be so great that when a worker leaves the plant on weekends or for vacation, and the stimulus for dilation ceases, the muscle tone will be too strong, and the blood vessels will become very narrow. This can lead to chest pain, or even a heart attack. Many workers have to take nitroglycerine pills when off the job in order to keep the vessels dilated. Their bodies actually become addicted to the nitro compounds.

During exposure the blood vessels become thicker and more muscular, so that the supply of oxygen to the blood vessel wall itself becomes insufficient. When there is not enough oxygen in a blood vessel wall, cholesterol accumulates. This is the beginning of *hardening of the arteries,* or *arteriosclerosis.* If the circulation through one of these vessels ceases, a heart attack results. If circulation through a blood vessel to the brain is blocked, a stroke results.

Small, repeated doses of these nitrates can lead to weakness, general depression, headache, and mental disorder.

Several other nitro compounds are listed, with their effects, in Table 27.

Amines. Some of these nitrogen-containing compounds and their effects are listed in Table 28.

Table 27. Some Nitro Compounds and Their Effects

NITROPARAFFINS	NITROOLEFINS	ALKYL NITRITES
nitromethane	nitrobutene	methyl nitrite
nitroethane	nitrononene	ethyl nitrite
1-nitropropane	nitrohexene	isoamyl nitrite
2-nitropropane		
1-nitrobutane		
2-nitrobutane		
Cause liver and kidney damage.	Effects occur at low concentrations.	High concentrations have narcotic effect.
Affect central nervous system.	Cause increase in rate of heartbeat and rapid breathing immediately upon exposure.	Can affect blood, making hemoglobin unable to carry oxygen.
Toxicity of compounds containing more than one nitro group is greater than that of single nitro group.	Affect nervous system; can lead to a staggering gait.	Decompose to highly toxic oxides of nitrogen gases.
Low concentrations can lead to nausea, headache, and diarrhea.	Can affect blood, making hemoglobin unable to carry oxygen.	In the body, nitrites are transformed to nitrosamines, which can cause cancer.
Can affect blood and lead to anemia, or change hemoglobin so it cannot transport oxygen.	Extremely toxic to lungs.	
Irritate upper respiratory tract. Can also lead to slowed heartbeat.		

Table 28. TOXIC PROPERTIES OF SOME AMINES

Aliphatic Amines	As a group the aliphatic amines form strongly alkaline solutions. The vapors are irritating to the mucous membrane linings of the throat and air tubes. The mists are irritating to the eyes and can even cause cornea damage. Some people develop skin allergies to the amines, and others develop skin irritation. The aliphatic amines are flammable.
Methyl amine	Strongly irritating to eyes and upper respiratory tract.
Dimethyl amine (DMA)	Very alkaline and an irritant.
Ethylamine	Strong eye irritant, can lead to cornea and liver damage from vapor; forms very alkaline water solution.
Propylamine	Vapor strongly irritating to eyes and upper respiratory tract.
Butylamine	Severely irritating to eyes and respiratory tract. Affects nervous system in animals.
Allyl amine	Intensely irritating vapor. Explosion risk.
Cyclohexamine	Has fishy smell and irritant mist. Direct skin contact can lead to irritation or development of allergy.
Ethylenediamine	Vapors irritating to eyes and can damage skin, eyes, and lining of throat and air tubes. Can lead to development of allergy.
Ethanolamine	Smells like amonia, but no toxic effects observed by scientists. Can irritate skin.
Ethylene imine + propylene imine	Severely irritating to skin after brief contact and can lead to blisters and burns. Vapors can damage eye membranes, and recovery may be slow. Irritating to breathing tubes and throat, where it can even cause ulceration. In lungs, can lead to bronchitis, pneumonia, and pulmonary edema. Vapors can lead to nausea and vomiting, headache, and dizziness. May also cause cancer. Vapors can be explosive in the presence of acid.
Morpholine	Irritating to eyes, skin, and upper respiratory tract. When heated, it decomposes to toxic nitrogen oxide gases, which have severe effects on lungs.

Hydroxylamine and salts	Extremely corrosive to skin and highly irritating to eyes. Repeated exposure causes contact dermatitis, which may not occur for 1–2 weeks to 5 or more years after initial contact. Causes rupture of red blood cells, which can lead to anemia. Also changes hemoglobin so it can no longer transport oxygen (methemoglobinemia). In animals, high concentrations cause convulsions and death.
Hydrazine and derivatives (phenylhydrazine)	Reactions may be accompanied by explosions. Strong odor may prevent acute exposures. Severely irritating to eyes and respiratory tract. Eye contact may damage cornea and impair vision. Can be absorbed through skin, and may damage brain, liver, kidneys, and blood-forming organs. Affect nervous system; can lead to tremors, excitement, and convulsions. Change hemoglobin to methemoglobin, so it can no longer transport oxygen. Can also rupture red blood cells and cause severe, sudden anemia. Can cause chemical hepatitis and even lead to chronic liver disease.

Sulfur-Containing Compounds

Mercaptans are derivatives of hydrogen sulfide and are very foul-smelling: the scent of a skunk is a mercaptan. Their odor is so offensive that an exposed worker may actually get sick to the stomach and vomit. They affect the hemoglobin in the blood so that it cannot transport oxygen properly. This may cause the victim to turn blue, especially the lips, whites of the eyes, and fingernails, but she or he may not actually feel ill and may even be euphoric. This condition, *cyanosis,* calls for immediate bed rest until it disappears. Repeated attacks are thought to lead to blood damage. The mercaptans also depress the central nervous system and can cause breathing to stop. High concentrations can knock a person out and also severely irritate the respiratory system, perhaps even leading to pulmonary edema. When the mercaptans are heated, they release **sulfur dioxide,** a very toxic gas. Some mercaptans are:

methyl mercaptan   dodecyl mercaptan
ethyl mercaptan   perchloromethyl mercaptan
n-butyl mercaptan

Sulfur monochloride is flammable. When heated, it can release **sulfur dioxide** and **hydrogen chloride,** toxic gases that irritate and burn the eyes, skin, air passages, and lungs. It should be handled in enclosed processes.

Sulfur dichloride, thionyl chloride, sulfuryl chloride, and **chlorosulfuric acid** are other chlorinated sulfur compounds with similar corrosive effects.

Aminothiazole interferes with thyroid gland function and may slow down the body's chemical processes. The thyroid may enlarge to form a goiter. Workers using this compound have suffered loss of appetite, nausea, and vomiting, and their skin and urine have become darkened. It is also irritating to the skin, causing itching, hives, or a rash. Pain in the joints and muscles may occur as an allergic reaction.

Dimethyl sulfate is extremely hazardous. It acts like a combination of sulfuric acid and methyl alcohol, the two substances it breaks down into in the body. Severe irritation of the eyes, throat, upper air passages, and lungs usually occurs 4 to 8 hours after the initial exposure. There may be swelling and obstruction of the lungs. Eye injury is severe and painful and may lead to blindness. If the victim survives an acute reaction, eye pain, light sensitivity, and lack of vision may last for a long time, and there may be permanent damage to the cornea. Prolonged low-level exposure causes less severe irritation of the eyes and respiratory tract.

Carbon disulfide is a volatile liquid used in the manufacture of viscose rayon. Both liquid and vapor are highly irritating to the skin, eyes, nose, and air passages. This local irritation, however, is overshadowed by the serious effects on the body after the chemical has been absorbed through the skin or the lungs. High concentrations rapidly affect the brain, causing loss of consciousness and even death. Lower concentrations may cause headache or giddiness and irritation of the lungs and stomach.

Prolonged, repeated exposures to moderately high levels affect several parts of the body. Brain damage results in mental abnormalities such as depression, euphoria, agitation, and hallucinations. Nerve injury can cause blindness when the optic nerve is involved (*optic neuritis*) or weakness of a leg or an arm when a peripheral nerve is inflamed (*peripheral neuritis*). There may also be symptoms of stomach ulcers, and heart, kidney, and liver damage.

Since World War II, levels of carbon disulfide in most viscose rayon factories have been low enough to prevent most of these illnesses. However, it has recently been suggested that after prolonged exposure—5 to 15 years—to low levels of the vapor, viscose rayon workers tend to develop more and earlier heart attacks and high blood pressure than people who do not work in these factories.

The threshold limit for carbon disulfide in the United States is 20 parts per million parts of air by volume, or 60 milligrams per cubic meter. In Czechoslovakia, the limit is 30 milligrams per cubic meter, but peak values of up to 150 milligrams per cubic meter are permitted for a short time, provided no average exposure for an 8-hour period is more than 30 milligrams per cubic meter. This means that a worker exposed to 150 milligrams per cubic meter for 96 minutes could not be exposed to any carbon disulfide for the remainder of the work shift. In the Soviet Union the ceiling value is 10 milligrams per cubic meter.

ACIDIC AND ALKALINE MATERIALS

Sodium hydroxide and **potassium hydroxide** are the most caustic of the alkalies. Whether in solid form, dusts, sprays, mists, or concentrated liquid solution, they are more corrosive to tissue than most acids because they combine with tissue proteins and fats to form gelatinous burns that are deep and painful. Even dilute solutions are very irritating. But after working in an atmosphere slightly contaminated with alkalies, a worker can become accustomed to their effects on the throat and respiratory tract, although tissue damage may still be occurring. The greatest hazard results from eye contact. Alkaline particles or solutions that splash into the eyes can cause blindness. Protective glasses that are effective at all angles must be worn. In case of any contact, the exposed skin or eye should be flushed immediately with large quantities of cold running water. An eye-bath station should be standard emergency safety equipment for any operation involving use of these materials.

Several related compounds such as **sodium carbonate, sodium peroxide, trisodium phosphate,** and **sodium silicate** are also moderately caustic and should be handled in the same way. **Ammonia** is discussed in the section on gases.

Calcium hydroxide (slaked lime, hydrated lime) and calcium oxide (lime, burnt lime, quicklime). Lime is produced by burning limestone in kilns. It combines with water to form calcium hydroxide, liberating large amounts of heat as well as carbon monoxide and carbon dioxide. Workers have been asphyxiated by this combination. Both calcium oxide and calcium hydroxide are caustic irritants to the skin, eyes, and upper airways. The skin irritation can be so bad that ulcers are formed. Inhaling calcium oxide dust can cause *chemical pneumonia,* a chemically induced inflammation of the lungs.

Lime also contains silica, so that many limestone quarry workers and processors may develop silicosis as well as chronic sore throat, bronchitis, and emphysema. Adequate industrial hygiene techniques are necessary to prevent these diseases.

Hydrochloric acid (hydrogen chloride, muriatic acid) has a detectable odor at the threshold limit. It is highly corrosive in solution, and the vapor is extremely irritating to all parts of the respiratory tract. It is usually so irritating to the upper air passages that people leave the vicinity to avoid further exposure. However, severe lung damage (pulmonary edema) may result from short high-level exposures. Prolonged exposure to moderately high concentrations of the gas can cause tooth erosion, ulcers of the mouth, skin, and gums, and perforation of the nasal septum, the cartilage separating the nostrils. **Hydrogen bromide** is similar in its effects.

Hydrogen fluoride is highly irritating and corrosive to the mucous membranes of the respiratory tract and to the skin. Usually, burns from hydrogen fluoride are extremely painful immediately. However, burns from dilute solutions are not immediately painful and if not treated properly can result in the fluoride ion penetrating the skin and causing painful, slow-healing ulcers. Mild burns should be flushed with large amounts of water, followed by swabbing with dilute ammonia solution (a 10% dilution of 28% aqueous ammonia) and then by bathing in water. Finally, an ointment composed of 20% magnesium oxide in a glycerine base should be applied. Obviously, these items should be standard safety equipment for any operation using hydrogen fluoride. If there is a severe burn, a calcium injection should be given to stop the fluoride from penetrating deep into the body.

The compounds of fluoride, **calcium fluoride, cryolite, aluminum fluoride, fluorosilicic acid,** and **sodium hexa-fluorosilicate,** can all be fatal if inhaled or eaten. In industry, they are more likely to cause chronic illness. Their irritating properties can cause bronchitis, chemical pneumonia, and even pulmonary edema. Repeated exposure to low doses leads to increased deposition of fluorides in the bones and ligaments. This condition, called *fluorosis,* has no symptoms at first, but the bones and teeth become dense, which shows up on X-rays. Continued exposure can lead to brittle bones, spontaneous fractures, and bony spurs on the edges of joints and in ligaments and tendons. These deposits can cause painful arthritis.

Fluoride can be measured in the urine. Levels above normal should be followed up by X-rays. Adequate engineering controls are necessary to prevent exposure to fluoride compounds.

Nitric acid is extremely corrosive and attacks the eyes, skin, and mucous membranes. The fumes contain **nitrogen dioxide,** which is highly toxic. Exposure to high concentrations can lead to serious lung disease. Low-level, repeated exposures can probably result in lung disease also.

Sulfuric acid (oil of vitriol) is quite volatile when concentrated and gives off **sulfur trioxide** gas and sulfuric acid mist, both strongly irritating to the respiratory tract. In fact, sulfur trioxide has asphyxiating properties. In solution, it is corrosive to the skin and teeth. Acute respiratory effects are similar to those of hydrochloric acid. Lung scarring (*pulmonary fibrosis*) and emphysema may follow acute reactions. Chronic effects of long-term exposure are not known.

METALS

Metals are shiny chemical elements that conduct electricity and can be formed into various shapes. They also form compounds with other elements such as oxygen and sulfur. Workers can be exposed either to the pure metal and its fumes or dusts during such processes as grinding or welding, or to the compounds. Some of these are not harmful, but many have extremely serious and even fatal effects. Since the industrial uses of metals are enormous, nearly all the metals are discussed in this section.

Metals affect the nervous system and cause nervous dis-

	Aluminum	Antimony	Arsenic	Cadmium	Chromium	Copper	Fluorine	Iron	Lead	Manganese	Mercury	Nickel	Silver	Tin	Zinc
Pulp, paper mills, paperboard, building paper, board mills					●	●			●		●	●			●
Organic chemicals, petrochemicals	●		●	●	●		●	●	●		●			●	●
Alkalies, chlorine, inorganic chemicals	●		●	●	●		●	●	●		●			●	●
Fertilizers	●		●	●	●	●	●	●	●	●	●	●			●
Petroleum refining	●		●	●	●	●	●	●	●			●			●
Basic steel works, foundries		●	●	●	●	●	●	●	●		●	●		●	●
Basic non-ferrous metals—works, foundries	●	●	●	●	●	●	●		●		●		●		●
Motor vehicles, aircraft—plating, finishing	●			●	●	●					●	●	●		
Flat glass, cement, asbestos products, etc.					●										
Textile mill products					●										
Leather tanning, finishing					●										
Steam generation power plants					●										●

Note: Plastic materials, synthetics, meat products, dairy products, fruits and vegetables, grain milling, beet sugar, beverages, and livestock feedlot industries have no heavy metal discharges.

Table 29. HEAVY METALS FOUND IN MAJOR INDUSTRIES

orders. They also attack and damage various organs of the body. Some metals cause *metal fume fever,* an illness with symptoms very similar to the flu: muscular aches and pains, fever, chills, weakness, nausea, burning throat, and dry cough. This illness is very brief, complete recovery usually occurring in 12 to 24 hours.

Aluminum in most forms is not toxic to the human body. The fumes produced in furnacing bauxite ores and in manufacturing corundum can cause *shaver's disease,* which resembles rapidly progressive silicosis. A similar illness may occur in workers exposed to flaked aluminum in the stamping of the metal.

Antimony. Industrial poisoning from antimony is not a clear-cut illness because the worker is usually exposed simultaneously to arsenic, present as an impurity in the metal. Reports of illness in workers, especially in mining and smelting, describe eye inflammation, nasal irritation, perforation of the nasal septum, chronic dermatitis ranging from mild rashes to eruptions resembling chicken pox, and muscle pain and weakness. Inflammation of the respiratory tract can lead to chemical pneumonia. Irritation of the digestive tract (*gastritis*) may cause loss of appetite, nausea, vomiting, diarrhea, and abdominal pain. Cases of heart disease have been related to antimony. The weakness and fatigue characteristic of the chronic poisoning may be due to anemia caused by antimony.

Most studies of antimony toxicity are of acute poisoning, and different levels of toxicity are found in industries where exposure levels vary. Because of these variable effects, medical checkups are necessary, including electrocardiograms and chest X-rays. A urinalysis can help determine the extent of poisoning.

Stilbene, a compound of antimony, is a highly toxic gas encountered in metallurgy, welding, soldering, and the etching of zinc. Acute poisoning causes headache, nausea, weakness, slow breathing, and weakened, irregular heart action.

Stilbene also breaks up the red blood cells and interferes with brain function.

Arsenic and some of its derivatives are used as fungicides and herbicides, as well as in medicine. They form part of various metal alloys, giving increased heat resistance and hardness. Acute poisoning usually results from suicide attempts in which the arsenic compound is eaten. Chronic poisoning is caused by exposure to fumes or dusts of pure arsenic or its compounds. These fumes and dusts can cause rashes and darkening of the skin, and can irritate the nose, leading to nosebleeds and perforation of the nasal septum. Eye irritation can lead to corneal damage. Arsenic compounds may cause heart disease; exposed persons often show abnormal electrocardiograms. Arsenic affects the nerves of the hands and feet, resulting in loss of sensation, pain, and a feeling of weakness. Exposed workers may also develop serious anemia. Arsenic causes skin cancers at area of local contact. Dust and fume exposure can lead to lung cancers, which appear after more than 20 years of exposure. Workers should allow *no* contact with arsenic and its compounds. There is a debate about whether it causes kidney disease.

Respirators used for protection should not be made of rubber, because serious skin irritation can result from contact with rubber in the presence of arsenic. Thorough, periodic medical examinations are necessary, including urinalysis to determine arsenic level.

Arsine is a by-product of the reaction of strong acids with arsenic impurities in many metals. An extremely toxic substance, it has a faint odor of garlic at 1 ppm, well above the legal limit. It combines with the red blood cells and ruptures them. This means there are not enough red blood cells to carry oxygen, which leads to heart and brain damage (*chemical asphyxia*). The ruptured cells release hemoglobin into the bloodstream, causing kidney damage and kidney failure. There is also direct damage by arsine to the liver.

The symptoms of acute arsine poisoning are weakness,

stomach pains, nausea, vomiting, red urine, and jaundice. Intensive medical care is necessary to prevent death. Chronic poisoning is a less severe version of acute poisoning. There may be headache, nausea, vomiting, and symptoms of anemia: weakness, shortness of breath, and mild jaundice. Long-term effects may be quite serious.

Barium. While the soluble compounds of barium are highly toxic if eaten, this occurs rarely in industry. The primary effect on health in industry is *baritosis,* which results from dust trapped in the lungs. The dust deposits show up in chest X-rays, but do not seem to cause changes in lung function. More serious reactions may occur from exposure to fine dusts of barium compounds during grinding operations. These compounds stimulate all the muscles and nerves, causing muscle spasms, intestinal spasms, and pain. The blood vessels also narrow and the heart beats irregularly. It may beat so rapidly that it can no longer pump effectively; a person in this condition may die.

Beryllium is used to harden alloys; exposure also occurs when the ores are processed. Beryllium lung disease is discussed in the section on dusts. This metal also affects the skin and liver. The skin may be inflamed or ulcerated by beryllium dust entering open sores, and the ulcers will not heal until the metal is removed. *Granulomas*—little lumps under the skin—may develop. Under a microscope these look the same as the lung inflammations observed in beryllium lung disease, and are thought to be an allergic response. The liver also may develop granulomas. Standard patch tests can be negative even in advanced cases. The only sure test is study of tissue removed from the affected area—obviously not a routine procedure.

Very careful regulation of exposure to beryllium is necessary. Strict standards do exist, but there is evidence that many industries are not meeting them. Workers should insist that all necessary engineering precautions be taken.

Bismuth compounds are insoluble in body tissues and

fluid. There are no reports of toxicity from industrial exposure.

Boron. The familiar compound boric acid is responsible for some accidental household poisonings of children. However, most of the boron salts are not important industrial poisons. An exception are the highly toxic **boranes,** used in high-encrgy fuels. **Diborane** mainly affects the lungs; exposure to high doses causes chest tightness, burning pain, and shortness of breath. Repeated low-dose exposures produce chronic lung disease resembling allergic bronchitis, with dry cough, wheezing, and shortness of breath. Repeated exposure also causes severe brain damage, with symptoms of dizziness, headache, fatigue, muscular weakness, and chills. Blood tests show evidence of kidney and liver damage. **Decaborane** is somewhat less toxic to the lungs, mainly causing nervous system disorders with dizziness, headache, nausea, and drowsiness. **Pentaborane** is almost as toxic as diborane and also attacks the nervous system, causing extreme nervousness, headache, dizziness, drowsiness, hiccough, nausea, painful muscle cramps, and spasms. Symptoms may not appear for 40 to 48 hours; then tremors, lack of coordination, and seizures may occur. Here, too, there is evidence of liver and kidney damage.

A self-breathing respirator or a respirator filled with silica gel provides complete protection, but should not have to be used for long periods or on a regular basis. Exposure must be eliminated by proper ventilation and engineering control.

Cadmium. Industrial poisoning can occur when cadmium-coated metals are fired or welded, releasing cadmium fumes, or when cadmium is present as an impurity in other metals. Because of its extreme toxicity, metals containing cadmium are required by law to be labeled. Management must be forced to comply with this law.

The earliest symptoms of acute cadmium poisoning are similar to those of metal fume fever: dry, burning throat, cough, chest tightness or pain (especially when inhaling), nausea, vomiting, chills, fever, and headache. Pneumonia or pulmonary edema may develop. There may be severe shortness of breath and a blue color in the eyes and lips due to lack of oxygen in the blood. Recovery may occur over a period of 1 to 2 weeks, but chronic emphysema often results.

The chronic effects of cadmium-fume poisoning may become evident long after exposure has ceased, and may progressively worsen even without further exposure. The effect on the lungs is a severe and unusual form of emphysema, with no preceding bronchitis or lung scarring. Apparently cadmium acts directly on the air-sac walls, causing them literally to disappear. The kidneys are damaged by repeated exposure. Kidney stones may develop, leading to kidney failure. Urine tests can be designed specifically to discover cadmium poisoning, but even if there is no urinary sign of cadmium, lung disease can still be present. There may be bone-marrow damage, causing anemia, or damage to the hard part of the bones, resulting in pain, weakening, and occasional spontaneous fractures. Finally, the nose is continually irritated and there can be a loss of the sense of smell.

It is disputed whether the legal limit for cadmium provides safety for exposed workers. One scientist has stated that it is impossible to predict a "safe" level because the cause-and-effect relationships between exposure and illness vary so much from case to case.

Cerium is toxic to animals, but there have been no reports of ill-health from industrial exposure.

Cesium is toxic when given to animals and causes nervous system disorders, but no cases of cesium illness in industry have been reported. However, when metallic cesium is exposed to moist air, it can ignite spontaneously, forming **cesium oxide,** which is extremely corrosive to the eyes, skin, and throat.

Chromium. The principal areas of injury by chromium compounds are the skin, the nasal membranes, and less frequently the voice box *(larynx)* and the lungs. The skin may become itchy and inflamed as an allergic reaction to **chromic acid** solution or its vapor, or to **chromates** found in paints, wool processing, or lithography. If the skin is broken, chromic acid or its salts may enter and cause deep, slow-healing ulcers. The nose lining is damaged by chromic acid and chromate mists and dusts. This can cause bleeding, scabs, dryness, and eventually ulceration, which may progress so far that the nasal septum is perforated. A worker who inhales large quantities of chromic acid vapor may develop chemical pneumonia from the severe lung irritation. The accompanying weakness, chest pain, shortness of breath, and cough may take as long as 6 months to disappear.

The most important medical effect of chromium compounds is lung cancer, which may develop 20 to 30 years after exposure. One study has shown that the death rate from lung cancer for exposed workers is 29 times that for average members of the population.

Chromium can be measured in blood, urine, and tissues, and is easily measured in air. Since the safety of the legal limit for a lifetime of exposure has not been proved, workers should see that the level they are exposed to is well below the legal limit.

Cobalt produces an allergic skin irritation, especially in workers exposed to it in alloys and in cemented tungsten carbide. These substances apparently roughen the skin, allowing the cobalt to enter and sensitize it. Workers in the tungsten carbide tool industry develop a lung disease, or pneumoconiosis, as an allergic response to cobalt dust. Scarring of the tissues surrounding the smaller air tubes and air sacs causes the lungs to become stiff, so that less air can be breathed in. While cobalt pneumoconiosis can be fatal, some victims improve or even recover completely.

Symptoms of asthma, such as chest tightness, wheezing,

and shortness of breath, have been found in workers exposed to cobalt during ore processing. As these workers were also exposed to a number of other chemicals including nickel, arsenic, ammonia, sulfur dioxide, and sulfuric acid mist, the particular causes of their symptoms have not yet been determined. However, the cobalt ores are suspected.

Cobalt acetate dust causes pain and tenderness in the stomach and pain and weakness in the arms and legs, followed after several days by vomiting of blood and bloody stools. Recovery is complete, but may take 3 to 4 weeks. The medical effects of the **cobalt carbonyls** are not known.

Many workers become sensitized to cobalt at concentrations lower than the legal limit. Engineering control is obviously necessary to prevent exposure.

Columbium (niobium) is used as an alloy with steel and with aluminum iron for magnets. It is toxic when given to animals, but no industrial toxicity has been described.

Copper is highly toxic when given to animals, but there are few reports of illness from industrial exposure. However, acute poisoning can result from exposure to the fumes of **copper oxide,** causing metal fume fever, also called *brass chills* or *brass-foundry worker's ague.* The symptoms result from irritation of the lungs and intestines. In severe cases there are chills, fever, cough, nausea, vomiting, diarrhea, stomach pains, thirst, and exhaustion. Recovery after 1 or 2 days is the rule.

Long-term exposure to mists or fumes of copper compounds causes chronic irritation of the skin, eyes, nose, and throat, which can lead to throat congestion and to ulceration and perforation of the nasal septum. The skin may develop an allergy to copper, resulting in an itchy, raw dermatitis. Eye contact causes swelling of the lids, and inflammation and even ulceration of eye membranes. The cornea, normally a thin clear membrane over the lens, may become thickened or damaged and interfere with vision. Copper can cause the lens to become opaque (commonly called a *cataract*).

There is disagreement about whether long-term exposure results in liver disease, kidney disease, and anemia. Engineering controls should be instituted to protect workers while scientists are drawing their conclusions. Examination of blood and urine for excess copper should be routine for all exposed workers.

Gallium is quite toxic when given to animals. When its radioactive form is used medically, it causes skin rash, itching, peeling, and intestinal upset. However, there are no documented industrial exposures showing toxicity.

Germanium has important physical properties that make it very useful in electronics, and is also used in alloys. Only **germanium hydride, germanium tetrachloride,** and **germanium tetrafluoride** are reported to be industrial health hazards. The halogenated germanium gases are irritating to the tissues lining the respiratory tract. **Germanium tetrahydride** causes destruction of red blood cells similar to but less severe than that caused by arsine. No industrial exposures resulting in this illness have been reported.

Indium is highly toxic when fed to or injected into animals. However, little experience with industrial exposure to indium fumes or dusts has been reported. Indium fumes from welding are believed to be hazardous.

Iron. Industrial exposures to iron occur in workers doing welding and cutting, metallizing, and processing, as well as mining. Iron salts such as **ferric chloride** and **ferric sesquichloride** cause skin allergies, and may irritate the respiratory tract when inhaled. Iron oxide fumes may cause metal fume fever. **Iron carbonyl** vapors can produce severe lung irritation. Inhalation of **iron oxide** dust leads to abnormal X-rays, but apparently causes no other symptoms.

Lanthanum. Workers exposed to lanthanum fumes from cored carbon-arc lamps suffer from headache and nausea, but no specific damage to body organs has been reported.

Lead. The dangers of lead poisoning are present both for workers who use lead in various industrial processes and for consumers who are exposed to it in paint and automobile exhaust. Inorganic lead, such as is found in paint or in grinding operations, usually enters the body when metallic lead dusts or fumes are inhaled. Organic lead, or **tetraethyl lead,** the anti-knock component in gasoline, is easily absorbed through the skin, as well as through the lungs.

The first symptoms of inorganic lead poisoning are inability to sleep, fatigue, and constipation. These symptoms may be so mild at first that the person is unaware of developing lead poisoning. If severe exposure is continued, anemia, colic, and neuritis will develop. The colic, which may be so intensely painful as to be incorrectly diagnosed as appendicitis, is thought to come from the effects of lead on the nerves of the intestines, not from its effects on the intestines themselves. Involvement of the brain causes headache, loss of appetite, weakness, and sometimes double vision. The nerves that extend to arms, legs, and all other parts of the body may become inflamed and painful and a fine tremor may develop. Finally, a person may lose his or her teeth and have sore gums. A blue "lead line" on the gums is evidence of lead poisoning.

Lead affects the blood-forming tissues in the bone marrow, producing anemia. All lead workers should be given regular blood examinations to check for blood disease. After many years of lead poisoning, a person may develop kidney complications that lead to high blood pressure and even to complete kidney failure. However, this kidney damage sometimes cannot even be found in urinalysis.

Children who eat leaded paints show many of the same signs and symptoms. The damage to their brains can be so great that they become permanently retarded.

Organic-lead poisoning produces different symptoms. Unlike inorganic lead, which is stored mainly in the bones, organic lead is stored mainly in the brain. Symptoms are mental disturbance, inability to sleep, and general anxiety. In acute exposures the victim can become delirious and die, though if he or she recovers, symptoms disappear in a few weeks.

One problem with lead is that the body stores it for extremely long periods. People who have not been exposed for years will still excrete lead from their bodies. All lead workers should have regular blood tests and urinalyses. The air should be sampled regularly for lead content. Medical treatments for lead poisoning exist, but are dangerous in themselves and their effectiveness is questionable. The answer is to improve the engineering design of operations so that *all* exposure is eliminated. Workers are exposed to lead in their daily non-work lives as well, and this will have a cumulative effect.

Lithium is almost always found compounded with other chemicals. **Lithium hydride** apparently most often causes illness in industry, as the dust is highly irritating to the eyes, nose, and respiratory passages and corrosive to the skin. Lithium is also toxic to the kidneys. A major danger in handling lithium hydride is that it easily releases large volumes of hydrogen gas, which may explode or burn spontaneously. Therefore the legal limit is 1/200 of the level that causes acute irritation or corrosive effects. Workers should ensure that this limit is observed in their workplace. Adequate ventilation is necessary in all lithium processes, including welding and brazing.

Magnesium is always found in small amounts in the body, where it produces energy for the body's chemical reactions. Only a few instances of ill-health have been reported from industrial exposure. **Magnesium oxide** fumes can cause metal fume fever, though this is rare, and complete recovery usually occurs in 12 to 24 hours. Magnesium ores or dusts are moderately irritating to the eyes, nose, and upper airways, and sometimes printers exposed to fine dusts have a chronic cough. The chronic effects of such exposure have not been studied. Some scientists have found that magnesium plant workers have more digestive disorders than normal. Excess magnesium can lead to *magnesium tetany,* where all the muscles contract against each other. Much higher concentrations than are found in the properly maintained workplace are necessary for this condition to occur.

There is a safety hazard involved in using finely divided magnesium dust, because it ignites easily and burns at a temperature above 2,000° F. It can cause severe burns.

Manganese. Manganese fumes, released near reduction furnaces and from manganese-coated welding rods, can cause metal fume fever. Workers exposed to finely divided manganese dioxide in mines or in the manufacture of permanganate often suffer irritation of the entire respiratory tract. As this weakens the body's defenses, they usually develop secondary infection such as bacterial pneumonia. In one study of manganese pneumonia there were at least 15 cases per 1,000 workers in 7 years, as compared with an average of 0.73 cases of ordinary pneumonia per 1,000 people not exposed to manganese.

Manganism is a severely crippling, permanently disabling disease of the nervous system that closely resembles Parkinson's disease. It occurs principally among miners exposed to the pyrolusite ore, but has also been found in workers involved in the arc-cutting of manganese steel, especially in unventilated areas. The earliest symptoms are indefinite: headache, lack of appetite, apathy, and weight loss. Later there may be uncontrollable laughter, euphoria, impulsiveness, inability to sleep followed by overpowering sleepiness, sexual excitement followed by impotence, leg cramps, and changes in speech patterns. In the final stage the face is rigid and masklike, the speech is thickened and infrequent, and walking and the use of the hands and arms are hampered or prevented entirely by extreme muscle rigidity.

Until recently there was no effective therapy for manganism, but the drug L-dopa, used in treatment of Parkinson's disease, has been found helpful in some cases. Urinalysis is not a reliable test, as severe cases may not show a high manganese level in the urine. Since manganism cases have been reported from prolonged exposure to levels lower than the legal limit, workers should see that levels significantly lower than this limit are provided and that unnecessary processes involving manganese are eliminated. Manganese workers should be examined regularly for signs of nervous system disorder, lung disease, and blood disease. Engineering control of manganese processes is essential. Welding operations, often a source of manganese dusts and fumes, must also be controlled.

Mercury is the only metal that is a liquid. Whenever it is left open to the atmosphere, mercury vapor will be present. As this is extremely toxic, mercury vessels or spills must never be left uncovered. Mercury salts such as **mercuric chloride** and organic forms such as **phenyl mercuric acetate** are also toxic.

Mercury is stored mainly in the kidneys, but its most striking effects are on the nervous system. A person with mercury poisoning can develop a slight tremor of the hands and no longer be able to write properly. Before that, there may be emotional problems such as anxiety, indecision, embarrassment, and depression, and also excessive blushing and sweating. Mercury poisoning can lead to speech disorders and loss of coordination. The victim may develop a staggering gait. Serious changes in mental ability and personality may occur. The expression "mad as a hatter" refers to the fact that most hatters used to become crazy after practicing their profession over many years. The mercury they used to soften the felt in the hats poisoned them and eventually drove them mad.

Mercury also affects the vision and the eye reflexes. It particularly affects the mouth and teeth, causing loose teeth and sore gums. The kidneys may also be damaged.

These serious and often irreversible effects make it imperative to prevent all exposure to mercury vapor and other forms of mercury. Engineering control must be used to eliminate open sources, and maintenance and housekeeping must ensure that no pipes leak vapor and no spills are left. There should be regular monitoring of the air for the mercury level. All exposed workers should have regular urinalyses to determine the mercury level in their bodies.

Molybdenum, when heated, forms **molybdenum hexacarbonyl, molybdenum hydroxide,** and **molybdenum trioxide,** which are all hazardous. The trioxide is irritating to the eyes, nose, and throat. Molybdenum poisoning affects the blood and can lead to anemia. Cattle that have a high concentration of molybdenum in their diet develop deformities of the leg joints.

Nickel. Skin contact with nickel salts causes *nickel itch,* an allergic dermatitis that begins with burning and itching of the hands and wrists, followed by bumps that break and release pus. Once sensitized, a person may react even to extremely small quantities of nickel. There is some evidence that metallic nickel dust causes cancer of the respiratory tract and nasal sinuses.

The most toxic compound is **nickel carbonyl,** produced in many welding and furnace operations. Acute poisoning results when the vapor is inhaled in concentrations as low as 0.5 percent, and the compound may also be absorbed through the skin. Symptoms begin with giddiness, headache, shortness of breath, and vomiting. These are relieved by fresh air, but 12 to 36 hours later respiratory inflammation recurs with fever, cough producing blood-stained sputum, shortness of breath, and low blood oxygen resulting in blue lips and fingers. Delirium, hallucinations, and muscular disorders may also occur. In severe cases death results from pulmonary edema. Long-term exposure to lower concentrations is reported to cause asthma and *Loeffler's pneumonia,* a chronic lung inflammation that occurs intermittently and is thought to be an allergic reaction. Cancer of the nasal sinuses and respiratory passages is reported as a result of long-term exposure, though its incidence appears to be decreasing.

Contact with nickel carbonyl fumes, and with other forms of nickel that produce dust and fumes, should be very limited, and adequate ventilation and other engineering controls should be provided. Skin contact should be avoided by carefully planned industrial procedures, and workers should be provided with protective clothing and facilities and time for frequent washing.

Osmium in its metallic form is harmless, but in the air at room temperature it slowly changes to **osmic acid** or **osmium tetroxide**. Both are extremely hazardous, causing severe irritation of the eyes, nose, and respiratory tract. Osmium tetroxide fumes produce stinging and tearing of the eyes and even blindness. Irritation can continue as long as 2 weeks after exposure. Lung symptoms range from asthma to a more severe illness resembling pneumonia. Osmium tetroxide also inflames the liver and kidneys. Skin irritation may result in a painful, slow-healing dermatitis.

Palladium has not been sufficiently reported on or studied.

Platinum. Metallic platinum has not been found toxic, but some platinum compounds cause an allergic response in the respiratory system. Initial symptoms resemble hay fever: sneezing, itching, and runny nose and eyes. This can progress to asthma, with chest tightness, shortness of breath, and wheezing. After many allergic episodes, X-rays may show a mild degree of lung scarring and emphysema. The skin also may show signs of allergy: itching, hives, or peeling.

Rhenium has not been sufficiently reported on or studied.

Rhodium has not been sufficiently reported on or studied.

Rubidium is used mainly in photoelectric cells. It is toxic to animals. It ignites spontaneously in air, forming **rubidium oxide,** which is very corrosive to the eyes, skin, and respiratory tract.

Ruthenium is used in alloys with platinum and palladium. **Ruthenium oxide** fumes are strongly irritating to the eyes, respiratory tract, and lungs.

Selenium. Acute poisoning can result from inhaling the fumes or vapors of selenium compounds. **Hydrogen selenide** gas, the most common cause of this poisoning, is discussed in the section on gases in this chapter. Other compounds are extremely irritating to the skin, eyes, and upper airways. Severe burns may result from skin contact. Chronic poisoning can occur from exposure to dusts or vapor, resulting in absorption through the skin. Symptoms are nausea, vomiting, nervousness, tremor, dizziness, fatigue, and a garlic smell on the breath. Liver damage has been found.

Silver. Absorption of silver into the tissues is called *argyria*. Silver is very irritating and can cause chronic bronchitis. Some cases of kidney damage and hardening of the arteries have been reported. Silver deposited in the skin causes to it become leaden grey. Silver deposits in the eyes are common and result in slight loss of night vision.

The medical tests for silver, such as blood analyses, are positive only during exposure. Silver can be seen in the skin with an ultraviolet lamp, and on the cornea with a simple test called a "slit lamp" examination.

Strontium has not been reported to cause illness in humans. However, its radioactive form can be dangerous. For discussion, see Chapter Six under "Ionizing Radiation."

Tantalum, in this country, has not been found to cause ill-health. A study of exposed Russian workers shows moderate lung scarring.

Tellurium fumes generated in silver, gold, and copper refining and alloy making can cause acute poisoning. The earliest symptoms, a garlic odor on the breath and a metallic taste in the mouth, may last 1 to 2 months after exposure has ceased. The TLV is set to avoid these symptoms. More severe intoxication results in nausea, loss of appetite, constipation or diarrhea, and sleepiness. Lack of sweating and a skin rash may be followed by skin infection if exposure is prolonged. Tellurium fumes released in welding can cause metal fume fever. Alcohol appears to make all symptoms worse. There is no information on the effects of long-term low-level exposure.

Adequate ventilation, mechanical grinding machines, use of pelletized tellurium, and protective clothing are necessary to prevent exposure. Urinalysis for tellurium can be made after acute exposure and as a routine check on workers repeatedly exposed.

Thallium, a highly toxic substance, has caused many accidental deaths and has been used in suicides. However, industrial exposures seem not to have been fatal. Thallium can enter the body by inhalation of the dust, skin contact, or contaminated foods. Acute exposure can lead to damage of the nerves and eyes. Other symptoms—fatigue, leg pains, hair loss, double vision, and an involuntary jerking motion —may result from nerve damage. Skin contact can lead to an allergic reaction.

Tin. Inorganic tin compounds are severely toxic and may cause death when eaten, but do not seem to have caused illness in industrial exposures. Tin dust and fumes deposited in the lungs cause changes in X-rays, but apparently no changes in lung function. Organic tin compounds are highly irritating and corrosive. Direct skin contact or contact through clothing can produce burns and severe itching. Less severe contact causes redness, itching, and peeling.

Titanium. Only two titanium compounds are reported to cause illness in industrial exposures. **Titanium dioxide** dust is moderately irritating to the upper respiratory passages, and the fumes, released from electric furnaces or during machining, may cause typical metal fume fever. Moderate lung scarring has been reported after many years of exposure to the dust; this is similar to silicosis and asbestosis, but less severe. **Titanium tetrachloride,** a volatile liquid, is corrosive and irritating to the eye membranes, skin, and respiratory tract. It can cause pulmonary edema and lung scarring.

Tungsten is mildly toxic to animals. It has caused severe

lung disease in workers in hard-metal production. Many of
its effects are similar to those of dusts that cause scarring and
stiffening of the lungs.

Uranium. Because radioactive uranium produces ionizing
radiation, its effects are discussed in Chapter Six under
"Ionizing Radiation." Uranium compounds are toxic to
animals and may be able to penetrate normal human skin.
The kidneys and liver are the principal sites of absorption
and often become diseased. **Uranium hexachloride** fumes
irritate the lungs and can cause severe chemical pneumonia.

Vanadium dusts are irritating to the respiratory passages,
often causing nosebleeds or continuous nasal congestion.
The throat becomes dry and sore. Repeated inhalation of
large quantities produces cough, asthma, shortness of breath,
bronchitis, and pneumonia. Asthma attacks and bronchitis
may continue for years after an acute attack of vanadium
pneumonia. It is claimed that emphysema can follow vanadium poisoning, though not all scientists agree on this.

Hives or an itchy, peeling rash may develop as an allergic
reaction to vanadium solution or dusts. The eyes may be
irritated, and often the tongue turns green. (Apparently the
green color occurs at lower exposure levels than are necessary for other symptoms to show up.) Workers have developed intestinal troubles with nausea, vomiting, and stomach
pains, or nervous symptoms such as tremors and mental depression. Finally, the heart may be affected; exposed workers show abnormal electrocardiograms.

Zinc (zinc oxide). Workers exposed to zinc oxide fumes
may develop metal fume fever. This illness is over in a few
hours, leaving the worker with an immunity to the fumes—
which, however, is lost after a day or two away from the job.
Zinc chromate, zinc sulfate, and **zinc cyanide** may cause
chronic skin inflammation. **Zinc chloride** is especially toxic
and corrosive and causes skin ulcers; its vapors irritate the
eyes and respiratory tract and can cause chemical pneumonia. **Zinc stearate powder** causes a disease similar to asbestosis, with lung scarring and air-sac destruction leading

to emphysema. Exposed workers have also developed anemia and intestinal troubles.

Zirconium is moderately toxic to animals, but no reports of industrial toxicity have been found. Some users of deodorants containing zirconium have developed an allergic reaction to it.

Threshold Limiting Values (TLVs) for Commonly Used Chemicals *
(C = ceiling, or maximum level)

SUBSTANCE	PARTS PER MILLION PARTS OF AIR (PPM)	MILLIGRAMS PER CUBIC METER OF AIR (MG/M^3)
Acetaldehyde	200	360
Acetic acid	10	25
Acetic anhydride	5	20
Acetone	1,000	2,400
Acetonitrile	40	70
Acetylene dichloride, see 1, 2-dichloroethylene		
Acetylene tetrabromide	1	14
Acrolein	0.1	0.25
Acrylamide—skin		0.3
Acrylonitrile—skin	20	45
Aldrin—skin		0.25
Allyl alcohol—skin	2	5
Allyl chloride	1	3
C Allylglycidyl ether (AGE)	10	45
Allyl propyl disulfide	2	12
2-aminoethanol, see Ethanol-amine		
2-aminopyridine	0.5	2
Ammonia	50	35
Ammonium sulfamate (ammate)		15
n-amyl acetate	100	525
sec-amyl acetate	125	650
Aniline—skin	5	19
Anisidine (o, p-isomers)—skin		0.5
Antimony and compounds (as Sb)		0.5
ANTU (alpha naphthyl thiourea)		0.3
Arsenic and compounds (as As)		0.5
Arsine	0.05	0.2
Azinphos-methyl—skin		0.2
Barium (soluble compounds)		0.5

* From the *Federal Register*, vol. 37, no. 202 (October 18, 1972), Tables G-1, G-2, G-3, pp. 22140–2.

SUBSTANCE	PARTS PER MILLION PARTS OF AIR (PPM)	MILLIGRAMS PER CUBIC METER OF AIR (MG/M^3)
para-benzoquinone, see Quinone		
Benzoyl peroxide		5
Benzyl chloride	1	5
Biphenyl, see Diphenyl		
Bisphenol A, see Diglycidyl ether		
Boron oxide		15
C Boron trifluoride	1	3
Bromine	0.1	0.7
Bromoform—skin	0.5	5
Butadiene (1, 3-butadiene)	1,100	2,200
Butanethiol, see Butyl mercaptan		
2-butanone	200	590
2-butoxy ethanol (butyl cello-solve)—skin	50	240
Butyl acetate (n-butyl acetate)	150	710
sec-butyl acetate	200	950
tert-butyl acetate	200	950
Butyl alcohol	100	300
sec-butyl alcohol	150	450
tert-butyl alcohol	100	300
C Butylamine—skin	5	15
C tert-butyl chromate (as CrO₃)—skin		0.1
n-butyl glycidyl ether (B G E)	50	270
Butyl mercaptan	10	35
para-tert-butyltoluene	10	60
Calcium arsenate		1
Calcium oxide		5
Camphor	2	
Carbaryl (Sevin®)		5
Carbon black		3.5
Carbon dioxide	5,000	9,000
Carbon monoxide	50	55
Chlordane—skin		0.5
Chlorinated camphene—skin		0.5
Chlorinated diphenyl oxide		0.5
Chlorine	1	3
Chlorine dioxide	0.1	0.3
C Chlorine trifluoride	0.1	0.4
C Chloroacetaldehyde	1	3
alpha-chloroacetophenone (phenacylchloride)	0.05	0.3
Chlorobenzene (monochloro-benzene)	75	350
ortho-chlorobenzylidene malononitrile (OCBM)	0.03	0.4
Chlorobromomethane	200	1,050

SUBSTANCE	PARTS PER MILLION PARTS OF AIR (PPM)	MILLIGRAMS PER CUBIC METER OF AIR (MG/M^3)
2-chloro-1,3-butadiene, see Chloroprene		
Chlorodiphenyl (42 percent chlorine)—skin		1
Chlorodiphenyl (54 percent chlorine)—skin		0.5
1-chloro, 2,3-epoxypropane, see Epichlorohydrin		
2-chloroethanol, see Ethylene chlorohydrin		
Chloroethylene, see Vinyl chloride		
C Chloroform (trichloromethane)	50	240
1-chloro-1-nitropropane	20	100
Chloropicrin	0.1	0.7
Chloroprene (2-chloro-1,3-butadiene)—skin	25	90
Chromium, soluble chromic, chromous salts as Cr		0.5
metal and salts		1
Coal tar pitch volatiles (benzene soluble fraction) anthracene, BaP, phenanthrene, acridine, chrysene, pyrene		0.2
Cobalt, metal fume and dust		0.1
Copper fume		0.1
dusts and mists		1
Cotton dust (raw)		1
Crag® herbicide		15
Cresol (all isomers)—skin	5	22
Crotonaldehyde	2	6
Cumene—skin	50	245
Cyanide (as CN)—skin		5
Cyclohexane	300	1,050
Cyclohexanol	50	200
Cyclohexanone	50	200
Cyclohexene	300	1,015
Cyclopentadiene	75	200
2, 4-D		10
DDT—skin		1
DDVP, see Dichlorvos		
Decaborane—skin	0.05	0.3
Demeton®—skin		0.1
Diacetone alcohol (4-hydroxy-4-methyl-2-pentanone)	50	240

SUBSTANCE	PARTS PER MILLION PARTS OF AIR (PPM)	MILLIGRAMS PER CUBIC METER OF AIR (MG/M^3)
1,2-diaminoethane, see Ethylenediamine		
Diazomethane	0.2	0.4
Diborane	0.1	0.1
Dibutylphthalate		5
C ortho-dichlorobenzene	50	300
para-dichlorobenzene	75	450
Dichlorodifluoromethane	1,000	4,950
1,3-dichloro-5,5-dimethyl hydantoin		0.2
1,1-dichloroethane	100	400
1,2-dichloroethylene	200	790
C Dichloroethyl ether—skin	15	90
Dichloromethane, see Methylene-chloride		
Dichloromonofluoromethane	1,000	4,200
C 1,1-dichloro-1-nitroethane	10	60
1,2-dichloropropane, see Propyl-enedichloride		
Dichlorotetrafluoroethane	1,000	7,000
Dichlorvos (DDVP)—skin		1
Dieldrin—skin		0.25
Diethylamine	25	75
Diethylamino ethanol—skin	10	50
Diethylether, see Ethyl ether		
Difluorodibromomethane	100	860
C Diglycidyl ether (DGE)	0.5	2.8
Dihydroxybenzene, see Hydro-quinone		
Diisobutyl ketone	50	290
Diisopropylamine—skin	5	20
Dimethoxymethane, see Methylal		
Dimethyl acetamide—skin	10	35
Dimethylamine	10	18
Dimethylaminobenzene, see Xylidene		
Dimethylaniline(N-dimethyl-aniline)—skin	5	25
Dimethylbenzene, see Xylene		
Dimethyl 1,2-dibromo-2,2-di-chloroethyl phosphate, (Dibrom)		3
Dimethylformamide—skin	10	30
2,6-dimethylheptanone, see Diiso-butyl ketone		

SUBSTANCE	PARTS PER MILLION PARTS OF AIR (PPM)	MILLIGRAMS PER CUBIC METER OF AIR (MG/M^3)
1,1-dimethylhydrazine—skin	0.5	1
Dimethylphthalate		5
Dimethylsulfate—skin	1	5
Dinitrobenzene (all isomers)— skin		1
Dinitro-ortho-cresol—skin		0.2
Dinitrotoluene—skin		1.5
Dioxane (diethylene dioxide)— skin	100	360
Diphenyl	0.2	1
Diphenylmethane diisocyanate, see Methylene bisphenyl isocyanate (MDI)		
Dipropylene glycol methyl ether —skin	100	600
Di-sec, octyl phthalate (Di-2-ethylhexylphthalate		5
Endrin—skin		0.1
Epichlorohydrin—skin	5	19
EPN—skin		0.5
1,2-epoxypropane, see Propylene-oxide		
2,3-epoxy-1-propanol, see Glycidol		
Ethanethiol, see Ethylmercaptan		
Ethanolamine	3	6
2-ethoxyethanol—skin	200	740
2-ethoxyethylacetate (cellosolve acetate)—skin	100	540
Ethyl acetate	400	1,400
Ethyl acrylate—skin	25	100
Ethyl alcohol (ethanol)	1,000	1,900
Ethylamine	10	18
Ethyl sec-amyl ketone (5-methyl-3-heptanone)	25	130
Ethyl benzene	100	435
Ethyl bromide	200	890
Ethyl butyl ketone (3-heptanone)	50	230
Ethyl chloride	1,000	2,600
Ethyl ether	400	1,200
Ethyl formate	100	300
C Ethyl mercaptan	10	25
Ethyl silicate	100	850
Ethylene chlorohydrin—skin	5	16
Ethylenediamine	10	25

SUBSTANCE	PARTS PER MILLION PARTS OF AIR (PPM)	MILLIGRAMS PER CUBIC METER OF AIR (MG/M^3)
Ethylene dibromide, see 1,2-di-bromoethane		
Ethylene dichloride, see 1,2-di-chloroethane		
C Ethylene glycol dinitrate and/or nitroglycerine—skin	a 0.2	1
Ethylene glycol monomethyl ether acetate, see Methyl cellosolve acetate		
Ethylene imine—skin	0.5	1
Ethylene oxide	50	90
Ethylidine chloride, see 1,1-di-chloroethane		
N-ethylmorpholine—skin	20	94
Ferbam		15
Ferrovanadium dust		1
Fluoride (as F)		2.5
Fluorine	0.1	0.2
Fluorotrichloromethane	1,000	5,600
Formic acid	5	9
Furfural—skin	5	20
Furfuryl alcohol	50	200
Glycidol (2,3-epoxy-1-propanol)	50	150
Glycol monoethyl ether, see 2-ethoxyethanol		
Guthion ®, see Azinphosmethyl		
Hafnium		0.5
Heptachlor—skin		0.5
Heptane (n-heptane)	500	2,000
Hexachloroethane—skin	1	10
Hexachloronaphthalene—skin		0.2
Hexane (n-hexane)	500	1,800
2-hexanone	100	410
Hexone (methyl isobutyl ketone)	100	410
sec-hexyl acetate	50	300
Hydrazine—skin	1	1.3
Hydrogen bromide	3	10
C Hydrogen chloride	5	7
Hydrogen cyanide—skin	10	11
Hydrogen peroxide (90%)	1	1.4
Hydrogen selenide	0.05	0.2
Hydroquinone		2
C Iodine	0.1	1

a An atmospheric concentration of not more than 0.02 ppm or personal protection may be necessary to avoid headache.

SUBSTANCE	PARTS PER MILLION PARTS OF AIR (PPM)	MILLIGRAMS PER CUBIC METER OF AIR (MG/M³)
Iron oxide fume		10
Isoamyl acetate	100	525
Isoamyl alcohol	100	360
Isobutyl acetate	150	700
Isobutyl alcohol	100	300
Isophorone	25	140
Isopropyl acetate	250	950
Isopropyl alcohol	400	980
Isopropylamine	5	12
Isopropylether	500	2,100
Isopropyl glycidyl ether (IGE)	50	240
Ketene	0.5	0.9
Lead arsenate		0.15
Lindane—skin		0.5
Lithium hydride		0.025
L.P.G. (liquified petroleum gas)	1,000	1,800
Magnesium oxide fume		15
Malathion—skin		15
Maleic anhydride	0.25	1
C Manganese		5
Mesityl oxide	25	100
Methanethiol, see Methyl mercaptan		
Methoxychlor		15
2-methoxyethanol), see Methyl cellosolve		
Methyl acetate	200	610
Methyl acetylene (propyne)	1,000	1,650
Methyl acetylene-propadiene mixture (MAPP)	1,000	1,800
Methyl acrylate—skin	10	35
Methylal (dimethoxymethane)	1,000	3,100
Methyl alcohol (methanol)	200	260
Methylamine	10	12
Methyl amyl alcohol, see Methyl isobutyl carbinol		
Methyl (n-amyl) ketone (2-heptanone)	100	465
C Methyl bromide—skin	20	80
Methyl butyl ketone, see 2-hexanone		
Methyl cellosolve—skin	25	80
Methyl cellosolve acetate—skin	25	120
Methyl chloroform	350	1,900
Methylcyclohexane	500	2,000
Methylcyclohexanol	100	470

SUBSTANCE	PARTS PER MILLION PARTS OF AIR (PPM)	MILLIGRAMS PER CUBIC METER OF AIR (MG/M^3)
ortho-methylcyclohexanone—skin	100	460
Methyl ethyl ketone (MEK), see 2-butanone		
Methyl formate	100	250
Methyl iodide—skin	5	28
Methyl isobutyl carbinol—skin	25	100
Methyl isobutyl ketone, see Hexone		
Methyl isocyanate—skin	0.02	0.05
C Methyl mercaptan	10	20
Methyl methacrylate	100	410
Methyl propyl ketone, see 2-pentanone		
C alpha-methyl styrene	100	480
C Methylene bisphenyl isocyanate (MDI)	0.02	0.2
Molybdenum:		
soluble compounds		5
insoluble compounds		15
Monomethyl aniline—skin	2	9
C Monomethyl hydrazine—skin	0.2	0.35
Morpholine—skin	20	70
Naphtha (coal tar)	100	400
Naphthalene	10	50
Nickel carbonyl	0.001	0.007
Nickel, metal and soluble cmpds, as Ni		1
Nicotine—skin		0.5
Nitric acid	2	5
Nitric oxide	25	30
para-nitroaniline—skin	1	6
Nitrobenzene—skin	1	5
para-nitrochlorobenzene—skin		1
Nitroethane	100	310
Nitrogen dioxide	5	9
Nitrogen trifluoride	10	29
Nitroglycerin—skin	0.2	2
Nitromethane	100	250
1-nitropropane	25	90
2-nitropropane	25	90
Nitrotoluene—skin	5	30
Nitrotrichloromethane, see Chloropicrin		
Octachloronaphthalene—skin		0.1
Octane	500	2,350
Oil mist, mineral		5

SUBSTANCE	PARTS PER MILLION PARTS OF AIR (PPM)	MILLIGRAMS PER CUBIC METER OF AIR (MG/M^3)
Osmium tetroxide		0.002
Oxalic acid		1
Oxygen difluoride	0.05	0.1
Ozone	0.1	0.2
Paraquat—skin		0.5
Parathion—skin		0.11
Pentaborane	0.005	0.01
Pentachloronaphthalene—skin		0.5
Pentachlorophenol—skin		0.5
Pentane	1,000	2,950
2-pentanone	200	700
Perchloromethyl mercaptan	0.1	0.8
Perchloryl fluoride	3	13.5
Petroleum distillates (naphtha)	500	2,000
Phenol—skin	5	19
para-phenylene diamine—skin		0.1
Phenyl ether (vapor)	1	7
Phenyl ether-biphenyl mixture (vapor)	1	7
Phenylethylene, see Styrene		
Phenyl glycidyl ether (PGE)	10	60
Phenylhydrazine—skin	5	22
Phosdrin (Mevinphos®)—skin		0.1
Phosgene (carbonyl chloride)	0.1	0.4
Phosphine	0.3	0.4
Phosphoric acid		1
Phosphorus (yellow)		0.1
Phosphorus pentachloride		1
Phosphorus pentasulfide		1
Phosphorus trichloride	0.5	3
Phthalic anhydride	2	12
Picric acid—skin		0.1
Pival ® (2-pivalyl-1,3-indandione)		0.1
Platinum (soluble salts) as Pt		0.002
Propargyl alcohol—skin	1	
Propane	1,000	1,800
n-propyl acetate	200	840
Propyl alcohol	200	500
n-propyl nitrate	25	110
Propylene dichloride	75	350
Propylene imine—skin	2	5
Propylene oxide	100	240
Propyne, see Methylacetylene		
Pyrethrum		5
Pyridine	5	15
Quinone	0.1	0.4

SUBSTANCE	PARTS PER MILLION PARTS OF AIR (PPM)	MILLIGRAMS PER CUBIC METER OF AIR (MG/M^3)
RDX—skin		1.5
Rhodium, metal fume and		
dusts, as Rh		0.1
soluble salts		0.001
Ronnel		10
Rotenone (commercial)		5
Selenium compounds (as Se)		0.2
Selenium hexafluoride	0.05	0.4
Silver, metal and soluble com-		
pounds		0.01
Sodium fluoroacetate (1080)—		
skin		0.05
Sodium hydroxide		2
Stilbene	0.1	0.5
Stoddard solvent	500	2,950
Strychnine		0.15
Sulfur dioxide	5	13
Sulfur hexafluoride	1,000	6,000
Sulfuric acid		1
Sulfur monochloride	1	6
Sulfur pentafluoride	0.025	0.25
Sulfuryl fluoride	5	20
Systox, see Demeton ®		
2,4,5T		10
Tantalum		5
TEDP—skin		0.2
Tellurium		0.1
Tellurium hexafluoride	0.02	0.2
TEPP—skin		0.05
C Terphenyls	1	9
1,1,1,2-tetrachloro-2,2-difluoro-		
ethane	500	4,170
1,1,2,2-tetrachloro-1,2-difluoro-		
ethane	500	4,170
1,1,2,2-tetrachloroethane—skin	5	35
Tetrachloroethylene, see Perchlo-		
roethylene		
Tetrachloromethane, see Carbon		
tetrachloride		
Tetrachloronaphthalene—skin		2
Tetraethyl lead (as Pb)—skin		0.075
Tetrahydrofuran	200	590
Tetramethyl lead (as Pb)—skin		0.07
Tetramethyl succinonitrile—skin	0.5	3
Tetranitromethane	1	8
Tetryl (2,4,6-trinitrophenyl-		
methylnitramine)—skin		1.5

SUBSTANCE	PARTS PER MILLION PARTS OF AIR (PPM)	MILLIGRAMS PER CUBIC METER OF AIR (MG/M^3)
Thallium (soluble compounds)— skin as Tl		0.1
Thiram		5
Tin (inorganic, except oxides)		2
Tin (organic cmpds)		0.1
Titanium dioxide		15
C Toluene-2,4-diisocyanate	0.02	0.14
ortho-toluidine—skin	5	22
Toxaphene, see Chlorinated camphene		
Tributyl phosphate		5
1,1,1-trichloroethane, sec Methyl chloroform		
1,1,2-trichloroethane—skin	10	45
Trichloromethane, see Chloroform		
Trichloronaphthalene—skin		5
1,2,3-trichloropropane	50	300
1,1,2-trichloro 1,2,2-trifluoroethane	1,000	7,600
Triethylamine	25	100
Trifluoromonobromomethane	1,000	6,100
2,4,6-trinitrophenol, see Picric acid		
2,4,6-trinitrophenylmethylnitramine, see Tetryl		
Trinitrotoluene—skin		1.5
Tri-ortho-cresyl phosphate		0.1
Triphenyl phosphate		3
Turpentine	100	560
Uranium (soluble compounds)		0.05
Uranium (insoluble compounds)		0.25
C Vanadium:		
V$_2$O$_5$ dust		0.5
V$_2$O$_5$ fume		0.1
Vinyl benzene, see Styrene		
C Vinyl chloride	500	1,300
Vinylcyanide, see Acrylonitrile		
Vinyl toluene	100	480
Warfarin		0.1
Xylene (xylol)	100	435
Xylidine—skin	5	25
Yttrium		1
Zinc chloride fume		1
Zinc oxide fume		5
Zirconium compounds (as Zr)		5

MATERIAL	8-HOUR TIME WEIGHTED AVERAGE	ACCEPTABLE CEILING CONCENTRATION	ACCEPTABLE MAXIMUM PEAK ABOVE THE ACCEPTABLE CEILING CONCENTRATION FOR AN 8-HOUR SHIFT.	
			CONCENTRATION	MAXIMUM DURATION
Benzene (Z37.4–1969)	10 ppm	25 ppm	50 ppm	10 minutes
Beryllium and beryllium compounds (Z37.29–1970)	2 mg/M³	5 mg/M³	25 mg/M³	30 minutes
Cadmium fume (Z37.5–1970)	0.1 mg/M³	3 mg/M³		
Cadmium dust (Z37.5–1970)	0.2 mg/M³	0.6 mg/M³	100 ppm	Do
Carbon disulfide (Z37.3–1968)	20 ppm	30 ppm	200 ppm	5 minutes in any 4 hours
Carbon tetrachloride (Z37.17–1967)	10 ppm	25 ppm		
Ethylene dibromide (Z37.31–1970)	20 ppm	30 ppm	50 ppm	5 minutes
Ethylene dichloride (Z37.21–1969)	50 ppm	100 ppm	200 ppm	5 minutes in any 3 hours
Formaldehyde (Z37.16–1967)	3 ppm	5 ppm	10 ppm	30 minutes
Hydrogen fluoride (Z37.28–1969)	do			
Fluoride as dust (Z37.28–1969)	2.5 mg/M³			
Lead and its inorganic compounds (Z37.11–1969)	0.2 mg/M³			
Methyl chloride (Z37.18–1969)	100 ppm	200 ppm	300 ppm	5 minutes in any 3 hours
Methylene chloride (Z37.3–1969)	500 ppm	1,000 ppm	2,000 ppm	5 minutes in any 2 hours

MATERIAL	8-HOUR TIME WEIGHTED AVERAGE	ACCEPTABLE CEILING CONCENTRATION	ACCEPTABLE MAXIMUM PEAK ABOVE THE ACCEPTABLE CEILING CONCENTRATION FOR AN 8-HOUR SHIFT.	
			CONCENTRATION	MAXIMUM DURATION
Organo (alkyl) mercury (Z37.30–1969)	0.01 mg/M^3	0.04 mg/M^3		
Styrene (Z37.15–1969)	100 ppm	200 ppm	600 ppm	5 minutes in any 3 hours
Trichloroethylene (Z37.19–1967)	do	do	300 ppm	5 minutes in any 2 hours
Tetrachloroethylene (Z37.22–1967)	do	do	do	5 minutes in any 3 hours
Toluene (Z37.12–1967)	200 ppm	300 ppm	500 ppm	10 minutes
Hydrogen sulfide (Z37.2–1966)		20 ppm	50 ppm	10 minutes once only if no other measurable exposure occurs
Mercury (Z37.8–1971) (Z37.7–1971)		1 mg/10M^3		
Chromic acid and chromates (Z37.7–1971)		do		

MINERAL DUSTS

SUBSTANCE	MILLIONS OF PARTICLES PER CUBIC FOOT OF AIR (MPPCF)	MILLIGRAMS PER CUBIC METER OF AIR (MG/M³)
Silica:		
Crystalline:		
Quartz (respirable)	$\dfrac{250}{\% SiO_2 + 5}$	$\dfrac{10 \text{ mg/M}^3}{\% SiO_2 + 2}$
Quartz (total dust)		$\dfrac{30 \text{ mg/M}^3}{\% S_3O_3 + 2}$
Cristobalite: Use ½ the value calculated from the count or mass formulae for quartz.		
Tridymite: Use ½ the value calculated from the formulae for quartz.		
Amorphous, including natural diatomaceous earth	20	$\dfrac{80 \text{ mg/M}^3}{\% SiO_3}$
Silicates (less than 1% crystalline silica).		
Mica	20	
Soapstone	20	
Talc (non-asbestos-form, containing less than 1% quartz)	20	
Talc (fibrous). Use asbestos limit		
Tremolite (see talc, fibrous)		
Portland cement	50	
Graphite (natural)	15	
Coal dust (respirable fraction less than 5% SiO_2)		2.4 mg/M³
For more than 5% SiO_2		or $\dfrac{10 \text{ mg/M}^3}{\% SiO_2 + 2}$
Inert or Nuisance Dust:		
Respirable fraction	15	5 mg/M³
Total dust	50	15 mg/M³

NOTE: Conversion factors—
mppcf$\times$35.3=million particles per cubic meter
=particles per c.c.

CHAPTER EIGHT

WELDING HAZARDS

cornea: the clear membrane that covers the outer part of the eye
IR: infrared radiation
ppm: concentration of a chemical in *parts per million*
pulmonary edema: lung disease in which the lungs fill with fluid
UV: ultraviolet radiation

THE TERM *welding* covers a variety of processes found in industry, including brazing, soldering, resistance welding, electric-arc welding, and oxyacetylene welding and cutting. While each of these methods involves different materials and equipment, they have one important feature in common: they all require the generation of large enough amounts of energy to bring about the melting or fusing of metals.

The liberation of this heat and energy into the workplace can cause chemical and physical reactions that do not normally take place at room temperature. These reactions release different kinds of radiation and many annoying and toxic chemicals in the form of fumes, dusts, and vapors. Thus welding, if not properly controlled, creates serious workplace hazards in almost all industries.

Figure 20 gives a pictorial summary of many of the hazards to be described in this chapter.

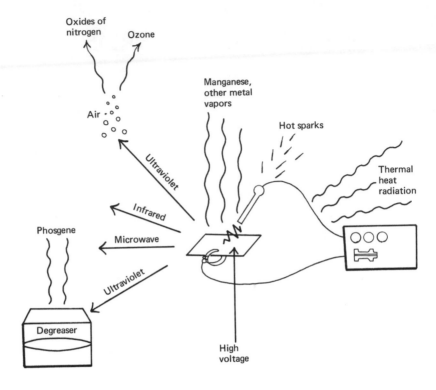

Figure 20. SOME WELDING HAZARDS

PHYSICAL HAZARDS

In electric-arc welding, an electricity generator is hooked up to two electrodes, one attached to the workpiece being welded and the other held in the hand of the welder. In the usual process, the welder "strikes an arc" by touching his electrode to the workpiece and causing a short circuit. The electricity pulses through, typically at up to 100 volts, stripping electrons from gases in the surrounding air, ionizing it, and thus producing the arc. Once formed, this region of ionized air can conduct electricity (normal air cannot), and the welder pulls his electrode back a few inches.

The heat released in the arc melts both the electrode tip and the metal around the joint being welded. Droplets of the liquefied electrode are propelled onto the workpiece as a fine metallic spray, which then cools and solidifies to form the weld.

The physical hazards involved are the presence of a live electrode carrying high voltage, and flying metallic sparks and molten metal. Many industrial accidents involve shocks and burns to welders and their helpers, due to exposed electrodes and poor electrical-cable splices. Some workers tend to shrug off the effect of the constant shower of sparks that comes from electric-arc welding, because it is such a common occurrence. Actually, these sparks consist of tiny shreds of extremely hot metal, sometimes still molten, which can be hotter than 1,000° F. even after flying 30 feet through the air. Sparks such as these cause painful burns if they land on exposed skin. Also, numerous industrial fires have been caused by welding sparks falling on flammable material nearby, or by overheated welding cables.

Radiation Hazards

Radiation hazards are discussed in greater detail in Chapter Six, and so will be mentioned only briefly here.

When a metal is heated to its melting point, it glows with a color that indicates its temperature. The cooler-melting metals glow red, hotter-melting ones are yellow, then white-hot. These colors are evidence that visible light is being radiated. Visible light, however, is only one part of the total radiation spectrum, because metals being welded also give off ultraviolet (UV), visible, and infrared (IR) waves, and often microwaves.

The major effect of **infrared and visible-light radiation** from welding is the intense heat, which can cause burns, headache, fatigue, and eye damage. Some of the IR is stopped by the upper layer of the skin, but part of the radiation penetrates to the living layer under the upper skin and can cause serious skin burns and persistent discoloration or pigmentation. Unfortunately, the eye has no outer absorbing layer, and can be severely damaged by the infrared rays. A common condition, *heat cataract,* in which the lens of the eye becomes opaque, is often found in people exposed to IR.

Ultraviolet radiation given off in welding is the same radiation as the portion of sunlight that causes sunburn. It can cause painful burns of the skin, and prolonged exposure can lead to skin tumors that occasionally become malignant. Inert-gas-shielded welding is a particularly strong source of UV.

The majority of electric-arc welders have at some time suffered from eye damage, which they call "arc eye." A recent survey of welders in 7 factories showed that nearly half have had pink-eye *(conjunctivitis)* and that over one-tenth suffered damage to the cornea. This may or may not cause permanent loss of vision, depending on how severe the irritation is.

CHEMICAL HAZARDS

When welding takes place, many toxic fumes and chemicals are released into the atmosphere. These consist not only of the vapors of metals being welded, but also of by-products of oxidation of gases in the air, lubricating and cleaning chemicals used in the welding process, and unstripped coatings on the workpiece, such as metallic-based paints.

These vapors can affect people in a variety of ways, both obvious and not so obvious. The gases released can either replace the normal oxygen supply in the air and cause suffocation, or can cause immediate acute poisoning. Acute poisoning causes great discomfort and illness and is easily recognized.

Welding also gives off fine dusts, which can act on the body over periods of months or years. Long-term chronic effects are often not attributed to welding or even recognized as occupational diseases, probably because they have not been studied as such by doctors and scientists.

Acute Effects on the Respiratory System

In oxyacetylene (OA) welding, oxygen gas is used to burn acetylene (C_2H_2), which burns easily because of its chemical structure. The chemical reaction that takes place between oxygen and acetylene gives off water and carbon dioxide:

Oxygen		Acetylene		Water		Carbon dioxide
$2\frac{1}{2}\ O_2$	$+$	C_2H_2	$\rightarrow$	H_2O	$+$	$2\ CO_2 +$ heat energy

The heat energy liberated by the reaction is used to melt or cut the metal. **Carbon dioxide** (CO_2) is a necessary by-product of the reaction, and escapes immediately into the surrounding air. Carbon dioxide occurs naturally in the atmosphere and is not itself a toxic gas. However, if the workspace is not well ventilated, carbon dioxide builds up in the atmosphere and cuts off a worker's normal supply of oxygen. When the concentration of carbon dioxide reaches 3 to 5 percent (the normal concentration in air is about 0.001 percent), a worker's breathing rate increases noticeably. If the buildup reaches 8 to 15 percent, not enough oxygen reaches the lungs and the worker suffers the usual symptoms of suffocation: headache, dizziness, nausea, vomiting, and eventually unconsciousness.

In a damp atmosphere, carbon dioxide can combine with water vapor and form **carbonic acid,** which is irritating to the eyes, skin, and mucous membranes.

Carbon monoxide, another important product of oxyacetylene welding, results from incomplete burning of the acetylene. It unites with the hemoglobin in the blood and blocks oxygen from reaching the tissues of the body. In carbon-dioxide-shielded arc welding, it has been found to block up to 15 percent of the oxygen transfer.

Acetylene itself can escape unburned from an oxyacetylene torch. In low concentrations, acetylene is a mild narcotic or intoxicant. When allowed to build up in great concentrations, it, like carbon dioxide, cuts off the oxygen supply and causes rapid breathing and air hunger, with loss of coordination. Commercial acetylene is contaminated with chemical impurities, such as 0.06 percent **phosphine, hydrogen sulfide, arsine,** and **carbon disulfide,** all extremely toxic, as well as carbon monoxide. Phosphine, besides being quite irritating to eyes, nose, and skin, acts as an anesthetic; it also causes kidney damage, lung irritation, and a variety of other illnesses. Arsine causes similar symptoms, and in addition destroys red blood cells.

Electric-arc welding releases enough energy into the atmosphere to change the nitrogen and oxygen normally found in air to the **oxides of nitrogen** and to **ozone.** Both nitrogen dioxide and ozone work to destroy enzymes within the body tissues.

Nitrogen dioxide is a brownish-red gas that is extremely irritating to eyes, nose, and throat. Continuous exposure can cause yellow staining of skin and teeth. Exposure to a high concentration causes immediate coughing and chest pain, and the lungs become irritated and fill with fluid. This reaction often occurs after the worker has gone home. In a small unventilated room, high concentrations of nitrogen dioxide can build up in only a few minutes. This has killed many oxyacetylene welders exposed to this kind of atmosphere. The changes from an acute exposure can also show up years later on X-rays, and can develop into scarring of the lungs. Long-term exposure to low concentrations can probably lead to chronic lung disease.

Ozone is a gas that is sometimes formed during electrical storms, and its sweet smell is familiar to everyone. The ozone produced during welding by ultraviolet radiation reacting with oxygen in the air has the same sweet odor. If this odor is noticed, welding should be halted immediately. Without ventilation, the ozone concentration in the air gets up to about 0.06 ppm for flux-covered electrodes, and up to 0.5 ppm for bare-wire, argon-shielded welding of aluminum. (The TLV is 0.1 ppm.) Like nitrogen dioxide, ozone is irritating to the eyes and mucous membranes. At concentrations above 0.05 ppm, it causes lung irritation and may cause fluid in the lungs, hemorrhage, shortness of breath, headache, and drowsiness.

One further by-product of welding is **phosgene gas,** which is produced when ultraviolet rays given off by welding decompose degreasing chemicals that may be nearby. A minute amount of phosgene can be deadly, but its effects, like those of nitrogen dioxide, may be delayed for hours. Welding should never be done within 200 feet of degreasing equipment. If phosgene gas is smelled, welding should be stopped immediately, and the room evacuated and ventilated with fresh air.

Welding fumes consist of vaporized metals, metallic oxides, and chemical by-products of the welding process. Examination of the fumes by X-ray and electron microscope, as well as by chemical analysis, shows that they consist of extremely fine dust particles containing various oxides of iron, the welded metals and their oxides, and silica particles.

Much welding involves coated electrodes, in which a surface-cleaning flux is built into the electrode and melts along with it. During welding with these electrodes, fumes from the coating are released into the air every second. These consist of 3 to 10 percent manganese and 10 to 17 percent fluorides, which is the flux. For carbon-dioxide-shielded arc welding, the greater the current intensity, the higher the percentage of manganese dioxide in the fumes. Whatever metals are in the welding rod are sure to be found in the surrounding vapor.

Nearly all the metallic dusts cause skin irritation, or *contact dermatitis,* and many people become sensitive to them after short exposures. **Brass** is usually an alloy of copper, zinc, lead, and tin, and its dust and slivers cause dermatitis. **Cadmium** also causes dermatitis and allergic hypersensitization. Chronic exposure to cadmium can cause yellow discoloration of the teeth. **Nickel, nickel carbonyl,** and **chromium** cause contact dermatitis. Contact with chromium can also cause skin ulcers and pink-eye. Inhaling chromium can cause a hole to form in the nasal septum, the cartilage separating the nostrils. **Manganese** dust is irritating to the upper respiratory tract, while **titanium** and its oxides are highly corrosive to skin and lungs. Most welders sooner or later deal with all of these metals and many develop occupational dermatitis, a response to skin exposure to all of these substances.

The use of fluoride-containing fluxes in coated welding rods releases **hydrogen fluoride** into the atmosphere. When this dissolves in the water in the skin and in the mucous membranes of the nose, throat, and lungs, it becomes **hydrofluoric acid,** a substance that is used to etch glass. Needless to say, it does the body no good. It causes burning of the lungs, chills, fever, painful breathing, and coughing. It also causes severe burns of the skin, besides having long-term effects on the bones and teeth through replacement of calcium, which makes them brittle. Workers who use low-hydrogen electrodes have elevated fluoride levels in their urine.

Since the metallic welding spray is essentially a dust, one expects to find symptoms of the usual dust diseases in welders.

Siderosis is a lung abnormality caused by deposits of **iron** in the small air sacs of the lung. It especially affects electric-arc welders working in confined spaces. Although it is a form of pneumoconiosis, doctors do not believe that it causes any disability. At high welding temperatures, iron also reacts with carbon monoxide to form **iron carbonyl, a** highly toxic vapor.

Exposure to metal vapors causes *metal fume fever,* which was discussed in Chapter Seven. Welding of **brass** (zinc-lead alloy) or **zinc** causes a variation of this known as "brass founder's ague" or "zinc chills," consisting of chills, fever, nausea, vomiting, muscular pain, dryness of mouth, headache, and fatigue.

Lead fumes cause typical lead-poisoning symptoms, with stomach cramps, headache, and muscular aches and pains.

Cadmium and **chromium** both cause severe problems when breathed. Cadmium is a common paint base, and is a filler metal used for brazing. It causes acute inflammation of the stomach lining *(gastroenteritis),* lung and chest pains, and such lung diseases as bronchitis, pneumonia, or pulmonary edema (filling of the lungs with fluid). Chromium poisoning causes bronchial asthma and may result in lung cancer after many years of exposure.

Nickel. The welding of nickel alloys produces **nickel carbonyl,** a highly toxic vapor that causes headache, dizziness, and disorders of the central nervous system, and can produce pulmonary edema and allergic bronchial asthma.

Manganese. Electric-arc welders are sometimes affected by the presence of manganese in the electrodes. Users of no. 4 manganese-coated electrodes have been found to have up to 6 times the normal manganese content in the blood, and 10 times the normal manganese level in the urine. Chronic manganese poisoning is a three-stage disease that first shows up as a minor headache, apathy, sexual impotence, and a diminished desire to talk. Speech disturbances and slowed reflexes gradually develop, and in the latter phases the victim falls frequently, has a high-stepping gait, and suffers from disability of the central nervous system. This disease is known as *manganism* and resembles Parkinson's disease.

CONTROL OF WELDING HAZARDS

Welding is a hazardous profession, and safety instructions should be a part of every welder's training. The American Welding Society has developed guidelines for safe welding practices, which have recently been incorporated into the federal standard for welding, under OSHA, and found in the *Federal Register* of regulations. A successful program of welding safety must be a three-part program, involving safe practices in the work area, personal protection, and adequate ventilation in conjunction with careful monitoring.

In Table 30 we have summarized the major hazards encountered in the most important welding operations. By referring to appropriate sections of this book for individual hazards such as types of radiation and chemical fumes and vapors, a union safety committee can easily pinpoint any areas of concern for the protection of the workers. The accompanying "Sample Welding Safety Survey Form" should also prove useful in familiarizing non-welders with the conditions in the welding areas of their plants.

Federal law makes it the employer's responsibility to see to it that spark shielding is adequate and that welding is not performed with poorly spliced or faulty equipment, or near flammable materials. Since the worker is the first person affected, however, he must make it part of his job to insist on enforcement of these safety rules.

Ordinary clothing is usually sufficient protection against ultraviolet and infrared radiation. The important point is to cover *all* exposed areas, including face, wrists, and hands. Asbestos welding gloves with gauntlets are ordinarily used to protect hands from electric shock and from heat and sparks.

Eye protection is by far the most critical problem. Welders' masks are fitted with tinted glass that can stop flying molten metal and that absorbs most of the radiation. In-

Table 30. HAZARDS OF WELDING PROCESSES

TYPE OF WELDING	HAZARDS
Shielded metal-arc	
1. low-alloy steel electrode	Oxides of nitrogen; iron oxide.
2. low-hydrogen electrode	Flux fumes, chiefly hydrogen fluoride and other fluorides. Oxides of nitrogen.
3. High-alloy and stainless steel electrode	Fumes may contain up to 6% chromates.
Submerged arc	Fluxes contain 2–5% fluorides, which are vaporized.
Gas metal-arc	Higher temperatures used in this process give rise to high ultraviolet and infrared radiation.
1. Argon, helium shielding	Ozone, which increases with increased argon flow. Decomposition of trichloro- and perchloroethylene to the extremely poisonous phosgene gas. The decomposition is accompanied by a disagreeable odor and can affect people 200 feet away. Welding should be stopped immediately if the odor is noticed.
2. Carbon dioxide shielding	In addition to the above, can produce carbon monoxide in high concentrations in the path of the fumes.
3. Nitrogen shielding	Production of nitrogen dioxide is especially hazardous.
4. Thorium-tungsten electrode	Thorium, a slightly radioactive substance, will be vaporized. In general, metal fumes from almost every metal being welded will be produced.

Table 30. HAZARDS OF WELDING PROCESSES (CONTINUED)

TYPE OF WELDING	HAZARDS
Gas welding, cutting and brazing	Acetylene used in this process is an asphyxiant that causes suffocation by decreasing the amount of oxygen available. Fluoride fluxes yield fluoride fumes. Some silver brazing alloys contain cadmium, which is *extremely* dangerous, causing severe damage and even death after short exposures.
Resistance	Metal fumes.
Electron beam	Produces X-rays and should be shielded. Handling of the welded pieces may disturb deposited fumes and expose workers to substances like beryllium.
Plasma arc	Excessive noise; all the other hazards produced by gas-metal-arc welding.

SAMPLE WELDING SAFETY SURVEY FORM

Name _____ Plant _____ Date _____

FILL OUT A SEPARATE SHEET FOR EACH DIFFERENT TYPE OF WELDING JOB
DONE IN YOUR PLANT

1. What type of welding operation is this? (Check one)

   Oxyacetylene _____

   Electric arc _____

   Resistance _____

   Other (fill in) _____

2. If oxyacetylene, how is the acetylene generated? If by gasoline engine,
   how is the exhaust ventilated?

3. If arc welding:
   a. What type of electrodes are used? Use trade name and model, if
      known.
   b. If a gas engine is used to generate electricity, how is the exhaust
      ventilated?

4. Where is this job usually performed? _____

   _____

   a. Is it always performed in the same location in the plane? _____
   b. Is it indoors or outdoors? _____
   c. If indoors:
      i. How are the welding fumes ventilated?
      ii. What protection is provided for other workers not involved in the
          actual welding operation? For example, goggles for other personnel,
          asbestos or canvas screening around the welder.
      iii. How is the job ventilated?

          Room ceiling fan _____

          Hood _____

          Exhaust vent directly over fumes _____

          Fan blowing directly on work _____

          Other _____

5. WORKER PROTECTION
   a. What training, if any, is given to inexperienced welders?
   b. What special clothing (such as aprons) is given to welders?
   c. Which of the following equipment is issued to welders? (Check off)

          Tinted goggles _____

          Face shield _____

          Face and shoulder shield _____

          Respirator _____

creasingly darker shades of glass are used for higher voltages. However, if people working near the welder are without eye protection, they can be seriously hurt. If possible, welding should be isolated from other jobs; if that cannot be done, workers near the operation must be warned and given eye protection.

Federal guidelines have been set for the amount of air turnover per minute needed for each welder in a room, and for the size of the room itself. For example, mechanical ventilation of 2,000 cubic feet or more per minute must be provided if welding is being done on metals in a room with less than 10,000 cubic feet of space per welder, or with a ceiling less than 16 feet high.

It is widely recognized that the best way to control welding fumes at the workbench is with a ventilating hood, which the law requires to draw better than 100 linear feet per minute. An even more satisfactory method is a hood with a suction extension that can be moved close to the work in order to draw off the most toxic fumes before they enter the atmosphere. A typical industrial design for such a hood is shown in Figure 21.

In order to assure safe working conditions, it is essential that welders have access to monitoring equipment, which their safety committee should be able to obtain and operate. The drawing ability of a hood can easily be measured with a *velometer,* or *air-flow meter.* Similar equipment can be used to check the ventilation of an entire room or suction system. The pollutants themselves can be detected and their concentrations in the workplace measured with instruments like the *universal tester* or the *explosimeter.* This is the best for the gas given off. Unfortunately, simple testing devices for the metallic vapors themselves are not widely available, and more elaborate testing procedures are probably needed for them.

When welding has to be done in a confined space such as a boiler, the welder must wear a respirator, because the

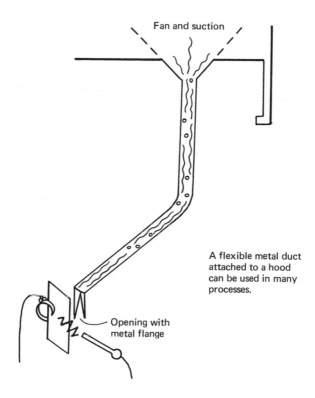

Fan and suction

A flexible metal duct
attached to a hood
can be used in many
processes.

Opening with
metal flange

Figure 21. A PORTABLE HOOD

fume concentration becomes deadly. He must have a helper
stationed outside with a safety line to be able to pull him
out should the situation become more dangerous. In addition, a test for explosive gases within a confined chamber
should be made before striking an arc. When welding is
done on pipes or containers that have held flammable material, the container should be cleaned or purged with inert gas.

CHAPTER NINE

CONTROLLING POLLUTION IN THE WORKPLACE

MANY OF THE HAZARDS discussed in this book are carried through the air, and controlling them is especially important. There are three basic methods of controlling or defending against airborne hazards such as dusts, fumes, vapors, and mists: (1) eliminating the hazard at its source, (2) proper ventilation, and (3) using protective devices. In general, the best means of control is to eliminate the hazard at its source; the second best, to ventilate a process or area; the last resort is to use respirators.

ELIMINATION OF THE HAZARD AT ITS SOURCE

Hazards can be generated by many industrial processes. Dusts come from handling bulk materials or from cutting, grinding, mining, or blasting. Welding, soldering, and heat-treating generate fumes. Since all these processes are so different, eliminating health hazards at the source is a problem that must be solved for each process. Rather than listing all the possible ways of changing processes to eliminate hazards, it is simpler to describe an over-all method of analyzing and changing conditions.

When you work with a process, you can almost always make suggestions for simplifying or improving it. The discussion that follows is designed to help you use your common-sense ideas about your work in order to make it safe. Of course, the best time to think about the safety hazards of industrial processes is when the process is first designed. Safety and health should be considered from the start, rather than added as an afterthought. But except for very well known hazards, such as radioactive substances, chlorine gas, and sulfuric acid, design for health and safety is usually added on to existing design when disastrous accidents have occurred, when the conditions become unbearable, or when the workers demand change.

Workers themselves should participate in decisions to change or design workplace equipment and practices, because they often have the most experience in the actual work and also have a personal interest in health and safety. Engineers who work in air-conditioned offices can easily overlook basic health factors.

Design for health and safety in any working situation should consider plant layout, the division of work, what materials are handled and in what manner, maintenance needs, the need for ventilation and possibly for respirators, and the content of work. For example, in considering layout, processes that produce health and safety hazards that cannot be eliminated should be isolated in one part of the plant and provided with special protective features. When considering materials, less toxic substances should be substituted for more toxic ones: toluene can be substituted for benzene as a solvent, zinc or barium can be used instead of lead in paint, steel shot instead of sand for blasting, and so on. The way materials are handled can be changed, too; for example, pellets or briquettes used instead of bulk material can eliminate dusts. Or else the process itself can be changed: metals may be joined by crimping instead of welding or soldering; the temperature, speed, or pressure of a process may be changed; and vats may be filled continuously or mechanically rather than manually or in batches. In any change of a process or work design, care should be taken that no new health or safety hazard is introduced.

If substitution is not possible, hazards can be controlled by totally enclosing the process or by shielding it. The workers can be given an enclosed space, like a booth, and operate machines with remote controls.

In all design changes, it is important to know the real meaning of the proposed changes. For example, a plant manager who was more interested in cutting down on the number of jobs rather than on hazards could redesign an entire process, saying that he was eliminating health and safety hazards, but really eliminating jobs instead. No one should be forced to give up a job just because he or she wants it to be safe.

VENTILATION

If it is not possible to eliminate or to control airborne hazards, the next best answer is to use ventilation to remove them. Ventilation systems are one of the most important yet most neglected parts of industrial plants. Ventilation is usually used as the first line of defense against airborne health hazards, but ventilation systems are often not maintained properly, even though they are as important to your health as the garbage and sewer systems in your community.

One of the major problems of ventilation systems is neglect. Because such systems are not directly related to production, there is not a great deal of incentive for maintenance. If a piece of production machinery breaks down, it is usually fixed immediately, but the ventilation is allowed to run down without a second thought. Even if a safety or maintenance engineer does look after a ventilation system, he frequently does not understand health hazards. Most engineers understand fans, cleaners, and air-flow problems but not lungs and poisons. On the other hand, if there is an industrial hygienist or company doctor around, he or she may understand the lungs but not the air-flow problems.

Finally, even though a ventilation system may be designed properly in the first place, it is often overextended. If additional hoods and ducts are added on without increasing the exhaust capacity of the fans and cleaners, they become ineffective. This chapter will try to help you overcome some of these problems by introducing some basic theory behind ventilation systems and suggesting ways in which you might inspect your own system.

There are two basic types of ventilation systems: *general* or *dilution ventilation,* and *local* or *exhaust ventilation.* The purpose of a general ventilation system is to distribute fresh air throughout the plant at a comfortable temperature and humidity. A general system consists of a series of blowers, inlets, outlets, ducts, and air-treating equipment. A general system is usually not very effective in controlling hazards, since it relies on fresh air merely to dilute harmful substances. A local or exhaust system, on the other hand, removes hazards from the workplace at the point where they are generated, and is thus more effective at controlling hazards. Each process that releases harmful airborne substances should have a local ventilation system so that those hazards can be removed immediately before they are inhaled.

A general ventilation system also replaces air removed by the exhaust system. If this *makeup air* were not supplied, exhaust hoods would find it harder to suck contaminated air out of the workplace because the suction forms a relative vacuum in the area being ventilated and leads to bothersome cross drafts and to pressure differences across doors and windows, making them difficult or dangerous to open. The need for makeup air is frequently overlooked, like many other aspects of ventilation systems that are taken for granted.

For general ventilation systems, there are several principles of good practice:

1. Air that is brought into the plant should go to work areas first and then to areas where there are harmful substances, so that workers can breathe as fresh air as possible.
2. Air should be evenly distributed throughout the plant so that cross drafts are avoided.
3. The inlet or general system should be located well away from the outlet for the exhaust system or from other sources of foul air.
4. Air should be treated (heated, cooled, humidified, dehumidified) for the greatest comfort and health of the workers.

A local ventilation system consists of four main elements: (1) one or more hoods, (2) a network of ducts, (3) air-cleaning equipment, and (4) a fan or blower. All of these elements operate in an environment where clean, fresh air replaces the withdrawn contaminated air. The main purpose of a local ventilation system is to remove contaminated air from the work area.

Hoods

The most important part of a local system is the design and placement of the hood, which actually captures and removes contaminants. In order to understand how hoods work, we must follow the path of an unwanted particle from its release into the air to its being captured by the hood and carried into the duct. We also need to know how particles get into the air, what happens to them once they are in the air, what air currents in the vicinity of the hood do to them, the speed and volume of air needed to carry the particles into the hood (measured in feet per minute), and how much contaminated air must be removed to provide adequate air for breathing (measured in cubic feet per minute).

Particles are released into the air by many different processes: grinding, blasting, sawing, polishing, evaporation, welding, spray painting, degreasing, and others. These processes give the particles an initial velocity, which may be very small, as from open vats and degreasing tanks; or moderate, as from low-speed conveyors, welding operations, and container filling, with initial velocities of about 100 feet per minute; or high, as from high-speed conveyors, blasting, and grinding, with initial velocities of up to 1,500 feet per minute.

Table 31. VELOCITY OF PARTICLES GENERATED
BY DIFFERENT PROCESSES

CONDITION OF CONTAMINANT DISPERSION	EXAMPLES	CONTROL VELOCITY AT POINT OF ORIGIN, FEET PER MINUTE
Released with essentially no velocity into still air	Evaporation from tanks, degreasing, plating	50–100
Released at low velocity into moderately still air	Container filling, low-speed conveyor transfers, welding	100–200
Active generation into zone of rapid air motion	Barrel filling, chute loading of conveyors, crushing, cool shakeout	200–500
Released at high velocity into zone of very rapid air motion	Grinding, abrasive blasting, tumbling, hot shakeout	500–2000

What happens to particles once they are released depends on this initial speed or velocity, the size of the particles, and air currents. Smaller particles tend to float in the air and are carried by air currents; larger ones tend to fall and are less affected by air currents and more by their own initial velocity.

An example is a small quartz crystal, 10 microns ($\frac{1}{2540}$ inch) in diameter, that is thrown with a high initial velocity, as from a grinding process. This crystal will travel only about $1\frac{1}{2}$ inches before it is overcome by air resistance. If it is not thrown from a grinder but just released from a height of 5 feet and allowed to fall in perfectly still air, it will take more than 30 seconds to reach the ground. This smaller particle actually floats in the air, and tends to go wherever the air it floats in goes. In contrast, if a larger quartz particle with a diameter of $\frac{1}{16}$ inch (or 1,590 microns) is thrown with the same initial velocity, it will travel about 150 feet before it is overcome by air resistance. If dropped from a height of 5 feet, it will reach the ground in less than half a second. Thus larger particles are affected more by their own initial speed than by air currents.

Gases and vapors behave like the smallest particles. They mix with air and float either up or down, depending on whether they are lighter or heavier than air. But regardless of their weight, the path they follow depends on air currents. It is important to realize that since the small particles, gases, and vapors float, they are easily inhaled with the air you breathe, go right past the normal filtering of your nose and throat, and penetrate deep into your lungs, where they can cause disease and disability. The particles that cause the most harm are less than 5 microns (1/5,080 inch) in diameter. These cannot be seen with the naked eye unless they are in very large concentrations. Air that looks clear is not necessarily clean and fit to breathe.

Air currents that occur naturally in the workplace are called *secondary air currents*. Secondary currents should either be eliminated or used to help move the unwanted substances away, or, if they go in the wrong direction, they should be overcome by generating currents that carry the particles in the right direction. Moving machinery parts, gas escaping from cylinders, or air from pneumatic devices all cause secondary air currents. Vibration, heat, the movements of workers, or cross drafts do also. Proper design and placement of hoods can use the currents to carry particles into the hood, as in spray painting toward the hood opening, or in placing the hood above a hot operation.

The air speed at which particles will be removed is called the *capture velocity*. Practical experience accumulated over many years is the best guide to determining the right capture velocity. Capture velocities for selected industrial operations are listed in Table 32. The capture velocity needs to be created at the point where the contamination is released and not at the hood opening. This means that the hood must have the right air flow and be in the right place in order to create the correct capture velocity.

Table 32. MINIMUM AIR FLOW NEEDED
FOR DUST COLLECTION

MATERIAL	FEET PER MINUTE (FPM)
Very fine, light dusts	2000
Fine, dry dusts and powders	3000
Average industrial dusts	3500
Coarse dusts	4000–4500
Heavy or moist dust loading	4500 and up

MATERIAL OPERATION, OR INDUSTRY	FPM
Abrasive blasting	3500–4000
Aluminum dust, coarse	4000
Asbestos carding	3000
Bakelite molding powder dust	2500
Barrel filling or dumping	3500–4000
Belt conveyors	3500
Bins and hoppers	3500
Brass turnings	4000
Bucket elevators	3500
Buffing and polishing Dry	3000–3500
Sticky	3500–4500
Cast iron boring dust	4000
Ceramics, general Glaze spraying	2500
Brushing	3500
Fettling	3500
Dry pan mixing	3500
Dry press	3500
Sagger filling	3500

MATERIAL OPERATION, OR INDUSTRY	FPM
Clay dust	3500
Coal (powdered) dust	4000
Cocoa dust	3000
Cork (ground) dust	2500
Cotton dust	3000
Crushers	3000 or higher
Flour dust	2500
Foundry, general	3500
Sand mixer	3500–4000
Shakeout	3500–4000
Swing grinding booth exhaust	3000
Tumbling mills	4000–5000
Grain dust	2500–3000
Grinding, general	3500–4500
Portable hand grinding	3500
Jute	
Dust	2500–3000
Lint	3000
Dust shaker waste	3200
Pickerstock	3000
Lead dust	4000
with small chips	5000
Leather dust	3500
Limestone dust	3500
Lint	2000
Magnesium dust, coarse	4000
Metal turnings	4000–5000
Packaging, weighing, etc.	3000
Downdraft grill	3500
Pharmaceutical coating pans	3000
Plastics dust (buffing)	3800
Plating	2000
Rubber dust	
Fine	2500
Coarse	4000
Screens	
Cylindrical	3500
Flat deck	3500
Silica dust	3500–4500
Soap dust	3000
Soapstone dust	3500
Soldering and tinning	2500
Spray painting	2000
Starch dust	3000
Stone cutting and finishing	3500
Tobacco dust	3500
Woodworking	
Wood flour, light dry sawdust and shavings	2500
Heavy shavings, damp sawdust	3500
Heavy wood chips, waste, green shavings	4000
Hog waste	3000
Wool	3000
Zinc oxide fume	2000

Generally speaking, the hood opening should be as close as possible to the point where the contaminant is released. This is because the *reach* of a hood, or the distance over which it can pull air into itself, is very short. For example, if air is drawn into a plain duct opening at a speed of 4,000 feet per minute (fpm) inside the duct, this draft will create an air velocity of only 400 fpm, or one-tenth as much, at a distance equal to only 1 diameter of the duct. In contrast, if air is *blown* out of the duct at the same speed—4,000 fpm— air will be moving at 400 fpm at a distance of 30 times the diameter of the duct. This difference is due to the fact that when air is drawn into the duct, it comes in from all directions around the opening, but when it is blown out, it goes in one direction only. Placing a flange or a tentlike hood at the opening of the duct will restrict the direction from which air can flow into the duct and thus somewhat increase its effective reach. In some applications, an additional blower can blow air into the hood opening, creating a "push-pull" effect.

Another major consideration in hood design involves how much air must be evacuated from the work area to reduce the concentration of the contaminant to safe levels. This volume depends on the amount of the contaminant released by the process and how dangerous it is. The best way to find out how fast the contaminant is produced is by taking air samples. When the pollutant comes from evaporating liquids, the rate of generation can be calculated by knowing the temperature, vapor pressure, and surface area of the liquid exposed.

If a contaminant is a fire or explosion hazard, ventilation will usually be installed. Ventilation for health is not the same as ventilation to prevent fire or explosion. Explosions occur when explosive gases, vapors, or dusts reach a concentration high enough for a spark, flame, or the right temperature and pressure to set off an explosion. Gases and vapors that are lighter than air will concentrate near the ceiling, and a hood will be placed there to remove the dangerous mixture. The opposite is true for substances heavier than air. Ventilation designed to prevent explosive mixtures will lower the concentration of dangerous substances to the *lower explosive limit* (LEL), or concentration below which an explosion will not occur. This LEL is not the same as the TLV or *threshold limiting value,* the legal limit at which damage to your health will occur, and could be higher or lower than the TLV.

When hoods are designed for potentially explosive mixtures, it is important to consider both the LEL and the TLV. Also, when air samples are taken for explosive levels, it makes sense to sample near the ceiling or floor, if that is where the hazards are. When sampling for health hazards, the samples should be taken at the level where the workers breathe.

In summary, there are four things to remember about good design of hoods:

1. The hood should be as close as possible to the source of the contaminant for fastest removal and to prevent spread to other areas.
2. Natural drafts should be used wherever possible to help the flow of contaminants into the hood. This also keeps contaminants from spreading.
3. Wherever possible, ventilated processes should be totally enclosed.
4. The air flow should not be through the breathing zone of the worker. For example, if a worker must lean over an open vat, a hood placed directly above the vat would draw the contaminants directly into his or her face, while a hood designed to draw vapors away to the side would not.

Ducts

Ducts carry the contaminants captured by the hood away from the workplace. Ducts must be made of corrosion-resistant materials and should be designed for fire safety. To keep the air flowing smoothly, ducts should be round, with as few bends as possible, and where there are bends, these should be gradual rather than sharp. When branch ducts join main ducts, the main ducts should enlarge gradually to allow for the increased air flow added by the branch, and the speed of air flow inside the duct should be high enough to keep the particles from settling. This *transport velocity* varies with different materials (see Table 32). The entire system should be balanced with enough suction to remove particles. Where the metal of the ducts is joined, the seams should be folded away from the direction of air flow to help prevent leaks. Ducts should be inspected periodically for leaks, settling, condensation, plugging, or damage from dents or corrosion. When a ventilated process is shut down, the ventilation system should be run a few minutes longer to evacuate contaminants thoroughly from the ducts so that they do not settle and possibly leak out and create secondary pollution problems.

Air Cleaners

Polluted air that is removed from the workplace by the hood is carried by the duct system to the air cleaner, which removes the pollutant from the air before blowing it outside. Air cleaners are not absolutely essential to maintaining a clean workplace, but without them, the neighborhood around the plant will be polluted. Neither workplaces nor the areas surrounding them should have to suffer from industrial processes.

There are several types of air cleaners. The most common is an ordinary filter, similar to the filter bags used on home vacuum cleaners. *Electrostatic precipitators* are the most efficient cleaners for removing very small particles, say, down to 1 micron in diameter. *Cyclones* are useful as "pre-cleaners" to remove large particles from contaminated air before it goes to more efficient cleaners or to the fan itself. *Scrubbers* are any type of cleaner that uses water or some other liquid to wash the air. *Settling chambers* are simple but inefficient cleaners that hold air in an enclosed space while dust particles settle out.

Fans

Fans are the part of the ventilating system that suck out the air. For this reason, it is very important to keep them in good working order. Fans should be installed with a straight duct to avoid interfering with the air flow. They should be mounted to keep the noise level down; this can be done by putting them outside or by placing them on rubber mountings. Mounting a fan directly on a steel floor will cause the steel to act as a sounding board. Fans should also be installed on the clean-air side of the air cleaner to help prevent corrosion and clogging. If explosive materials are

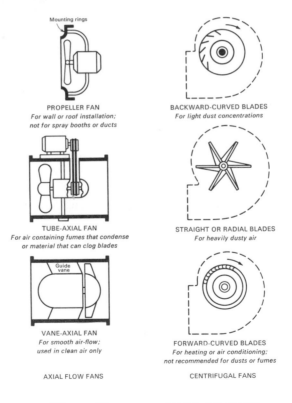

Mounting rings

PROPELLER FAN
For wall or roof installation;
not for spray booths or ducts

BACKWARD-CURVED BLADES
For light dust concentrations

TUBE-AXIAL FAN
For air containing fumes that condense
or material that can clog blades

STRAIGHT OR RADIAL BLADES
For heavily dusty air

Guide vane

VANE-AXIAL FAN
For smooth air-flow;
used in clean air only

FORWARD-CURVED BLADES
For heating or air conditioning;
not recommended for dusts or fumes

AXIAL FLOW FANS

CENTRIFUGAL FANS

Figure 22. SOME TYPES OF FANS

being evacuated, fire- and spark-resistant materials should be used and the motor should be fully shielded to avoid the possibility of sparks that might ignite the mixture.

Choosing the right fan is very important, and it is necessary to think about how much air needs to be moved and what the contaminants are. Once the type of fan is selected, it is easy to consult with manufacturers' published ratings and to select the particular fan required. Fans are rated according to tests established by the Air Moving and Air Conditioning Association.

More detailed information about fans can be obtained from the Air Moving and Air Conditioning Association, from the *Industrial Ventilation Manual* (twelfth edition, 1972) published by the American Conference of Governmental Industrial Hygienists, or by consulting the American Society of Heating, Refrigerating, and Air Conditioning Engineers *Guide and Data Book,* 1969, Chapter 4.

Testing the Ventilating System

The best test of any ventilating system is whether or not the air in the workplace is fit to breathe. This means taking an air sample. Once you know that the ventilating system is supplying air that is clean, you don't have to keep on taking samples but can just check on the system itself, unless the process it is ventilating is changed.

Inspecting a ventilation system is like diagnosing a patient. You have to (1) gather the records and specifications of the system, (2) interview the people who work around or on the system, (3) look at the system, (4) take measurements, collect data, and make a diagnosis.

Gathering Records. The basic documents to get for inspecting the system are a blueprint or plan of the system showing location and capacity of the hoods, the plan of the ducts, location and description of the air cleaner, location and capacity of the fan, and location of maintenance and inspection points. There should also be a standard record-keeping form. Records of past inspections and maintenance work, as well as of any changes in the system such as the addition of ducts or hoods, should also be obtained. These documents record the design specifications of the system. They form the basis by which an inspector will decide whether or nor the system is measuring up to its capacity or not. All of them are usually kept by the plant safety or maintenance engineer. If they are not readily available at the plant, they may usually be obtained from whoever installed the system in the first place—a ventilation contractor or sheet-metal contractor. If the management does not have them, that is an initial indication of a poorly maintained system. In any inspection, careful records should be kept of all steps taken and measurements made so that future inspections can have this record as a reference.

Other documents of interest include copies of government regulations, codes, and standards of good practice. The federal government, through OSHA, has published standards in the *Federal Register* (see references) for various industrial processes. State and local codes also frequently establish criteria for minimal good conduct. Standards of good practice are published by the American Conference of Governmental Industrial Hygientists (*Industrial Ventilation: A Manual of Recommended Practice*), the American National Standards Institute (ANSI Z9.2-1971, Z33.1-1961) the National Fire Protection Association (NFPA no. 204-1968, no. 90A-1971) and the National Safety Council (Data Sheet 428-1963). These are recognized standards accepted by most industrial managements. All these documents give information to help you evaluate the functioning of a local ventilation system.

Interviewing People. Another way to evaluate the system is to talk to people who work on or around it and who usually know a lot about it. They should be asked when the ventilation system is turned on and off and by whom, how often the system is inspected or shut down for maintenance, and how well the system works in general. Their personal knowledge may conflict with official records. Try to find out if the materials or processes have been changed from the time the system was originally installed. Be sure to talk to people from every shift, as there may be differences from one shift to another. People who work with a system usually have a feel for where it might be weak. Investigating a ventilation system is like doing detective work: a good ventilation detective asks all the neighbors whether the system has many problems, what they are, and even whether it works at all.

Looking at the system. The system may easily be examined at the same time people are interviewed. The actual works should be compared with blueprints, and maintenance and inspection stations should be located. Any damage, signs of corrosion, leakage, loose fittings, or blockage, or any sign of physical harm should be noted on the inspection sheet. The air cleaner should be examined to see if it is plugged up. The fan should be inspected for signs of corrosion or imbalance. The ducts should be tapped with a metal object to see how they sound; a dull thud may indicate that particles have settled in the duct, while a ringing sound indicates a clear duct. The inspector should also check that makeup air has ample opportunity to enter the work area without creating bothersome cross drafts that may add to worker discomfort or disturb the operation of hoods.

Taking measurements. Actually measuring the capacity of the system is the heart of an inspection. It is the most time-consuming part and the stage where the greatest care and imagination are required. Measurements of air velocity, suction (static pressure), power consumption by the fan motor, and speed of the motor are all useful. A smoke tube that chemically generates smoke may also be useful for actually tracing the flow of air into the hoods.

Ventilation systems break down for any number of reasons, the most common being neglect and improper maintenance. The parts most commonly neglected are the fan and the air cleaner. Filters must be inspected and cleaned regularly, unless they are cleaned automatically, and fan motors cleaned and lubricated. Failure to do this results in pressure losses at the filter and reduced fan capacity. Adding hoods without at the same time increasing the capacity of the fan or air cleaners results in reduced suction at all hoods and an overloaded filter.

Repairing the system may require anything from simple routine maintenance to major overhaul and redesign. Sometimes the ventilating system just has to be turned on. At any rate, the real test of good function is whether the workplace is safe and comfortable.

In pressing management for improvements in a ventilation system, it is always important to keep in mind that the final measure of success is not whether the system meets its design specifications but whether the air in the workplace is fit to breathe.

PROTECTIVE DEVICES—RESPIRATORS

If there is *no* way of eliminating or controlling atmospheric hazards, then some sort of personal protective equipment is required. Since many hazards enter the body through breathing, respiratory protective devices are common. However, some hazardous substances may also enter in other ways, such as through the skin. In these cases, additional protection, such as special clothing, is needed. In this section we will discuss only respiratory protective devices.

Different types of hazards—lack of oxygen, toxic gases and vapors, or particles—require different types of respiratory protection. There are various types of air filters for use with respirators. In order to be protected adequately from uncontrollable hazards, you have to know the hazard and its toxicity.

In determining the right respirator to use, hazards are ordinarily divided into the following categories:

1. Oxygen deficiency

Single-Filter Respirator

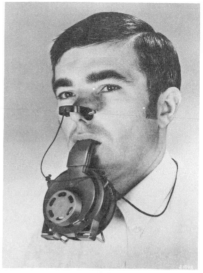

Pocket Respirator

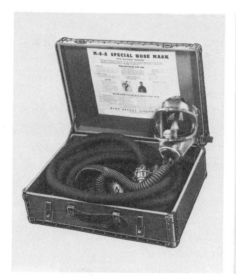

Special Hose Mask

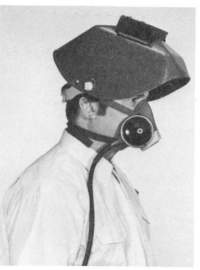

Demand-Type
Compressed-Air Mask

Figure 23. FOUR RESPIRATORS

2. Gas and vapor contaminants
   a. immediately harmful to life or health
   b. not immediately harmful to life or health
3. Particulate contaminants (dust, fog, fume, mist, smoke, and spray)
   a. immediately harmful
   b. not immediately harmful
4. Combination of gas, vapor, and particulate contaminant
   a. immediately harmful
   b. not immediately harmful

First, let's consider oxygen deficiency. Normal air contains about 21 percent oxygen, and air that contains 16 percent or less oxygen is immediately harmful to life and health. At higher altitudes or under other conditions of relatively low air pressure, a lack of oxygen is even more dangerous. A person breathing air with insufficient oxygen may begin to breathe faster, have a faster pulse, become dizzy or drowsy, have muscle cramps, become unconscious, and maybe die. Continuously working in such an atmosphere can give a worker a jaundiced appearance and may even lead to kidney and liver damage.

Oxygen deficiency may occur in any enclosed and unventilated space, such as tanks, wells, mines, holds of ships, or cellars; in any area where the oxygen may be absorbed or consumed, as by certain chemical reactions, in burning buildings, or in places where organic material is decaying; or where the oxygen may be displaced by other gases. Whenever a person is in an area where there is not enough oxygen, regardless of whatever other contaminants are in the air, he or she must have a supply of fresh air or oxygen.

Gases, vapors, and particulate contaminants may be divided into two groups, those that are immediately dangerous to life or health and those that are not. No person should be required to work continuously under conditions that are immediately harmful to life or health, but if such conditions sometimes occur, a respiratory protective device should be chosen with great care since the worker's life depends on it.

The determination of whether a hazard is immediately harmful is open to some interpretation and depends on the toxicity of the substance and its concentration. Some gases, such as phosgene, chlorine, fluorine, and hydrogen sulfide, even in very small concentrations, are extremely dangerous and are considered immediately harmful. Others become immediately dangerous at high concentrations, such as sulfur dioxide at concentrations of 400 ppm (TLV – 5 ppm). Asbestos, though it causes no immediate ill effects, can cause cancer 20 years after exposure. The same is true of radioactive dusts. If there is any doubt as to whether a gas or vapor is immediately harmful, it should be considered harmful unless it can be shown that it is not. For work in areas containing immediately harmful substances, the worker should have a supply of fresh air or oxygen for full protection. Devices that merely filter out contaminants may be adequate, but only when used with great care. In general, air-cleaning devices are not reliable protection because of leaks and inefficiency.

There are many gases, vapors, and particles that are not immediately harmful to life or health and that may require an equally wide variety of respiratory protective devices. These include air-supply systems, or air-filtering devices (gas masks, half masks, or mouthpiece devices) with the right type of filter.

Respirators may be divided into two general categories, air-supply systems and air-cleaning devices. An air-supply system provides air or oxygen to breathe that is independent of the air in which the person works, while an air-cleaning device merely removes harmful gases, vapors, or particles from the air being breathed.

Air-Supply Devices

The **self-contained breathing apparatus (SCBA)** is any portable device that gives a personal oxygen supply. This may be an oxygen or air tank (which must never be interchanged) or a breathing device that uses the exhaled breath of the wearer to release oxygen. These respirators are available in different styles: a full hood, a full-mask facepiece, or a half mask covering only the nose and mouth. They may be self-contained and have no contact with the outside air, or they may be open circuits and expel the exhaled air to the outside air. There are two types of respirators. One, the *demand type,* supplies oxygen into the mask only while the person breathes. The second, the *pressure-demand type,* maintains a positive pressure in the facepiece. The demand type may allow leakage from the contaminated outside air into the mask, which can't happen in the pressure-demand type.

There are problems with self-contained breathing systems. First, they are limited by the amount of air a worker can carry as well as by the existing air pressure. (If the air pressure is doubled, the service life of the device is cut in half.) How much extra effort a person has to make while using a mask is important, as breathing through a mask can place extra strain on the heart, which may be especially dangerous for some people.

Some devices are good for only a few minutes, while others can be used up to an hour or more. Each SCBA should have a warning device that indicates when its service life is about to expire. SCBAs are bulky and heavy, and require extensive training for use. They are awkward to use in confined spaces for long periods of time.

Air-line respirators supply high-pressure air (maximum 125 pounds per square liter) to a pressure-reduction device worn by the worker, then to a mask. Compressed air may be supplied either from cylinders or from an air compressor.

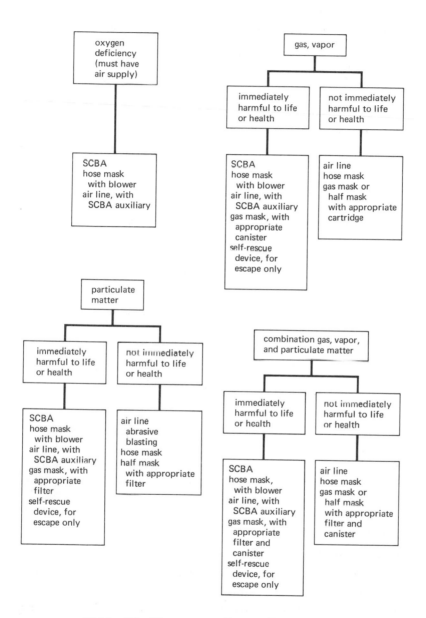

Table 33. USING THE RIGHT RESPIRATOR

Oxygen must never be used with an air line. The air line may be up to 250 feet long, and should be firmly attached to the wearer to prevent accidental disconnection. Air lines may be used with a full hood or suit, a full or half mask.

Air-line respirators are uncomfortable because they use bulky equipment, and the air line may become dirty, disconnected, cut, tangled, or pinched. Since the worker is connected to a lifeline, his or her motion is really limited. If the air comes from air compressors, it must be filtered. When the air compressors are internally lubricated, they should not be allowed to overheat because they might produce carbon monoxide. Air-line respirators are also noisy when used with a full hood or suit. They should not be used in immediately harmful areas, as there would be no protection if the line failed.

Hose-mask respirators are similar to air-line respirators and have the same problems. They supply the air through a large-diameter hose. However, hose masks may be used in air that contains immediately harmful substances, since if the blower should fail, the worker would still have enough protection to escape though a slight negative pressure would occur in the mask, threatening leakage. For full protection, the worker should also wear an auxiliary SCBA for emergency use.

Supplied-air suits provide protection against airborne hazards that may attack the skin. They should not be used in air that is immediately harmful, except when used with an auxiliary SCBA for emergencies. They may be used with either an air line or a hose.

Air-Purifying Devices

Since air-purifying devices simply clean the air and do not replace missing oxygen, they cannot be used in oxygen-deficient air. They should be used only in air that presents immediate hazards to life and health, and should be used with great care. For gases and vapors and for particles having a TLV less than 0.1 milligrams per cubic meter, the maximum concentration for which the cleaner provides protection is specified on the label.

Gas- and vapor-removing respirators may use either a full face mask, a half mask, or a mouthpiece respirator. They have a cartridge or canister containing a substance that removes certain gases or vapors by adsorption—that is, by forcing them to adhere to its surface—or by chemical reaction. Different canisters are useful against particular gases, such as chlorine; against a single class of gases, such as organic vapors; or against a combination of two or more gases and vapors. Canisters and cartridges must be labeled and color-coded for protection against particular gases or vapors (see Table 34). It is important to use the right canister for the right contaminant.

Canisters or cartridges are useful for a limited period of time, depending on the size and type of canister, the concentration of the contaminant, and the activity of the wearer. They also have a limited "shelf life," which means they can only be stored for a certain period before becoming chemically inactive. Canisters will usually heat up when they are functioning properly. If a canister gets very hot, this usually indicates that there is a high concentration of gas or vapor present, and the worker should leave the area immediately. If a canister is not functioning properly, the worker may become dizzy, have a headache, have difficulty seeing, or have a bad taste in the mouth. Again, the worker should leave the area and get some fresh air. A canister for protection against carbon monoxide has an indicator or timer that shows when it is no longer providing protection. Other canisters have window indicators for particular hazards. You should insist on using those canisters with window indicators, because these give a more reliable indication of the canister's effectiveness. Once a canister has been used, it should not be used again.

Gas- and vapor-removing cartridges and canisters provide

Table 34. COLOR ASSIGNED TO CANISTER OR CARTRIDGE

ATMOSPHERIC CONTAMINANT(S) TO BE PROTECTED AGAINST	COLOR ASSIGNED
Acid gases	White
Organic vapors	Black
Ammonia gas	Green (bright)
Carbon monoxide gas	Blue
Acid gases and organic vapors	Yellow
Acid gases, ammonia, and organic vapors	Brown
Acid gases, ammonia, carbon monoxide, and organic vapors	Red
Other vapors and gases not listed above.	Olive
Radioactive materials (except tritium and noble gases)	Purple (magenta)
Dusts, fumes, and mists (other than radioactive materials)	Orange

NOTE 1: A purple (magenta) stripe shall be used to identify radioactive materials in combination with any vapor or gas.

NOTE 2: An orange stripe shall be used to identify dusts, fumes, and mists in combination with any vapor or gas.

NOTE 3: Where labels only are colored to agree with this table, the canister or cartridge body shall be aluminum or gray in color.

NOTE 4: The user shall refer to the label wording to determine the type and degree of protection the canister or cartridge will afford.

no protection against particulate contaminants unless it specifically says so on the label. Canister masks should not be used in air that is immediately harmful to life or health. Some authorities, such as the American National Standards Institute and the National Safety Council, say that gas masks with the appropriate canister may be used in air that is immediately harmful since the wearer can smell or taste the contaminant when the mask is not functioning properly. However, once you can smell or taste a contaminant, particularly one that presents immediate hazards to your life or health, it may already have done some damage to your health. Therefore, gas masks should not be used in air that presents such immediate hazards. If it is necessary to enter such an environment, air-line respirators should be supplied.

Particle-removing respirators use filters in cartridges or canisters to remove dust, fog, fumes, mist, spray, or smoke. These may be attached to full face masks, half masks, or mouthpiece respirators. Filters are designed for use against particular types of particles (silica dust, coal dust, asbestos) or classes of particles (dusts, fumes, mists) and should be clearly labeled. They may be fixed to gas- or vapor-removing canisters to provide protection against particles as well. Particle-removing cleaners provide no protection against gases or vapors and require replacing or cleaning when breathing becomes difficult. If particles can be seen *inside* the respirator, it is clogged or does not fit properly, and is not providing protection.

Thermal-protection respirators protect against very hot air, where temperatures range up to 300° F. For work in such environments, additional protection—special clothing, hoods, etc.—is also necessary for full protection.

The **mouthpiece** or **pocket respirator** is for either continual or emergency use. It is held in the mouth and is provided with a nose clip to prevent inhalation through the nose. When not in use, it can be carried either around your neck or in a pocket. It can be fitted with a variety of air cleaners for protection against dusts, vapors, or gases. Its major use is for protection in emergencies, and it can be put on in a few seconds.

Maintenance and Care of Respirators

In order to ensure long and reliable service life for respirators, they should be regularly inspected and cleaned and properly stored. Instructions on the care and storage of respirators are frequently supplied with each device, and some manufacturers provide a regular maintenance service for their products. However, many respirators are supplied to workers with only a minimum of supplies for maintenance and repair, and sometimes suggested maintenance procedures are completely neglected. Proper maintenance and care varies with the type of respirator and its use—whether for constant, intermittent, or emergency applications.

There are some general guidelines. Respirators should be inspected before and after each use. Filters, canisters, valves, air-pressure regulators, connections, and rubber parts should be examined to check for clogging, leakage, or deterioration. Clogged or used air cleaners should be replaced or cleaned. Disposable filters should be thrown away, not cleaned. Rubber facepieces should be massaged so that they remain pliable and soft, and examined for cracks or other signs of deterioration. If respirators have been used in a toxic environment, they should be inspected only after they have been cleaned and the toxic substances removed. Each inspection should be recorded on a standardized form. Any damaged parts should be replaced by parts supplied by the manufacturer.

Respirators should be regularly cleaned and disinfected, particularly after each use. The following procedure is recommended:

1. Remove any filters, cartridges, or canisters.
2. Wash facepiece and breathing tube in cleaner-disin-

fectant or detergent solution. Use a hand brush to remove dirt.
3. Rinse completely in warm water.
4. Air-dry in a clean area.
5. Clean other respirator parts as recommended by the manufacturer.
6. Inspect valves, headstraps, and other parts; replace with new parts if defective.
7. Insert new filters, cartridges, or canisters, making sure there is a tight seal.
8. Place in a plastic bag or container for storage.

Effective cleaning agents vary because respirators are exposed to different chemicals, but adequate cleaning can usually be done with a quaternary ammonium solution, which is an effective disinfectant. An alternative cleaning procedure is to wash respirators in a detergent solution; rinse them in a hypochlorite solution (50 parts per million—ppm—chlorine) for two minutes, in an aqueous iodine solution (50 ppm iodine) for two minutes, or in a quaternary ammonium solution (200 ppm); then rinse thoroughly. Thorough rinsing is important, as the ammonium solution may cause dermatitis, and the hypochlorite or iodine solutions may deteriorate the rubber parts.

Following inspection and cleaning, respirators should be properly stored. They usually come in some sort of carrying case or plastic bag, which should be kept to store the respirator and protect it from dust and from being crushed. When stored, respirators should not be exposed to excessive heat, as from a steam pipe or radiator, direct sunlight, moisture, or dusty or otherwise dirty air. Unless fixed in a protective case, they should not be stored in a toolbox or similar place, as they may become crushed and distorted, making an airtight fit difficult.

Emergency respirators should be inspected at least every 30 days and after every use. They should be clearly labeled and stored in a well-protected but easily accessible place.

General Weaknesses of Respirators

If respirators are effective, they are bulky or uncomfortable. Attempts to remove the bulk and discomfort is usually done only by sacrificing effectiveness.

The most common and most persistent problem with all respirators is leakage around the edges of the mask. This occurs because it is impossible to design one type of mask to provide an airtight fit for the many different shapes and sizes of faces. Any respirator with negative pressure inside the mask will tend to leak. Men with beards or sideburns cannot get an airtight fit. (Several lower-court decisions have upheld the authority of fire chiefs to prohibit firemen from growing beards and sideburns because of this airtightness problem.) Any movement of the face or any talking can frequently cause leaks. People with one or both dentures missing may also have problems with negative-pressure masks. Usually leakage is overcome by varying the sizes of respirators to ensure a snug fit, by providing seals with moldable shapes, by providing the mask with a positive pressure so that any leakage is out rather than in, or by using a full hood with positive pressure. Attempts to overcome leakage may add problems. Tightening the fit frequently presses the mask so tightly against the face that pain spots develop, which are very uncomfortable over any period of time. Once a mask is airtight, the wearer's face may become hot and sweaty, particularly if he or she is working in a hot environment. The American National Standards Institute has ranked respirators according to their relative degree of airtightness.

Respirators are hard to talk through. Several devices have been developed to overcome this problem, such as a speaking diaphragm and different types of microphone-like devices. All of these distort the voice.

Respirators are difficult to breathe through, particularly those which use particulate filters that become clogged and those which rely on the wearer to draw air into the mask.

If the worker must do considerable labor, this resistance to breathing may become quite bothersome; it may heat up the air as it passes through the filter and thus dry out the throat. This resistance has other effects, such as increasing the pulse and slightly increasing the blood pressure. These can be serious in a worker with circulatory or heart problems, or in one who is subject to many other stresses as well.

Full face masks are also hard to see through. If the wearer must wear glasses, the temple bar must pass between the face and the mask, making an airtight fit impossible. Several devices attempt to overcome this, such as fixing the correct lenses inside the mask, or fixing them to the wearer's face. Full face masks restrict vision to the side and up and down. Double-window face masks restrict depth perception and make it difficult to see things close up. These masks also interfere with the sense of balance and contribute to a general sense of ill-being.

The various types of respirators are grouped below by the ANSI according to the increasing amount of inward leakage one might expect during use.

GROUP 1

Supplied-air suit
Pressure-demand full-facepiece open-circuit and air-line devices
Pressure-type full-facepiece closed-circuit self-contained breathing apparatus
Continuous-flow full-facepiece air-line respirator
Air-line respirator with loose-fitting hood gathered around the waist
Hose-mask with blower and full-facepiece mask

GROUP 2

Demand-type full-facepiece open-circuit self-contained breathing apparatus
Demand-type full-facepiece air-line respirator
Pressure-demand half-mask air-line respirator

GROUP 3

Continuous-flow half-mask air-line respirator
Demand-type half-mask air-line respirator
Air-line respirator with loose-fitting hood gathered around the neck

GROUP 4
   Air-purifying respirator with blower and with full-facepiece or hood
      gathered around the waist

GROUP 5
   Air-purifying respirator with blower and with half-mask facepiece
      or hood gathered around the neck

GROUP 6
   Hose mask without blower
   Air-purifying full-facepiece respirator without blower

GROUP 7
   Air-purifying half-mask respirator without blower
   Mouthpiece respirator

Full face masks also tend to fog up, though there are anti-fog masks and compounds that can be used.

In general, then, respirators are bothersome. They restrict motion and are cumbersome and inefficient.

Respirators have other weaknesses as well, which extend beyond their technical weaknesses. Even if the technical problems could be overcome, these other problems would persist.

Respirators are frequently used as a substitute for positive engineering controls of airborne contaminants. They are thought to cost less, while in reality, the cost of compensating for airborne contaminants is shifted from management and placed on the men and women who must wear the respirators. Respirators may be introduced as a positive protection against airborne hazards rather than as the mere aid or emergency protection they should be. Though OSHA regulations clearly state that respirators should be used only when other means of protection are unfeasible, management lobby groups interpret it differently. They claim that under some circumstances, which really can be decided on a case-by-case basis until there is more experience with OSHA, their responsibility to protect workers from health hazards is met through furnishing personal respirators. Furthermore, management efforts are aimed at establishing their right to discipline workers for not wearing respirators. Such a stance shifts the responsibility for the workers' health away from the people who created the problem in the first place and places it on the people who are most harmed by it, the workers themselves, who are honestly laboring for their livelihood.

Another related problem is that once respirators or any personal protective devices are supplied to workers, there is a tendency on management's part to allow hazards to exceed the TLVs. Their reasoning is that the respirator protects individual workers from airborne hazards, and that once they are introduced workers are protected regardless of the concentration of the hazard. Such reasoning sounds like, "Since we have put this armor on you, we can drop boulders on your head."

There are operations that require the use of respirators. However, the conditions for the reasonable application of respirators that most industrial managers would select would vary widely from those other people would select. There are many areas open to considerable interpretation. The intent of this book in general and of this chapter in particular is to provide people who work in hazardous environments with sufficient information so they may gain more control over factors that affect their day-to-day health in the workplace.

Laws and Regulations Covering the Use of Respirators

The laws and regulations that cover the use of respirators are simple and straightforward and convey a simple message: respirators are to be used only when other means of control fail. The OSHA regulations are published in the *Federal Register*, May 29, 1971:

> In the control of those occupational diseases caused by breathing air contaminated with harmful dusts, fogs, fumes, mists, gases, smokes, sprays, or vapors, *the primary objective shall be to prevent atmospheric contamination.* This shall

be accomplished as far as feasible by accepted engineering control measures (for example, enclosure or confinement of the operation, general and local ventilation, and substitution of less toxic materials). When effective engineering controls are not feasible, or while they are being instituted, appropriate respirators shall be used pursuant to the following requirements. (Sec. 1910.134, (a), (1).)

In order for respirators to be fully accepted, they must be approved by an appropriate government agency. If a respirator is not approved, it is not only violating regulations but also probably not providing adequate protection. If it is approved, that is still no real guarantee that it is protecting you, because of all the problems discussed above. The latest listing of approved respirators is published by the Bureau of Mines in their Information Circular no. 8559, *Respiratory Protective Devices Approved by the Bureau of Mines as of May 24, 1972* and in a supplemental listing by the National Institute of Occupational Safety and Health dated March 1, 1973. (These may be obtained from the Testing and Certification Laboratory, NIOSH, Morgantown, West Virginia 26505.) The respirators approved in these publications are listed according to their approval schedule, which categorizes them according to type—SCBA, gas-mask, supplied-air, particulate-matter, or chemical-cartridge respirator—and lists them with the name of the manufacturer.

The tests that the Bureau of Mines uses to determine whether or not a respirator is acceptable are listed in the *Federal Register*, March 25, 1972. These tests might serve as guidelines for tests you might use in evaluating your own respirator. This publication may also be obtained from the Bureau of Mines.

Standard use of respirators is defined by the American National Standards Institute (ANSI) in its publication Z88.2-1969, *Practices for Respiratory Protection*.[1] This stan-

1 This may be purchased from ANSI, 1430 Broadway, New York, N.Y. 10018, for $4.50.

dard is recognized as authoritative by most industrial managers.

A manual giving more complete technical information concerning respirators is the *Respiratory Protective Devices Manual,* published by the Joint Respiratory Committee of the American Industrial Hygiene Association (AIHA) and the American Conference of Governmental Industrial Hygienists (ACGIH).[2] This manual is old, having been published in 1963, and it does not take into account recent developments either in respirator technology or in laws and regulations. No new edition is planned. For all its weaknesses, it is the only complete technical manual of its kind.

What You Can Expect from a Respirator Program in Your Plant

In order for respirators to be effective, it is essential that the people who are to wear them be instructed fully concerning their use and limitations and that they be fitted properly to ensure the most efficient use. Minimum training and evaluation instruction should include the following:

1. Instruction in the nature of the hazard, its concentration toxicity and biologic effects. This should include a discussion of what may happen if the respirator is not used.

2. An explanation of more positive steps being taken to either eliminate the hazard or control it so that respirators will not be necessary. A timetable should be produced showing when the hazard will be brought under control.

3. An explanation of why the particular respirator in question is the correct one to use. Sometimes, in a token gesture of worker participation in decision making, workers are given a choice from among several respirators of the same type.

---

[2] This may be purchased for $8.50 from the Joint Respiratory Committee, AIHA/ACGIH.

4. A discussion of the respirator's capabilities and weaknesses.
5. Instruction, including field instruction, in the proper use of the respirator. This should include a thorough discussion of emergency procedures if they are necessary.
6. Proper fitting and testing of the respirator for effectiveness and airtightness.
7. A regular program of medical testing to ensure that the wearers of respirators are getting adequate protection.
8. Continuous monitoring of the air that workers are breathing. This can be done with a stationary monitoring device or one worn by the worker on his or her lapel.

There are relatively simple tests of the airtightness and effectiveness of respirators. One of the most direct is to close off the inlet valves and breathe in. If there are no leaks, a negative pressure can be maintained inside the mask, and the mask may even collapse slightly. Alternatively, close off the outlet valves and breathe out. A slight positive pressure will develop if the mask is airtight. Another simple test, immediately after taking the face mask off, is to examine your face carefully to see if there are any traces of dust that could result from inward leakage. A slightly more elaborate test is to install an organic-vapor cartridge in the respirator and wave a cotton swab soaked in isoamyl acetate (banana oil) around the perimeter of the mask. If the person wearing the mask smells bananas, the mask leaks. During the test, the person should move and talk as he or she would during work. If the person does not smell bananas, the mask is safe. The organic-vapor cartridge should be replaced with whatever cartridge is appropriate for that application. More elaborate tests are described in the *Federal Register* mentioned above.

REFERENCES

Ventilation

American National Standards Institute. *Fundamentals Governing the Design and Operation of Local Exhaust Systems,* American National Standards Publication no. Z9.2-1971. New York, 1971.

American Society of Heating, Refrigerating, and Air-conditioning Engineers. *Guide and Data Book: Applications* (1971), *Equipment* (1969), and *Systems* (1970). New York.

Committee on Industrial Ventilation, American Conference of Governmental Industrial Hygienists. *Industrial Ventilation: A Manual of Recommended Practice,* 12th ed. Cincinnati, Ohio, 1971.

W. C. L. Hemeon. *Plant and Process Ventilation.* New York: Industrial Press, 1955.

National Safety Council. *Checking Performance of Local Exhaust Systems,* Data Sheet no. 428, revised. Chicago, 1963.

Respirators

American National Standards Institute. *Practices for Respiratory Protection,* American National Standards Publication no. Z88.2-1969. New York, 1969.

Joint Respiratory Committee: American Industrial Hygiene Association and American Conference of Governmental Industrial Hygienists. *Respiratory Protective Devices Manual.* Detroit, Mich., 1963.

National Fire Protection Association. *Breathing Apparatus for the Fire Service,* rev. ed. Boston, 1971.

U.S., Department of the Interior, Bureau of Mines. *Respiratory Protective Devices Approved by the Bureau of Mines as of May 24, 1972.* Information Circular no. 8559. Washington, D. C., 1973. Supplemental listing by the National Institute of Occupational Safety and Health, March 1, 1973.

U.S., Department of the Interior, Bureau of Mines. "Respiratory Protective Devices; Tests for Permissibility; Fees," *Federal Register,* March 25, 1972.

NAMES AND ADDRESSES OF COMPANIES TO WHICH
APPROVALS ON CURRENTLY ACTIVE RESPIRATORY
PROTECTIVE DEVICES HAVE BEEN GRANTED

Acme Products, Scott Aviation, Division of "Automatic" Sprinkler Corp. of America, 1201 Kalamazoo St., South Haven, Mich. 49090

American Optical Corp., Southbridge, Mass. 01550

Aro Corp. (formerly The Firewel Co.) 400 Enterprise St., Bryan, O. 43506

Bausch & Lomb, Inc., P.O. Box 478, Rochester, N.Y. 14602

Binks Manufacturing Co., 3114 Carroll Ave., Chicago, Ill. 60612

E. D. Bullard Co., 2680 Bridgeway, Sausalito, Calif. 94965

Cesco Safety Products, 2727 West Roscoe St., Chicago, Ill. 60018

H. S. Cover, Station A, So. Bend, Ind. 46614

Davis Emergency Equipment Division of "Automatic" Sprinkler Corp.
of America, 45 Halleck St., Newark, N.J. 07104

The DeVilbiss Co., Toledo, Ohio 43601

Glendale Optical Co., 130 Crossways Park Drive, Woodbury, N.Y.
11797

Globe Safety Products, Inc., 125 Sunrise Place, Dayton, Ohio 45407

Metallizing Engineering Co., Inc., 1101 Prospect Ave., Westbury, N.Y.
11590

Mine Safety Appliances Co., 201 North Braddock Ave., Pittsburgh, Pa.
15208

Mine Safety Appliances Co., Ltd., Queenslie Industrial Estates, Glas-
gow E. 3, Scotland

Minnesota Mining and Manufacturing Co., 2501 Hudson Rd., St.
Paul, Minn. 55101

Norris Industries (formerly The Fyr-Fyter Co.) Fire and Safety Equipment Div., P. O. Box 2750, Newark, N.J. 07114

Pangborn Corp., Hagerstown, Md. 21740

Protector Pty., Ltd., P. O. Box 76, Sidney Mail Exchange, N.S.W.,
Australia

Pulmosan Safety Equipment Corp., 30-48 Linden Pl., Flushing, N.Y.
11354

Rite Hardware Manufacturing Co., 4429 San Fernando Rd., Glendale,
Calif. 91209

Safeline Products, P.O. Box 550, Putnam, Conn. 06260

Scott Aviation, Division of "Automatic" Sprinkler Corp. of America,
Lancaster, N.Y. 14086

U. S. Divers Co., 3323 West Warner Ave., Santa Ana, Calif. 92700

W. W. Sly Manufacturing Co., 4735 Train Ave., Cleveland, Ohio
44102

Welsh Manufacturing Co., 9 Magnolia St., Providence, R.I. 02909

Willson Products Div., ESB Incorporated, P.O. Box 622, Reading, Pa.
19603

CHAPTER TEN

MEASUREMENT AND MONITORING

In order to evaluate the extent of exposure to various hazards, it is necessary to monitor the environment. This monitoring may require a variety of instruments, some of which can be quite costly. It is the purpose of this chapter to acquaint workers with the various kinds of equipment available and some of the basic principles of air sampling and monitoring. It is obvious that no one is going to learn all there is to know about air sampling from this discussion, nor how to analyze totally his or her own workplace. Rather, it will provide general information on the techniques of monitoring, so that workers will know the kinds of demands to formulate when seeking to have their workplace monitored, and will also know whether the programs being instituted are effective.

Too often a government inspector or plant-safety engineer tries to make the technique of monitoring seem difficult. After reading this chapter, it is hoped that workers will realize that monitoring the workplace is really no more difficult than most of the jobs they do every day. And of course, once a monitoring program is set up, a worker can easily sample and analyze the air in the plant as part of the regular duties of his or her job.

All monitoring is done with two really simple questions in mind:

1. What pollutants are in the air?
2. How much of each pollutant is there?

To answer the first question often takes no more than knowing what work is being performed and what chemicals are being used. For instance, if highly toxic benzene is being used, then the monitoring should be done with equipment that measures the concentration of benzene in the air. When a complex process such as welding is being done, the result is a mixture of hazardous fumes and radiation, and tests can be performed for each of these.

When more exotic or unknown materials are involved, identification and analysis can become a real scientific challenge requiring a great deal of professional skill and experience. Here, we will stick to the measurement of common dust, vapor, radiation, and noise pollutants that frequently show up in the work atmosphere, and try to answer the question how to find out how much of a particular substance there is.

Basics of Air Sampling

Before we go over the various kinds of instruments and analyses used for testing of dusts and gases, the basic techniques of sample taking should be understood. All instruments generally consist of a pump, a meter, and a collecting device put together in a *sampling train*. The main parts are shown in the block diagram in Figure 24.

The sampling head or nozzle should be the right shape to capture a good fraction (at least 20 percent) of a fast-moving stream of air. Behind the sample head is a probe, which feeds the air sample into a collector. The sample collectors are usually small bottles called *impingers,* which may contain chemicals or substances like charcoal to trap the pollutants in the air that is traveling through them.

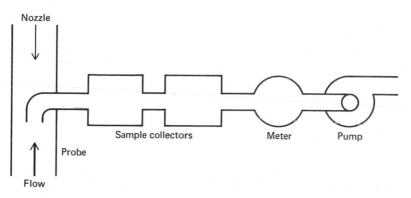

Figure 24. A SAMPLING TRAIN

The function of the meter in the sampling train is to keep track of how much air the pump is moving. In order to determine the concentration of hazards present, it is necessary to analyze a sample to see how much chemical was caught, and then to know how much air has flowed through the collector and left this chemical. The meter keeps track of the volume of air that flows through, so that a concentration can be calculated.

Finally, there is a pump that keeps a good steady flow through the whole train. Most pumps have regulating knobs to control the exact rate of air flow. Most are also small enough to be carried and set up easily at any location in a plant.

Choosing a representative sample. Two of the most crucial questions in any study of the amount of pollutant to which a worker is exposed are where to sample and what kind of sample to take. It does little good to know that the corridor is safe if the workers are in a room off the corridor. Location within the workroom is also important. Assuming that the main way pollutants enter the body is by breathing, samples should be taken at the workers' breathing level, near the nose and mouth. A meter placed near the ceiling or floor may be good for testing for explosives, but it does not give a good indication of how much pollutant is being breathed in.

Two specific kinds of samples can be taken, depending on the purpose of monitoring. The instantaneous or "grab" sample gives the hazard level, as the name implies, at the instant the sampling is done. It is a quick test to perform, and is usually done to test for immediate danger from a toxic gas such as carbon monoxide, or on routine plant surveys.

The second sampling method involves pumping air through a sampling container. This can take from minutes to hours, and can even be done continuously, so that hazard levels are checked and recorded daily. This type of air sampling gives average exposures over a work day.

The importance of representative sampling is illustrated by the famous story of the five blind men who ran into an elephant and tried to figure out what it was. One grabbed its leg and thought the elephant was like a tree, another grabbed its tail and thought it was like a snake, a third grabbed its trunk and thought it was like a rope, and so on. By just sampling non-representative parts, none of them had the true picture of an elephant.

The elephant problem can be extended to industrial processes. For example, if there is an operation that is done once an hour, so that the concentration goes up and down on an hourly schedule, sampling for 1 hour may be sufficient. A 1½-hour sample would be much worse, since it might include the same operation twice in less than a 2-hour period. While various formulas for calculating the average concentration exist, some common sense and planning will go a long way toward ensuring that a representative sample is chosen.

Calculating the concentration. Suppose the results of an air sampling for carbon monoxide show:

1 hour at 50 parts per million (ppm)
2 hours at 40 ppm
5 hours at 10 ppm

To get the average concentration for an 8-hour day, multiply all the concentrations found by the number of hours at that concentration, add them up, and divide the total by 8 hours. No matter how long a worker's day is, one always has to divide by 8 because the legal limits are all based on 8-hour days.

$$\frac{(1 \text{ hour} \times 50 \text{ ppm}) + (2 \text{ hours} \times 40 \text{ ppm}) + (5 \text{ hours} \times 10 \text{ ppm})}{8 \text{ hours/ day}} = 22.5 \text{ ppm average daily exposure}$$

Sampling for Dusts

The hazards that arise from dusts were discussed previously. There are two problems in monitoring dusts. First, the particles must be collected, and then they must be counted and sized because it is necessary to know how much of the dust is small enough to get into the lungs without being stopped by the nose and throat. A dust particle about 5 microns (1/5000 inch) in diameter is the largest that will get into the lungs. Second, it is necessary to know what the type of dust is. Some particles, such as silica, asbestos, or beryllium dusts, are extremely deadly. Others, such as iron or aluminum dust, require much higher dose levels to cause serious damage.

The type of instrument used for collecting dusts depends on whether the instrument is to be portable or not, and on whether an instantaneous or an average reading is needed. Methods for collecting dusts include the use of electrostatic force to attract the particle, and the use of impingers, in which a stream of dust-containing air is directed into water or onto a sticky surface, and then the dust is separated out. A typical impinger is pictured in Figure 25. The impinger assembly, with pump and several flasks and necessary accessories, tends to cost in the neighborhood of $200 to $250.

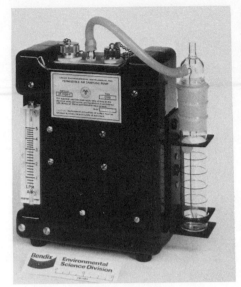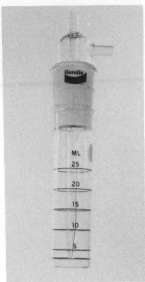

Figure 25. AN IMPINGER ASSEMBLY

Once the right trapping substance is placed in the impinger, all one has to do is turn the pump on and adjust the air flow to keep it constant. A little bobbing ball indicates the air flow. After the sample has been bubbled through the impinger, a portion of the sample of known size—say, 10 milliliters—is normally taken and evaporated. The dust particles can then be weighed, or counted under a microscope. This may also be useful for identifying the type of particle and for determining the size distribution.

Instead of using an impinger, the pump can draw the air through a filter. The filter can then be weighed and the amount of dust determined, or else the filter can be dissolved and the dust analyzed directly. Most impinger-pump setups also come with filters. This filter can also be used for counting bacteria, spores of fungi, and so forth.

A more expensive dust collector is the *electrostatic precipitator,* which places particles in a high electric field, giving them an electric charge. They then migrate to the walls, which are also charged, where they then can be collected. This method works well with fumes, which are extremely fine particles suspended in air, as well as with dusts. The precipitator can be used for short-term or long-term sampling, and the sampling head is portable. The instrument costs over $500.

There are other dust-sampling devices available. They can be found in the catalogues of the various companies listed at the end of this chapter. Also listed are several manuals that give more complete instructions on the use of this apparatus.

Sometimes it is desirable to obtain a sequence of samples automatically, over a period of hours. The sample is still collected in a midget impinger or filter, and a mechanical sequencer automatically advances the sampling to the next impinger or filter. For example, a *paper-tape sampler* collects samples in a spot on a paper tape that advances at regular intervals. It can continue unattended for days. Often the paper is treated so that it turns color when exposed to a particular chemical. The darkness of the spot indicates the extent of reaction, and can be measured with another instrument to determine total contamination. A paper-tape sampler is shown in Figure 26.

All the instruments discussed so far take samples over a period of time. A *konimeter,* which is used by the Bureau of Mines, takes 26 instantaneous or "grab" samples. With the aid of a spring, 1.5, 5.0, or 10.0 milliliters of air are impinged onto a jelly-covered sample area. The dust sticks, and the number of particles can be counted. In this way, changes in the amounts of dust can be easily determined. The dust can be counted in a relatively inexpensive microscope. This instrument is not efficient in very dusty atmospheres.

Evaluation of dust samples. Once the sample has been collected, two quantities must be determined: the total number of particles, and the number of particles of each size.

Figure 26. A PAPER-TAPE SAMPLER

The techniques for doing this are not really difficult. The problem of evaluating dust samples falls into three parts: (1) How many particles are there? (2) How big are the particles? (3) What kind of particles are they? The limit on the number of particles counted is set by the smallest particles that can be seen in the microscope. Counting the particles often means simply putting the filter paper under a microscope and counting the spots one by one. This can be extremely tedious, as well as inaccurate. Many clever techniques for treating the samples and attachments for the microscope are available to simplify the tests, but the details go beyond the scope of this chapter. The final information desired is simply the number of particles divided by the volume of air from which the particles were trapped. This result is the dust concentration, which is usually expressed in either milligrams per cubic meter of air (mg/m^3) or parts per million (ppm).

The second step is to identify the dust. Sometimes this is not a problem, because the type of dust in a given location is known. If an unknown dust is being sampled, the identifi-

cation techniques are very difficult, and such samples will
have to be sent to a professional laboratory for analysis. This
is usually an expensive undertaking, and the company should
pay the cost. Workers should see that the samples are sent to
a trustworthy laboratory so that they will get an unbiased
report.

Air Sampling for Toxic Gases

In this section we are primarily concerned with obtaining
samples of vapors such as those which evaporate from solvents or are given off in welding, but which do not exist in
particles, and which will go directly to the lungs if breathed
in.

Methods of collecting vapors do not differ much, in prin-
ciple, from methods of collecting dusts. However, some dust-
collecting methods will not work with vapors. Filters cannot
be used, since their pores obviously cannot capture a gas.
However, filter paper loaded with a chemical to react with
the gas can be used. There are many standard chemical
methods of evaluating the concentration of chemicals, and
many principles involved in dust sampling apply to sampling chemical vapors as well.

As in dust collecting, the air to be sampled is pumped
through a collecting device. The rate at which the air is
pumped through must be measured, in volume per time
(liters per minute or cubic feet per minute). This is multiplied by time of sampling to get the total sample size, in
liters or cubic feet. The *concentration* of the contaminant
is the total amount of the contaminant divided by the sample volume, and this concentration is the number to be
compared with the threshold limiting value (TLV) or other
standard used for deciding whether the air is "safe."

Devices for sampling and collecting gases are similar to
devices for sampling and collecting dusts. A pump is used
to move the air through a trap for the material to be sampled. The rate at which the air moves is determined with

Table 35. SOME GASES THAT CAN BE DETECTED
BY CONTINUOUS MONITORING DEVICES

GAS	ALARM CONCENTRATION
Sulfur dioxide	0.01–10 ppm +
Carbon monoxide	0.50 ppm
Chlorine	0.2 ppm
Total sulfur (hydrogen sulfide, sulfur dioxide, sulfur trioxide)	0.01–10 ppm
Oxides of nitrogen	0.10 ppm
Mercury	
Ozone	0.01–10 ppm
Tetraethyl lead	0.500 ppb +
Oxygen	
Carbon dioxide	
Combustible gases	
Hydrogen cyanide	0.10 ppm
Acid and alkaline gases	
Nickel carbonyl	0.250 ppb
Halogenated hydrocarbons	
Phosgene	0.5 ppm
Ammonia	0.500 ppb
Hydrazine	0.2 ppm
Fluorine	0.2 ppm
Hydrogen fluoride	0.2 ppm

ppm = parts per million   ppb = parts per billion

a flow meter. After collection, one has a sample either of air or of the pollutant already separated from the air. The sample is then analyzed, and, from a knowledge of the quantity of pollutant and of the amount of air pumped through, the concentration is determined.

Air pumps and midget impingers are the standard method for drawing an air sample. Sampling tapes, when used for toxic chemicals, do not trap particles but are impregnated with chemicals that react with the contaminant being monitored. Generally, the color that results is evaluated to give

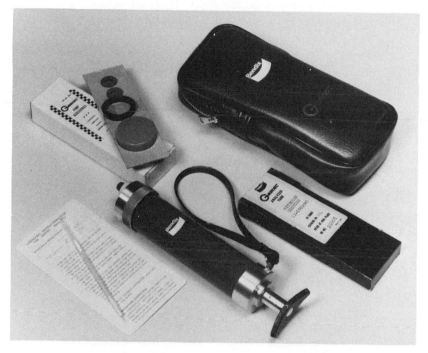

Figure 27. A UNIVERSAL TESTER

the quantity of contaminant. This is compared with the quantity of air sampled, to determine the concentration of the contaminant. Similar considerations apply to filters of all kinds.

There are monitoring devices available for specific gases such as carbon monoxide or chlorine. These devices can be used as permanent, continuous industrial monitors, which sound an alarm once a certain concentration level has been exceeded. Some gases for which alarms are available are listed in Table 35.

A useful device for routine surveys of the workplace is the *universal tester pump,* shown in Figure 27. This pump comes with tubes filled with reagents that react with specific chemicals. For example, if a tube that reacts with mercury is placed in the pump and air containing mercury is drawn through it, the chemical in the tube will change color or else a stain of a certain length will appear. The length of the stain or the depth of the color can be compared to a calibration chart that comes with the tubes, and the concentration determined.

The entire kit costs about $85, and tubes for about 120 chemicals are available. Some of them are listed in Table 36. The tubes cost between $0.50 and $1.25 each. The drawback to the method is that only instantaneous samples can be taken, not average readings, and sometimes the tubes are unreliable. However, the universal test is an excellent, inexpensive way to get a good idea of the levels of many pollutants in the workplace.

Table 36. CHEMICAL TUBES FOR THE UNIVERSAL TESTER

TUBE NO.	ANALYZER TUBE	MEASURING RANGE
81	Acetic acid	2–40 ppm
151	Acetone	0.01–0.8%
		0.8–2%
171	Acetylene	0.2–2.0%
		2.0–4%
		0.1–0.2
191	Acrylonitrile	10–500 ppm
3H	Ammonia—high range	1–16%
		16–32%
		0.2–1%
3M	Ammonia—middle range	50–500 ppm
		500–1,000 ppm
3L	Ammonia—low range	2–30 ppm
		30–60 ppm
		1–2 ppm
121	Benzene	10–200 ppm
174	Butadiene	50–1,400 ppm
104	Butane	50–1,400 ppm
142	Butyl acetate	0.01–0.8%
2H	Carbon dioxide—high range	1.0–10%
		10–20%
		0.5–1.0%

TUBE NO.	ANALYZER TUBE	MEASURING RANGE
2L	Carbon dioxide—low range	0.2–3.0% 3.0–6%
2LL	Carbon dioxide—extra low range	300–5,000 ppm
13	Carbon disulfide	5–50 ppm
1H	Carbon monoxide—high range	0.2–5.0% 5.0–10% 0.1–0.2%
1L	Carbon monoxide—low range	50–1,000 ppm 1,000–2,000 ppm 10–50 ppm
1La	Carbon monoxide—low range	25–600 ppm 600–1,200 ppm 8–150 ppm
1LL	Carbon monoxide—extra low range	5–50 ppm
8L	Chlorine—low range	1–30 ppm 30–60 ppm 0.5–1 ppm
103	Cyclohexane	0.03–0.6% 0.6–1.2% 0.015–0.03%
141	Ethyl acetate	0.04–1.5%
112	Ethyl alcohol (ethanol)	0.1–2.5% 2.5–5% 0.05–0.1%
125	Ethylbenzene	20–250 ppm 250–500 ppm 10–20 ppm
172	Ethylene	50–800 ppm
163	Ethylene oxide	0.1–3.0%
161	Ethyl ether (ether)	0.04–1.0%
91	Formaldehyde	2–20 ppm
101	Gasoline	0.03–0.6% 0.6–1.2% 0.015–0.03%
102H	n-hexane—high range	0.03–0.6% 0.6–1.2% 0.015–.03%
102L	n-hexane—low range	50–1,200 ppm
14L	Hydrogen chloride—low range	2–20 ppm 20–40 ppm 0.2–2 ppm
12H	Hydrogen cyanide—high range	0.05–2%
12L	Hydrogen cyanide—low range	5–50 ppm 2.5–20 ppm

TUBE NO.	ANALYZER TUBE	MEASURING RANGE
4H	Hydrogen sulfide—high range	100–1,600 ppm 1,600–3,200 ppm 10–100 ppm
4HH	Hydrogen sulfide—extra high range	0.1–2.0%
4L	Hydrogen sulfide—low range	10–120 ppm 120–240 ppm 1–10 ppm
4LL	Hydrogen sulfide—extra low range	5–60 ppm 1–12 ppm 0.5–6 ppm
113	Isopropyl alcohol (isopropanol)	0.04–2.5% 2.5–5% 0.02–0.04
111	Methyl alcohol (methanol)	0.02–1.5% 1.5–3% 0.01–0.02%
136	Methyl bromide	10–100 ppm 100–200 ppm
135	Methyl chloroform	100–500 ppm
152	Methyl ethyl ketone (MEK)	0.02–0.6% 0.1–2.0%
153	Methyl isobutyl ketone (MIBK)	0.01–0.6%
15L	Nitric acid vapor—low range	2–20 ppm 20–40 ppm 0.2–2 ppm
9L	Nitrogen dioxide—low range	2–100 ppm
10	Nitrogen oxides	10–300 ppm (NO) 5–300 ppm (NO_2)
11L	Nitrogen oxides—low range	2–100 ppm
133	Perchloroethylene	50–200 ppm 5–50 ppm
16	Phosgene	0.1–10 ppm
7	Phosphine	50–500 ppm 500–1,000 ppm 5–50 ppm
124	Styrene	20–500 ppm 500–1,000 ppm 10–20 ppm
5H	Sulfur dioxide—high range	0.5–4.0% 4.0–8% 0.05–0.5%
5M	Sulfur dioxide—middle range	100–1,800 ppm 1,800–3,600 ppm 10–100 ppm

TUBE NO.	ANALYZER TUBE	MEASURING RANGE
5L	Sulfur dioxide—low range	5–100 ppm
		100–200 ppm
		1–5 ppm
122	Toluene	20–250 ppm
		250–500 ppm
		10–20 ppm
132H	Trichloroethylene—high range	50–200 ppm
		5–50 ppm
132L	Trichloroethylene—low range	2–25 ppm
		25–50 ppm
		1–2 ppm
131	Vinyl chloride	0.05–1.0%
		1.0–2%
		0.025–0.05%
6	Water vapor	2–18 mg/1
		18–32 mg/1
		1 2 mg/1
100	LP gas (propylene)	0.02–0.8%
123	Xylene	20–250 ppm
		250–500 ppm
		10–20 ppm
18L	Ozone	.05–0.3 ppm
		0.1–0.6 ppm
		0.5–3.0 ppm

Another important monitor is one that monitors combustible gases and vapors. These are for the prevention of explosions and can be portable or stationary and hooked up to alarms. The monitors are set to determine the lower explosive limit (LEL) of combustible materials. They are generally used before a welder strikes an arc or in areas where there is a possibility of explosion.

It is important to realize that a gas or vapor can be toxic below its lower explosive limit, so that an explosion monitor cannot be used in place of monitoring devices for toxicity levels. Also, when monitoring for combustibles, the monitoring device is usually placed near the ceiling if the substances are lighter than air, and near the floor if they are heavier than air. When monitoring for toxic vapors and gases, the air samples should be taken at the level of the workers' nose and mouth, where the chemicals are being inhaled.

Explosimeters are essential devices for ensuring the safety of workers using flammable and explosive materials. They should not be considered substitutes for air sampling for toxic levels of industrial pollutants.

Measurement of Physical Hazards

We have discussed some of the methods of evaluating the environment for chemical pollutants. It is also possible to measure the levels of physical hazards such as noise and radiation in the workplace, using the methods described below.

Noise. The noise survey meter consists of a microphone, which converts the sound into an electrical signal, and an electronic device to sense the strength of the electrical signal and transmit it to a meter, from which the noise level can be read. It can be calibrated from a known signal. A typical noise meter is shown in Figure 14 on page 114. Noise survey meters are available from a number of firms. General Radio produces one that, with the calibrator, costs about $650. This meter records an "A" scale reading, corresponding to the legal standards. The "A" scale tunes the pickup to those frequencies picked up by the human ear.

Taking a reading involves simply turning on the meter and holding the microphone at a right angle to the path of the noise. The noise level is then read off the meter. Noise meters typically take several seconds to take a reading, so the ordinary survey meter cannot be used to determine the level of sharp, impulsive noises. Especially fast-responding equipment for these measurements is available as an attachment for the noise meter.

There is an inexpensive meter available from Radio Shack that does not give "A" scale readings but can give a fairly good estimate of noise levels. It costs about $40 and can at least give an idea of the noise level of the environment.

Just as a weighted average (see above, p. 331) was needed for determining whether the standard for a chemical pollutant was exceeded, so too a weighted average for noise exposure is needed. In a noise survey, you find the noise level at each operation, and the length of time a worker is exposed to the noise. Then the noise level is multiplied by the number of hours spent at each area, and the total added and divided by 8 hours. This gives the weighted average, which can then be compared to the standard.

There are noise dosimeters that measure the time that the noise is in excess of 85, 90, or 95 dB(A), so that an average can be obtained automatically.

Ionizing radiation and its effects on the body have been described in Chapter Six. Whenever ionizing radiation is used, careful monitoring is absolutely essential. The government is supposed to inspect all sites using radioactivity, and licensed health physicists must see to it that all precautions are being taken.

Radiation can be detected by its property of ionizing or creating electric charges in materials. Survey meters for ionizing radiation can read either in milliroentgens or in counts per minute. These give an approximate idea of whether there is immediate danger from the radiation. The most common meters are principally sensitive to beta rays. Other equipment is needed for alpha and gamma radiation.

For safety purposes, personal dosimeters are also very important. These are portable and normally worn on the clothing. One type is the film badge, shown in Figure 28. Ionizing radiation will fog film, and every week or month or whatever length of time is appropriate, the film can be developed and the extent of fogging determined. This gives the radiation exposure. Another type of personal meter is the pocket electroscope. This has a small electric charge, which is discharged by ionizing radiation. The instrument is about the size and shape of a fountain pen, and can be read by the person wearing it.

If there is danger of whole-body radiation, whole-body counters can be used. These are non-portable and have

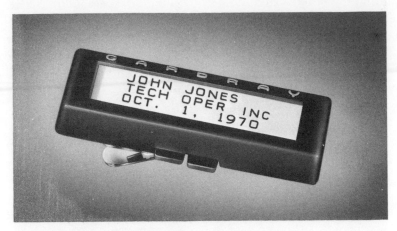

Figure 28. A FILM BADGE

counters that determine radiation coming from hands, feet, and the body as a whole. They should be located at an easily accessible spot in the workplace, so that the workers can step into them in order to have such radiation measured.

REFERENCES

U.S. Department of Health, Education, and Welfare, *The Industrial Environment, Its Evaluation and Control,* Public Health Service Publication no. 614 RS (Washington, D.C.: Government Printing Office, 1970), $1.50.
Morris Jacobs, *The Analytical Toxicology of Industrial Inorganic Poisons* (New York: Interscience Publishers, 1967).

SOME SOURCES OF MONITORING EQUIPMENT

Catalogues and information on equipment can be obtained from:
Mine Safety Appliances Company, Pittsburgh, Pa. 15208
Matheson Scientific, 1850 Greenleaf Avenue, Elk Grove Village, Ill. 60007
General Radio Company, 300 Baker Avenue, Concord, Mass. 01742

CHAPTER ELEVEN

KEEPING HEALTH RECORDS

AT THE BEGINNING of this book we made a very important distinction between *acute* health hazards and *chronic* or long-term hazards in industry. The medical effects of acute hazards, such as leaking chlorine or deafening noise, are relatively simple to detect and deal with. Long-term health hazards are much more difficult to diagnose and control. Their disabling or deadly effects on the human body develop gradually over the years, so that the worker is not fully conscious of the deterioration of his or her health. The problems themselves are often vague and non-specific and, like industrial hearing loss, often resemble the ordinary effects of aging. Since work-related chronic disease is almost never recognized as such by medical or labor authorities, a person who goes to a doctor with such an ailment usually cannot expect to have it properly diagnosed as an occupational disorder, and may never find out that it is related to his or her work. Symptoms of specific problems such as asbestosis or lung disease from chlorine exposure and other chemicals may be identical to or aggravated by non-work-related health hazards such as smoking or air pollution.

One result is that researchers who may be genuinely interested in studying these diseases are frustrated in their efforts because the very facts and figures they need are not available—not available, because no one is likely to collect statistics or medical records for an unrecognized disease.

Because chronic disease is so often treated as a non-disease, millions of American workers sacrifice their health on the job without knowing it and are not eligible for workmen's compensation or disability benefits. Many people never survive to enjoy their pensions, if they have any to begin with.

Our purpose in this chapter is to show how workers can collect the vital data necessary for gaining recognition for some of their own industry-specific illnesses. However, this is a political as well as a scientific struggle, as the black-lung miners' groups have found out, for it is very difficult to move medical and governmental authorities even after the data have been analyzed and the reports published. These health problems will be effectively recognized and dealt with only through a cooperative effort between unions, willing scientific researchers, and other concerned organizations.

How Is a Long-Term Toxic Substance Recognized?

The average person is exposed to many bodily insults every day: cigarette smoke, street and factory noise, air pollution, sonic booms. To try to isolate just one factor of daily life that contributes to over-all health loss is extremely difficult. Consequently, there are only a few substances for which long-term health effects are adequately known. To establish a chronic effect of one particular substance as a medical fact requires at least three kinds of scientific evidence.

1. **Population studies.** In this field of science, known as *epidemiology,* large numbers of people are selected for study on the basis of their exposure or non-exposure to the substance. For example, in studying the health effects of smoking, the average health of a large group of smokers is compared with the health of a similar group of non-smokers to see whether there is a meaningful difference. The most im-

portant and difficult consideration is to select subjects who are as much alike as possible except for the fact that some smoke and some do not. The group not exposed to the substance, such as tobacco smoke, are called the *controls*.

2. Animal experiments. Many animals, such as rats and monkeys, have a body chemistry similar to that of humans, but a much shorter life-span. Often, low-level exposure to chemicals will cause disease in a rat or monkey in weeks or months, whereas the effects of the same level of exposure would take years to show up in a human being. By using laboratory animals in careful experiments, researchers can sometimes identify the exact effect of toxic chemicals or dusts on specific body organs.

3. Clinical and autopsy studies. If many workers in one factory are observed to suffer from or die from one particular disease more often than other people, this is a good indication that a single factor in the plant environment is responsible. Unfortunately, this kind of information is rarely collected, because people go to widely scattered clinics or doctors. The only people with access to such data are company doctors, who rarely publicize their observations.

Autopsies are a method of identifying the specific cause of death. For example, an insulation worker who may have died from asbestosis should be autopsied to see if asbestos particles are embedded in his lungs. Unfortunately, autopsy studies are not required, and even when done, are generally not used to help establish proof of occupational disease.

In the rest of this chapter, we will concentrate on the methods of epidemiology, or statistics of large populations, so that workers can learn how to develop the necessary medical information. Smoking and asbestos are two examples of how statistical proofs were used to establish long-term health effects.

How Do We Know That Smoking Causes Lung Cancer?

Smoking is a good example, even though it is not strictly speaking an occupational health issue, because typical chemical hazards in industry follow exactly the same pattern: the smoker is exposed daily to low doses of a chemical or a mixture of chemicals. Long-term continuation of smoking, for many people, causes a variety of diseases such as lung cancer, emphysema, and heart disease. That this fact was disbelieved for many years and is still questioned by tobacco companies says something about the power of industry money over medical research.

During the last four decades, dozens of research teams across the world have interviewed hospital patients, lung cancer victims, and sufferers from emphysema and heart disease. They have collected statistics on cause of death for thousands of people. It would be impossible to summarize here the different types of studies or the different diseases looked at, so we will choose one.

A group of California researchers interviewed patients in in eleven hospitals. They chose 518 patients who had been admitted and diagnosed as having lung cancer, and 518 who were in the hospital for non-cancerous diseases. The people in the latter group, the controls, were matched one by one with the cancer group by race, sex, and age. All other factors being equal, one would expect that if smoking were unrelated to lung cancer the researchers would find roughly the same percentage of smokers in each group. The actual results of the study are presented in Table 37.

The statistician's way of describing the results would be to say that the control group represents the average public. In the population at large, we find that 42.7 percent, less than half, of the people are heavy smokers. But three-quarters of the lung cancer patients were observed to be heavy smokers. If you flip a coin 518 times, you expect it to come up heads about half the time. Of course, it is rarely

Table 37. PERCENTAGE OF LUNG CANCER PATIENTS
AND CONTROLS WHO WERE NON-SMOKERS,
AND WHO WERE HEAVY SMOKERS

LUNG CANCER PATIENTS		CONTROLS	
Non-smokers	Heavy smokers *	Non-smokers	Heavy smokers *
3.7	74.1	10.8	42.7

* 2 packs a day or more

exactly half, but it is certainly very close to it. The odds are extremely great against getting 75 percent heads in that many coin flips. Exactly the same odds hold against finding 75 percent smokers in a random group of people. Therefore, the lung cancer group is not "random," that is, there is some factor causing their statistics to be different from those of the controls. That factor has been determined to be cigarette smoke.

The study described is typical of the many which were cited in the Surgeon General's Report on Smoking and Health. It is important to understand that the proof of cause and effect for smoking and lung cancer was based, not on a single study, but on the accumulation of data on literally millions of people, *plus* the agreement of that evidence with laboratory findings in animal experiments and clinical observations and autopsies of smokers.

How Do We Know That Asbestos Is a Deadly Dust?

Asbestos is found in practically every industry, and its uses have been multiplying rapidly in the past 15 years or so. Here is another case where long-term health effects have been established: there is clear-cut, overwhelming evidence that breathing asbestos dust is extremely hazardous to health, and that these health effects take about 20 to 30 years to show up in humans.

One classic study was done by Dr. Irving Selikoff of the Mount Sinai School of Medicine in New York and his associates.[1] They obtained from the asbestos workers' union (International Association of Heat and Frost Insulators and Asbestos Workers) the work and medical records for all 632 men who were members of New York City and Newark locals on December 31, 1942. These men were then studied to determine how many had died and what they had died of over the past 20 years. This information was compared with information about how great their exposure to asbestos dust had been while they worked. At the end of 20 years, the statistics were evaluated and studied to find out how much the life-and-death expectancy of these workers, all of whom had been exposed to asbestos for 20 years or more, differed from the average life-and-death rate of the American population at large. They compared the number and cause of death of the 632 men with the *expected* rates, based on age, race, and sex. The results are shown in Table 38.

These results leave no doubt about how deadly a material asbestos is. The asbestos worker can be expected to increase his risk of lung cancer and other cancers of the respiratory tract by nearly fifteen times, his risk of stomach cancer by five times, and his risk of other kinds of cancer by double. In addition, asbestos workers have a high death rate from two extremely rare diseases: asbestosis, and mesothelioma, cancer of the lining of the chest or abdomen. These two diseases, which are almost never observed in people not exposed to asbestos, are truly industry-specific.

1 E. C. Hammond, I. J. Selikoff, and J. Churg, "Neoplasia Among Insulation Workers in the United States with Special Reference to Intra-abdominal Neoplasia," *Annals of the New Academy of Sciences*, vol. 132 (1965), pp. 519–25.

Table 38. OBSERVED AND EXPECTED NUMBER OF
DEATHS AMONG 632 ASBESTOS WORKERS
EXPOSED TO ASBESTOS DUST 20 YEARS
OR LONGER, 1943–1962

CAUSE OF DEATH	NUMBER OF DEATHS
Total cancer, all sites	
Observed	95
Expected	36.5
Cancer of lung and pleura	
Observed	45
Expected	6.6
Cancer of stomach, colon, and rectum	
Observed	29
Expected	9.4
Cancer of all other sites combined	
Observed	21
Expected	20.5
Asbestosis	
Observed	12
Expected	Less than 1
Total, all causes of death	
Observed	255
Expected	203.5

Keeping Medical Records

Some specific steps a group can take to win recognition
and control of the hazards in their own industry are these:

1. They can keep accurate records of local members, including birthdate, number of years on the job, type
   of work performed, any specific hazards they are
   exposed to, state of health, and cause and date of
   death if the member dies. The health records should
   be updated yearly. If possible, these should include

results of hearing tests (audiometry), lung-function tests, blood-pressure readings, and blood tests, if these are done. Medical tests on workers entering a local can establish a baseline by indicating how healthy they were when they began working.

2. If a specific aspect of the workplace is suspected, such as noise, dusts, or fumes, it should be measured regularly, and these records kept along with the medical records.

3. Any significant findings should be publicized as much as possible to bring the results to the attention of other workers. Publicity should be directed to the news media if possible, as well as to the trade and union publications. This proved to be extremely important in the fight against black lung.

4. Evaluation of studies and general advice on methods should be sought from good technical people, such as schools of environmental or occupational medicine. These departments will often assign one or more researchers to a specific problem of industry if the records are complete enough.

Two kinds of records have proved valuable in dealing with the problems of the workplace and establishing a baseline for normal health patterns in a given industry. The first is the health records of individual workers, and the second is a hazard record for the workplace itself.

Individuals can compare how their own health has fared over the years as compared with the general population. This comparison benefits the workers as a group because it tells them if their average collective exposure to hazards is significant, and it can give support to an individual's compensation claim or grievance in the case of chronic exposure to hazards.

It is very important to record when a worker retires or dies, and to keep the file. The file of any worker terminating employment should also be kept. The retirements, deaths, and terminations can all be saved in an alphabetical inactive file, and a rehired worker's record can be put back in the active list. These files should be kept in safe storage as long as possible. Union locals can try to pool their information with other locals in the same industry, to see if industry-wide trends are developing. Sample health-employment record forms are shown here. In this manner, a union can keep track of the sickness and death of its members.

HEALTH-EMPLOYMENT RECORD
(Note—When this sheet is full, attach a new sheet)

Name of Worker _____
　　　　　　　　Last　　　　　First　　　　Middle or Maiden

Current Address _____

Date of Birth _____ Present Age _____ Social Security No. _____

Male ___ Female ___

PREVIOUS EMPLOYMENT
What jobs have you had before, and how long did you hold each? (Put most recent first.)
1.
2.
3.

Were you ever overcome by chemicals or other hazardous conditions on any of these jobs? Yes ___ No ___ If yes, explain briefly what happened.
Have you ever worked in a mine?　　What kind?

Have you ever worked with asbestos?　When?　How long?
Have you ever worked with silica dust?　When?　How long?

SMOKING
Please fill out a Respiratory Function Questionnaire. If your smoking habits have changed recently, please fill out a new one.

PRESENT EMPLOYMENT
Date started _____
List each job held (title of job and duties) and date begun.
1.
2.
3.

List any absences due to work-related disabilities. What was the cause of the disability?

Have you been out sick for more than 7 consecutive days? What was the cause?

Have you noticed any change in your over-all health in the past year? Please describe.

POST-EMPLOYMENT (For secretary of local)

Date retired _____

Date and cause of death _____

RESULTS OF SPECIAL TESTS PERFORMED
   A. Lung function test. (Fill in or else attach spirogram.)

   Date_____ Results: FEV_1_____ . FVC_____ Other_____

   Date_____ Results: _____

   Date_____ Results: _____

   B. Audiometry (hearing) tests.
   Date _____ Results: Frequency _____ ; dB_____ .

   Date _____ Results: _____

   Date _____ Results: _____
   C. Blood

   D. Other

RESPIRATORY FUNCTION QUESTIONNAIRE

NAME

HEIGHT: AGE:

SMOKING HISTORY
   1. Do you smoke?   _____

   If YES: a. Do you inhale?   _____

             b. How many cigarettes a day   _____
                  with filters   _____
                  without filters   _____
             c. Ounces of tobacco a week   _____

             d. How old were you when you started
                  smoking regularly?   _____

   If NO: a. If not smoking now, have you ever
                  smoked?   _____
             b. Have you ever smoked as much as 1
                  cigarette a day for as long as a year?   _____
             c. How old were you when you started
                  smoking regularly?   _____

    d. How old were you when you last gave
       up smoking?   _____
    e. How much were you smoking at the
       time you stopped?   _____
    f. Why did you give up smoking?

Doctor_____   Unwell_____   Finance_____   Other_____

COUGH

1. Do you cough when you get up or first thing in the morning? _____
2. Do you cough like this on most days for as much as 3
months during the year? _____
3. Do you cough during the rest of the day? _____
4. Do you cough like this on most days for as much as 3
months during the year? _____

PHLEGM

1. Do you bring up phlegm when you get up or first thing in
the morning? _____
2. Do you bring up phlegm like this on most days for as much
as 3 months during the year? _____
3. Do you bring up phlegm during the rest of the day? _____
4. Do you bring up phlegm like this on most days for at least
3 months of the year? _____

BREATHLESSNESS

Do you have to walk slower than other people on level ground? _____

WHEEZING

Do you ever have wheezing or whistle in your chest when you
you don't have a cold? _____

WEATHER

Does the weather affect your chest? _____

MEDICAL HISTORY

In the last 3 years have you had a chest illness that kept you
off your job for at least a week? _____
What was it diagnosed as? _____
Have you ever had a heart attack?   When? _____

WORKING CONDITIONS

1. Do you work in a dusty area? _____
2. How much time do you spend in this dusty area? _____
3. Do you work with any materials that irritate your nose, lungs or
chest? _____   Which materials? _____
4. How long do you work with these materials? _____
5. How long have you worked on this job? _____
6. Have you worked on any other jobs with above conditions? _____
7. How long? _____

Any relevant comments on your working conditions? _____

Keeping Tabs on Your Own Workplace

An extremely effective approach to health and safety in the workplace has been developed by Dan Berman, a researcher in this area, and Teamsters Local 688, St. Louis. They have put together a problem-oriented booklet evaluating hazards in their plant, the Crane Corporation, which manufactures steel pipes and fittings. This type of study can be applied to almost any plant. In the summer of 1971 the 198 members of the local were asked to fill out a poll that asked, "Which of the following conditions or substances on the job bother you most?" Ninety-two men responded, and gave the following results:

PROBLEM	NUMBER OF MENTIONS
1. Ventilation	60
2. Fork-lift truck exhaust	59
3. Smoke or dust	59
4. Noise	56
5. Heat	48
6. Fumes	45
7. Faulty wiring	16
8. Graphite dust	14
9. Paints	6
10. Lubricating oils	5
	378

These results immediately pinpointed the main problems in the workers' opinion, such as ventilation and air quality. By conducting similar polls in the various departments (metallurgical laboratory, paint and grinding department, X-ray, press, welding), problems in each area were quickly identified. Once the main areas were known, hazard levels were tested to find out just how high the levels were. For instance, a sound-level meter gave a reading of 110 to 115 decibels in the flux-reclaiming part of the welding department, which is far in excess of safe levels as well as in violation of the 90-decibel legal limit.

The second half of the local's book is a series of "Hazardous Substances Inventory" sheets, one for each chemical, dust, or gaseous hazard found in the plant. Using a looseleaf notebook format, these pages are filled out and entered one at a time, as each new hazard is discovered and researched. Each form also has space to describe the methods for handling the specific problem. Thus the whole project is a continuing one, and the book becomes a running record not only of problems but also of solutions. Other groups can use this method. On the following pages we reproduce the inventory form used for carbon monoxide.

We should stress that it is not necessary to wait 20 years for all the facts to be in. If all the workers in a factory are tested for decreased lung function or for hearing loss, these results should be compared right away with the corresponding statistics for normal or unexposed people. If significant differences show up, even when a medical statistician would insist on much longer-term studies, the results can be used, with good scientific grounds, in safety grievances, arbitrations, and workmen's compensation hearings.

Often such collected evidence can make a difference in the outcome of a hearing, and if it does, it sets a precedent for other workers to quote so that they too can win better working conditions. These figures can also be of obvious use in collective bargaining, contract negotiations, and grievance procedures, where the subject of health and safety can be brought up. Knowledge of the health effects in the factory can be used as a lever to gain the right to monitor and control the level of exposure to hazardous substances on the job.

Any collected medical statistics can prove important in helping to set standards for exposure levels so that workers across the country can eventually benefit from the efforts of the unions that do the initial research.

HAZARDOUS SUBSTANCES INVENTORY, Teamsters Local 688, St. Louis (Form A-3)

1. Company name and address where substance is used:___Crane, St. Louis_____

2. Department where substance is used:_Press, welding, from fork-lifts_____

3. Trade name (include identifying numbers):___-----_____

4. Name and address of substance's manufacturer:_____-----_____

5. Description of substance: Colorless, odorless gas
   melting point207°C flashpoint-----boiling point:_-191°C_____

6. Chemical composition of substance by per cent:_-----_Is produced by many_
   operations, including driving, pressing hot steel, welding, fork-lift
   trucks, and other industrial processes, and burning of coke_____

7. Safety and health information on label (quote or attach label to
   this form):_-----_(generally a by-product of other processes)_____

8. Complaints of workers handling this substance:_headache, sleepiness, dizziness,
   confusion, depending on the concentration of the gas._____
   Long-term, low-level exposure has been linked to heart disease and
   other problems. Often used to commit suicide from car exhausts.

9. Toxicology (possible bad effects):_Carbon monoxide takes the place of oxygen
   in the blood and reduces its oxygen-supplying capacity, causing
   asphyxiation and death in high enough quantities. It is extremely
   dangerous, especially in high quantities._____

ITEMS 10 through 13: HAZARDS OF SUBSTANCE: For items 10-13 fill in whether
hazard of substance is "none," "slight," "moderate," "high," or "unknown."

10. Health Hazard Rating: Degree of threat of substance from

	contact	breathing	swallowing
local* damage, one large exposure:	none	none	none
local damage, low daily exposure:	none	high	none
systemic**damage, one large exposure:	none	high	none
systemic damage, low daily exposure:	none	high	none

*local damage occurs at the point of contact, such as skin, eyes,
 nose, mouth, throat, and lungs
**systemic damage occurs through absorption of the substance into the
 body

11. Fire hazard and fire-fighting: <u>Dangerous in high concentrations and when</u> <u>exposed to flame. No fire hazard in Crane.</u>

12. Explosion hazard: <u>none</u>

13. Disaster hazard: <u>In extremely high concentrations it can kill. In most</u> <u>plants it is a lingering hazard.</u>

14. Storage and handling: <u>Carbon monoxide hazards can be removed by good</u> <u>ventilation, in industrial operations.</u>

15. Ventilation control: <u>Necessary to prevent build-up of the gas</u>

16. Threshold limit value (TLV): <u>50 parts per million</u>

17. Hazard information source (include date of letter of inquiry): <u>Sax, 1968;</u> <u>APMIO, 1971</u>

18. Source of in-plant information (write initials): _____

19. This form (A-3) reviewed before filing by: <u>DMB</u> date: <u>13 Aug. '71</u>

IF MORE SPACE IS REQUIRED TO REPLY TO ABOVE QUESTIONS, PLEASE
INCLUDE THE INFORMATION ON THE BACK OF THIS FORM WITH THE
QUESTION NUMBER YOU ARE REFERRING TO.

CHAPTER TWELVE
WHAT IS TO BE DONE

EVERY CITIZEN in the United States is guaranteed the rights to life, liberty, and the pursuit of happiness. Working under dirty, dangerous, and unhealthy conditions is an infringement of these basic rights. Through long years of struggle and effort, workers have established the right to organize themselves and collectively bargain with their employers over wages, benefits, and *conditions of employment.* This right to collective bargaining and joint or concerted action to achieve improvements in their lives is a valuable tool that working people have. It is with this tool that the changes needed to make the workplace safe and healthful can best be achieved. Of course, the federal and state laws may also be utilized as means of bringing about change.

Achieving Change Through Government
Although the Occupational Safety and Health Act of 1970 guarantees every worker, in a union or not, the right to file a complaint without being discriminated against by the employer, in reality it is the rare individual who, standing alone, unsupported by an organization of her or his fellow workers, will use this right. Despite the guarantee against discrimination, the fear of reprisal will keep individual workers from using OSHA. In addition, when dealing with a huge government structure, the most effective way to cut

through red tape and utilize the benefits of the law is through another structure, a union. Those unions that have been trying to enforce the law have found that when the complaints of a local group are filed by the international union, and the international follows through on every step of the way and is prepared to fight for enforcement, then the law can be used most effectively. How to file a complaint and your exact rights under the law are laid out in detail by the AFL-CIO and many international unions.

But as we have said throughout this book, although the Occupational Safety and Health Act guarantees many rights and improvements to American workers, each standard that becomes part of the law must be fought over, each rule contested. Most important, sufficient funds must be appropriated for inspectors, testing facilities, research, and so on, so that enforcement of the act is possible. Because there are so many roadblocks to effective use of the law, workers cannot give up trying to utilize it. Since working people pay the taxes that carry the cost of running the government, it is their right to expect the government to protect them in their workplaces. If the government is swayed by the influence of the business community, then workers must see to it that their unions exert a counter influence. United, their strength cannot be overlooked, especially since without their labor no businesses would run at all. Of course, the unions can carry out these tasks only if workers actively participate in union affairs by such methods as frequent local meetings, strong and representative shop-steward programs, and educational activities and material for the members.

Using Your Rights as Workers

It did not take the Occupational Safety and Health Act to give workers the right to control their working conditions. It has long been the practice of many unions to negotiate these conditions. Electrical workers, construction workers, tunnel workers, boilermakers, firemen, and others have negotiated contract clauses that specify the way a job must be done in order to be safe, what equipment must be supplied, and so on. The work of these unions in improving job safety through negotiations has been excellent in many cases. Most of the contract clauses do not concern themselves with health effects at all, since it was not until recently that many of the long-term health effects of the chemical and physical hazards of the workplace were recognized. Now that they have been, contract negotiations must start to deal with them as well as with safety hazards.

Such negotiations have already been successfully carried out in some cases. For instance, the local of the United Paperworkers International Union that represented the workers at Johns-Manville, which has a severe asbestos problem, bargained for and won the right to see company medical records. The International Rubber Workers have negotiated a large joint union-management research fund to investigate the effects of their working conditions on the workers in the rubber industry. The union now employs a full-time industrial hygienist to guide it in its programs.

Besides being used for research into the medical effects of various working conditions, a negotiated fund can be used to find new engineering methods and process changes to eliminate hazards. It can also be used for regular medical screening programs, worker education and training, and so on. There are other contractual benefits that should be sought. For instance, workers should demand the right to monitor the environment, or to participate in the company's monitoring program. The union should have an effective safety committee to study health situations and to recommend specific changes and improvements, which must then be instituted. Regular medical examinations, including specific tests for the particular hazards of the job, should be provided by the company. The union should hire a physician to receive all the medical reports, so that a picture of the health status of the entire work force can be obtained to see if any trends of ill-health are developing that can be traced to particular working conditions. By maintaining these records, workers can also establish baseline statistics on individuals to see what effects working in these plants has had over the years.

The following is a sample of contract language that has been negotiated by the Oil, Chemical and Atomic Workers International Union with AMOCO (American Oil Company). Similar language has been negotiated by OCAW with all the oil companies in which it represents workers. Although workers in various industries may have different problems than chemical and refinery workers, the basic principles behind these clauses still apply.

1. There shall be established a joint labor-management Health and Safety Committee, consisting of equal Union and Company representatives, but not less than two nor more than four each.

2. The Company will, from time to time, retain at its expense qualified industrial health consultants, subject to the approval of the International Union President or his designee, to undertake industrial health research surveys, as decided upon by the Committee, to determine if any health hazards exist in the work place.

3. Such research surveys will include measurements of the exposures in the work place, the results of which will be submitted in writing to the Company, the International Union President, and the joint Committee by the research consultant, and the results will also relate the findings to existing recognized standards.

4. The Company shall pay for the appropriate physical examination and medical tests at a frequency and extent determined from time to time by the joint Committee.

5. The Union agrees that each research report shall be treated as privileged and confidential to the extent that disclosure of information in the nature of trade secrets

will not be made without the prior written approval of the Company.

6. At a mutually established time, subsequent to the receipt of research survey reports, the joint Committee will meet for the purpose of reviewing such reports and to determine whether corrective measures are necessary in light of the industrial consultant's findings and to determine the means of implementing such corrective measures.

7. Within sixty (60) days following the execution of this agreement and on each successive October 1 thereafter, the Company will furnish to the Union all available information on the morbidity and mortality experience of its employees.

8. The joint Committee shall meet as often as necessary, but not less than once each month, at a regularly scheduled time and place, for the purpose of jointly considering, inspecting, investigating and reviewing health and safety conditions and practices and investigating accidents, and for the purpose of jointly and effectively making constructive recommendations with respect thereto, including but not limited to the implementation of corrective measures to eliminate unhealthy and unsafe conditions and practices and to improve existing health and safety conditions and practices. All matters considered and handled by the Committee shall be reduced to writing, and joint minutes of all meetings of the Committee shall be made and maintained, and copies thereof shall be furnished to the International Union President. Time spent in connection with the work of the Committee by Union representatives, including walk-around time spent in relation to inspections and investigations, shall be considered and compensated for as their regularly assigned work.

9. In addition to the foregoing, Company intends to continue its existing industrial hygiene program as administered by company personnel.

10. Any dispute arising with respect to the interpretation or application of the provisions hereof shall be subject

to the grievance and arbitration procedures set forth in
the Agreement.

In addition to contract negotiations, workers have the
right to use the grievance procedure to grieve any condition of employment. Since potential health hazards are a
condition of employment, they are subject to the grievance
procedure. As an example, let us consider how to use this
procedure if the plant has a noise problem.

The union should file a grievance stating that the level
of noise within the workplace is in excess of the accepted
standard, and that it is affecting the health and well-being
of the workers in the area. The company can make a number of responses to the grievance. It can say that the noise
is not in excess. Then the union can demand to know what
criterion the company used to determine this. If it was by
sound-level meter, the union should demand to be present
at the calibration of the meter and the measurement of the
sound. If the company used no sound-level meter, the union
can demand that it obtain one to determine the sound level.
If the company refuses to provide the necessary equipment
to measure the possible deviation from the standard, the
union is then in a good position to go to arbitration and
win, citing the fact that the company has acted in an un-
scientific and capricious manner since there are ways to
measure sound. The local involved should call on its international, the AFL-CIO, or any other supporting groups for
substantiation and testimony at the arbitration.

If the company does provide the sound-level meter and
the tests show that there is excessive noise, the company may
then require the use of earmuffs. The union at this point
should demand an engineering survey to see if the noise can
be eliminated before the protective devices are used. This
right is guaranteed under the Occupational Safety and
Health Act. If the company refuses the survey, the same
steps that were detailed above can be carried out for the

earmuff question. The same type of tactic can be used in other areas of unsafe and unhealthy working conditions. It is likely that many successful changes in working conditions can be brought about through the active and proper use of the grievance procedure.

Using the grievance procedure means that there must be enough stewards to really represent the people in the plant, and that the stewards must be trained to do their job well. They should have regular meetings to exchange information and ideas. Using the grievance procedure for health and safety also requires that there be people in the plant who understand the potential hazards of their work areas. Such workers, who can be called health and safety stewards, can gather health and safety material on all the things used in their areas. The material can be taken from this book, and from labels, government publications, and other sources. The international can be asked to help provide the training and information. The safety stewards can be either elected or appointed, and can all serve together on a plant-wide safety committee. In the event of hazardous conditions, the safety steward can be the aggrieved worker and the area shop steward can file the grievance, setting the grievance procedure in motion. The grievance should be carefully planned so that it can be used effectively, as in the noise-problem example cited above.

Of course, sometimes contract negotiations, grievances, and complaints under the federal health and safety act cannot move management to institute changes and improvements. In that case, when all the workers in the plant are convinced that the problems are truly serious and are willing to fight for what they believe in, they can use the same tactics that have been used time and again by working people to achieve dignity and proper benefits for their labor. This is happening right now in many assembly-line plants. Workers in the United Auto Workers Union, through strikes, are demanding changes in the assembly line, which is a most dehumanizing way to work. This is their right, since their contract specifies that they may strike over health and safety. We can be sure that until workers in other plants are willing to wage this battle in a similar fashion, no real changes will take place. Workers themselves will have to see to it that no one has to pay any price in health, welfare, or dignity for trying to earn a decent living.

Some Exposures to Health Hazards Listed by Occupation

ABRASIVE WHEEL MAKERS AND
OPERATORS
  Abrasive dusts
    Chromium emery (lead)
    Silicon carbide
    Silicon dioxide
    Titanium
    Zirconium
  Bonding agents
    Formaldehyde-based resins
    Phenolic resins
    Polyester resins
    Rubber adhesives
    Shellac
ACTORS
  Cobalt and compounds
ADHESIVE MAKERS AND USERS
See also Airplane dope makers
    and users; Glue makers and
    users; Gum makers; Rubber
    cement makers and users
  Benzene
  Chromium compounds
  Dioxane (diethylene ether)
  Ethyl silicate
  Ethylenediamine
  Fluorides
  Ketones: acetone, butanone
  Methyl alcohol
  Plastics
    Amino resins: urea-formalde-
      hyde resins, melamine-
      formaldehyde resins
    Diisocyanate resins
    Epoxy resins
    Styrene
    Tolylene diisocyanate (TDI)
  Pyridine
  Xylene
  Zinc compounds
AGRICULTURAL WORKERS
See Farmers and agricultural
    workers

AIRCRAFT WORKERS AND
REPAIRMEN
  Chlorinated solvents
  Chromates
  Chromic acid
  Cutting fluids
  Cyanides
  Dichromates
  Glass fibers
  Hydraulic fluids: dichromates
  Hydrogen cyanide
  Hydrogen fluoride
  Lubricants
  Nitric acid
  Oils
  Paints and thinners
  Plastics, including tolylene di-
    isocyanate (TDI)
  Radiation: ultraviolet, X-ray
  Resins, including epoxy
  Rubber antioxidants and ac-
    celerators
  Solvents
  Vibrating tools
  Welding fumes
AIR CREWMEN AND PILOTS
  Carbon monoxide
  Decreased air pressure
  Microwaves
AIRPLANE DOPE MAKERS AND
USERS
See also Adhesive makers and
    users; Glue makers and users
  Butanone
  n-butyl acetate
  Cellosolves
  Diacetone alcohol
  Ethyl acetate
  Perchloroethylene
  n-propyl alcohol

ALKALOID PROCESSORS
  Ethylene dichloride
  Ethyl ether
  Isopropyl alcohol
  Methylene chloride
AMMONIA MAKERS
  Ammonia
  Calcium cyanamide
  Carbon monoxide
  Cerium
  Natural gas
  Osmium (in synthetic ammonia)
AMMUNITION MAKERS
See also Explosives workers
  Aluminum and compounds
  Cadmium
  Phosphorus
  Tetryl
ANTIFREEZE MAKERS AND USERS
  Amyl alcohol
  Diacetone alcohol
  Dichromates
  Ethyl alcohol
  Ethylene glycol
  Isopropyl alcohol
  Methyl alcohol
ASPHALT MAKERS AND WORKERS
See also Coal tar and pitch workers; Construction workers; Road workers and builders
  Acridine
  Coal tar and fractions (polycylic hydrocarbons)
  Copper and compounds
  ortho-dichlorobenzene
AUTO WORKERS AND REPAIRMEN
  Abrasive dusts
  Antifreeze fluids: dichromates

Brake fluids: bisphenol A, hydroquinone
  Carbon monoxide
  Cutting fluids
  Epoxy resins
  Gasoline
  Graphite
  Lead
  Lubricants
  Metal cleaners, including oxalic acid
  Oils
  Paint thinners: turpentine
  Paints
  Phthalic anhydride, polyester resins
  Plastics
  Rubber antioxidants and accelerators
  Soldering fluxes
  Solvents: chlorinated hydrocarbons, methyl alcohol
AVIATION MECHANICS
  Chlorinated solvents
  Fuels
  Hydraulic fluids
  Lubricants
  Oils (as aluminum-oxidation inhibitors)
  Zinc chromate (as aluminum-oxidation inhibitors)
BACTERIOCIDE MAKERS AND USERS
See also Detergent makers; Soap makers
  Fluorides
  Mercury and compounds
  Ozone
  Silver and compounds
  Tin and compounds
BAKERS
See also Food processors and handlers

Candle wax
Carbon dioxide
Disinfectants: formaldehyde,
    parabens
Dusts: flour, spices, sugar
Flavorings, natural and synthe-
    tic
Flour improvers
Fungus infections: monilia
Heat
Infrared radiation

BARBERS AND HAIRDRESSERS

Ammonium thioglycolate
Benzene
Cosmetics: talc
Depilatories: thioglycolic acid
Detergents (synthetic): hexa-
    chlorophene
Dyes: cobalt, resorcin, styrene
Hair tonics: lanolin, mercuric
    chloride, beta-naphthol
Infections: bacteria, fungi
Lacquer removers
Nail lacquers
Perfumes
Propellents
Soaps
Ultraviolent light
Vibrating machines
Wave solutions

BATTERY MAKERS AND
WORKERS

See also Electrical and electronics
    workers; plastic makers: resin
    makers
Amyl acetate (storage)
Antimony and compounds
    (storage)
Benzene (dry)
Cadmium (storage)
Carbolic acid
Chromium compounds (dry)

Coal tar and fractions (poly-
    cyclic hydrocarbons; dry)
Copper and compounds
Glass fibers
Graphite (dry)
Hydrogen chloride
Lead
Manganese compounds
Mercury
Nickel and compounds (stor-
    age)
Phenol (dry)
Picric acid
Plastics: epoxy resins, harden-
    ers
Silver and compounds
Sulfuric acid (storage)
Zinc chloride (dry)

BEARINGS MAKERS AND
WORKERS

Barium and compounds (for
    packing)
Fluorocarbons
Osmium and compounds
Phenolic resins
Rubber accelerators
Silver and compounds

BLACKSMITHS

Hydrogen cyanide
Infrared radiation

BLEACH MAKERS AND USERS

Arsine (in bleaching powder)
Chlorine compounds
Chloramine-T
Chromium compounds
Hydrofluoric acid
Hydrogen peroxide
Nitric acid
Optical brighteners
    Coumarins
    Stilbene
    Triazine
Oxalic acid
Peroxy salts

Potassium hydroxide
Sodium hydroxide
Sodium silicate
Sulfur compounds

BOATBUILDERS
See also Construction workers
Asbestos
Coal tar and fractions (polycyclic hydrocarbons)
Glass fibers
Plastics

BOILER CLEANERS, OPERATORS, AND SCALERS
Arsensic (operators)
Barium and compounds (operators)
Carbon dioxide
Carbon monoxide
Chromium compounds (scalers)
Hydrazine
Hydrogen chloride (scalers)

BOILERMAKERS
See also Construction workers
Cement and concrete: dichromates, epoxy resins
Glass fibers
Ultraviolet light
Welding fumes

BOOKBINDERS
See also Glue makers and users; Ink makers; Lithographers; Paper workers; Printers; Shellac makers and users; Typographers, electrotypers, and stenotypers
Acrolein
Amyl acetate
Asbestos
Formaldehyde
Glues (plastics)
Inks (acrylic monomer, cobalt, methyl salicylate, resin, rubber, carbon black)

Lead
Methyl alcohol
Paper dust
Shellac

BRAKE FLUID MAKERS AND USERS (HYDRAULIC FLUID)
Bisphenol A
n-butyl alcohol
Cellosolves
Diacetone alcohol
Ethylene glycol
Hydrazine
Hydroquinone
Phosphorus oxychloride
n-propyl alcohol
Tri-ortho-cresyl phosphate (TOCP)

BRAKE LINING MAKERS
Asbestos
Graphite
Plastics: phenolic resins
Xylene

BRASS MAKERS
See Metal workers: alloy makers

BREWERS AND BREWERY WORKERS, FERMENTATION
Carbon dioxide
Carbon monoxide
Hydrogen fluoride
Hydrogen sulfide
Sulfur dioxide

BRICK MASONS
See also Construction workers
Calcium oxide (lime)
Cement
Coal tar and fractions (polycyclic hydrocarbons)
Cold
Moisture
Parasitic skin infections
Plastics: epoxy resins
Preservers: ethyl silicate
Sunlight

BRICKMAKERS, BRICK BURNERS
Carbon dioxide
Carbon monoxide
Coal tar and fractions (poly-
cyclic hydrocarbons)
Lead compounds
Manganese compounds
Preservers: ethyl silicate
BRICK, REFACTORY
Beryllium
Chromates
Ethyl silicate
Graphite
Quartz
Silica
Titanium compounds
Zirconium
BROOM AND BRUSH MAKERS
Bleaches (chlorine)
Coal tar and fractions (poly-
cyclic hydrocarbons)
Colophony resin
Dyes
Ethyl acetate
Glues (adhesives)
Infections
Bacteria: anthrax
Parasites
Lead
Mercury (for carbon brushes)
Methyl alcohol
Organic dusts
Plastics
Shellac; varnish
Woods
BUTCHERS, SLAUGHTERHOUSE
WORKERS, AND MEAT
HANDLERS
See also Food processors
Antibiotics
Cold
Detergents (synthetic)
Hydrogen sulfide

Infections
Bacteria: brucellosis, anthrax,
tularemia
Fungi
CABINETMAKERS AND
CARPENTERS
See also Adhesive makers and
users; Construction workers;
Furniture workers; Glue
makers and users: Lacquer
makers and users; Shellac
makers and users; Stain mak-
ers and users; Varnish mak-
ers and users; Wax makers
and users; Wood preserva-
tive makers and users
Arsenic
Asbestos fibers
Bleaches: oxalic acid, chlora-
mine T
Coal tar and fractions
Ethylene dibromide
Glues
Acacia
Camphor
Formalin
Insulation: asbestos, fibrous
glass
Lacquers
Oils
Plastics
Resin
Shellac
Solvents
Sulfur monochloride
Stains: alizarin, cobalt
Synthetic resins: phenolic res-
ins, amino resins
Varnish; varnish removers
Woods
CABLE SPLICERS, COATERS, AND
WORKERS
See also Linemen
Antimony

Chlorinated diphenyls and
  naphthalenes
Coal tar and fractions
Dyes
Epoxy resin
Hydrogen sulfide
Solvents

CAISSON WORKERS AND
TUNNEL BUILDERS
  Carbon dioxide
  Hydrogen sulfide
  Increased air pressure

CAMPHOR MAKERS,
PROCESSORS, AND USERS
  Amyl acetate
  Aniline and derivatives
  1,2-dichloroethylene
  Ethylene dichloride
  Turpentine

CANDLEMAKERS
  Ammonium compounds
  Borax
  Boric acid
  Calcium oxide
  Carbon dioxide (in making tal-
    low)
  Chlorine
  Chromates
  Hydrochloride acid
  Potassium nitrate
  Sodium hydroxide
  Stearic acid
  Waxes

CARPET MAKERS
See also Textile workers
  Alizarin
  Aniline dyes
  Bacteria infection: anthrax
  Bleach (chlorine)
  Cleaners
    Formaldehyde
    para-dichlorobenzene
    Oils

Detergents
Fungicides
Glues (adhesives)
Insecticides
Loom oils
Soaps
Solvents: turpentine

CASE HARDENERS
  Ammonia
  Heat
  Hydrogen cyanide
  Oils
  Polycyclic hydrocarbons
  Sodium salts of cyanide dichro-
    mate and nitrate

CELLOPHANE FILM MAKERS
  Acrylonitrile
  Carbon disulfide
  Cobalt
  Dimethylamine
  Ethylene glycol
  Ethylene oxide
  Fluorocarbons
  Hydrogen sulfide
  Sulfuric acid
  Sodium hydroxide

CELLULOID MAKERS
  Dinitrobenzene
  Ethylene dibromide
  Ketones
  Naphthalene
  Oxalic acid

CEMENT MAKERS AND
WORKERS
  Calcium oxide
  Chromium compounds
  Cobalt compounds
  Dioxane
  Epoxy resins (for patching)
  Ethyl silicate (as preservative)
  Fluorides
  Phosphine
  Portland cement dust

Selenium and compounds
Zinc compounds (in magnesium oxide cement)

CEMENTED CARBIDE MAKERS AND WORKERS
Cobalt
Nickel and compounds
Titanium and compounds
Tungsten

CERAMIC MAKERS AND WORKERS
See also Enamel makers and workers; Pottery makers and workers
Acetylene
Aluminum and compounds
Antimony and compounds
Arsenic
Barium and compounds
Beryllium and compounds
Bismuth and compounds
Cadmium
Calcium oxide
Ceramic molding
Chromium and compounds
Cobalt and compounds
Fluorides
Freon
Hydroquinone
Lead
Manganese compounds
Mercury and compounds
Molybdenum and compounds
Nickel compounds
Oxalic acid
Phosphoric acid
Platinum and compounds
Selenium compounds
Silver and compounds
Tellurium compounds
Thorium and compounds
Tin and compounds
Uranium and compounds

Vanadium and compounds
X-rays
Zinc compounds

CLEANING COMPOUND MAKERS
ortho-dichlorobenzene
Ethyl alcohol
Ethylene dichloride
Ketones

COAL TAR AND PITCH WORKERS: ROOFERS AND ROAD WORKERS
Anthracene oil (acridine)
Ammonia
Aniline and derivatives
Benzene
Coal tar and fractions (polycyclic hydrocarbons)
Cresol
Graphite
Lead (roofers)
Naphthalene
Phenol
Skin infections
Sunlight

COBBLERS AND CEMENTERS OF RUBBER SHOES
See also Leather workers
Adhesives
Benzene
Carbon disulfide
Coal tar products
Glues
Methyl alcohol
Naphthalene
Polish

COKE MAKERS
See Metal workers: coke makers

COKE OVEN WORKERS
See Metal workers: coke oven workers

COMPOSITORS
See also Ink makers; Lithographers; Paper workers; Pho-

toengravers; Photographic
chemical makers and users;
Printers
Alkalies
Aniline and derivatives
Inks
Metals
Solvents

CONSTRUCTION WORKERS
See also Boatbuilders; Boilermak-
ers; Brick masons; Cabinet-
makers and carpenters; Plas-
terers; Road workers and
builders
Adhesives
Asbestos
Cement
Coal tar fractions (creosote,
etc.)
Cold
Epoxy resins
Gasoline
Glass, fibers
Lead (in riveting, tile makers)
Oils
Paints, paint removers
Pitch
Plastics
Solvents
Sunlight
Tolylene diisocyanate (TDI)
Welding fumes

COSMETICS MAKERS
See also Perfume makers
Aluminum
Asbestos
Beeswax
Bismuth and compounds
Carnuba wax
Cellosolves
Cobalt compounds
Dioxane

Ethyl alcohol
Ethylene glycol
Fruit and vegetable oils and
acids
Hexachlorophene
Isopropyl alcohol
Ketones
Lanolin
Parabens
n-propyl alcohol
Talc
Titanium compounds
Zinc compounds

DEMOLITION WORKERS
See also Construction workers
Lead
Trinitrotoluene (TNT)

DENTAL PRODUCTS MAKERS,
DENTAL TECHNICIANS, AND
DENTISTS
Anesthetics: ethyl chloride, ni-
trous oxide
Antibiotics
Cadmium (in amalgam)
Disinfectants (aromatics)
Germanium (in alloys)
Lead (in alloys)
Mercury (in alloys)
Methylene chloride
Natural oils
Eugenol
Menthol
Peppermint
Wintergreen
Phosphoric acid
Plastics: acrylic resins
Platinum (in alloys)
Soaps
X-rays
Zinc compounds (in cement)

DENTIFRICE MAKERS
Fluorides
Zinc compounds

DEODORANT MAKERS AND
USERS
See also Cosmetics makers; Drug
　makers
　Bismuth compounds
　Chloride of lime (calcium chlo-
　　ride, calcium hydroxide, cal-
　　cium hydrochloride)
　Cresol
　ortho-dichlorobenzene
　para-dichlorobenzene
　Dioxane
　Formaldehyde
　Hexachlorophene
　Zinc compounds
　Zirconium compounds
DEPILATORY MAKERS AND
USERS
　Beeswax
　Hydrogen sulfide
　Rosin
　Thallium compounds
　Thioglycolic acid
DEGREASERS
See also Metal degreasers; Sol-
　vent makers and users
　Benzene
　Carbon disulfide
　Carbon tetrachloride
　Dichloroethyl ethers
　Diethylene tetramine
　Dioxane
　Methylene chloride
　Perchloroethylene
　Sodium and potassium hydrox-
　　ides
　Stripping agents: coal tar frac-
　　tions
　Trichloroethylene
DETERGENT MAKERS
See also Soap makers
　Benzene
　n-butyl alcohol

Dioxane
Ethyl alcohol
Ethylene oxide
Naphtha (polycyclic hydrocar-
　bons
Oxalic acid
Perchloroethylene
Phosphoric acid
Silica
Sodium and potassium hydrox-
　ide
Sodium silicate
Sulfuric acid
Toluene
DIESEL ENGINE WORKERS, OP-
ERATORS, REPAIRMEN
　Acrolein
　Carbon monoxide
　Chromium compounds
　Coal tar and fractions (poly-
　　cyclic hydrocarbons)
　Sulfur dioxide
DISINFECTANT MAKERS AND
USERS
See also Detergent makers
　Acetaldehyde
　Aniline and derivatives
　Barium and compounds
　Benzyl chloride (germicide)
　Bismuth and compounds
　Carbon dioxide
　Chloride of lime (calcium chlo-
　　ride, calcium hydroxide, cal-
　　cium hypochlorite)
　Chlorine
　Coal tar fractions
　Cresol
　Ethyl alcohol
　Ethylene oxide
　Fluoride compounds
　Formaldehyde
　Furfural (furfuraldehyde)
　Hydrogen cyanide

Hydrogen peroxide
Iodine
Mercury and compounds
Methyl silicate
Nickel and compounds
Paradichlorobenzene
Phenol
Phthalic anhydride
Picric acid
Pine oil
n-propyl alcohol
Sulfur dioxide
Surfactants
Trichloroethylene
Zinc compounds

DOCK WORKERS
Asbestos
Carbon monoxide
Cold
Fumigants
  Acrylonitrile
  Chloropicrin
  Formaldehyde
Fungicides: phenylmercuric acetate, beta-naphthol
Infections: bacteria, parasites
Insecticides
Plant products
Tetramethylthiuram disulfide

DRUG MAKERS
Acetaldehyde
Acetonitrile
Acrolein
Allyl alcohol
Ammonia
Amyl alcohol
Aniline and derivatives
Arsenic
Barium and compounds
Benzene
Benzyl chloride
Bismuth and compounds
Bromine

n-butylamine
Carbon dioxide
Chlorinated hydrocarbons
Chromium compounds
Cobalt and compounds
Diacetone alcohol
Dimethyl formamide
Dimethyl sulfate
Ethyl acetate
Ethyl bromide
Ethyl chloride
Ethylene chlorohydrin
Ethylene dibromide
Ethylene glycol
Ethyl ether
Formaldehyde
Freons
Hexamethylenetetramine
Hydrogen bromide
Hydrogen chloride
Hydrogen peroxide
Hydroquinone
Isopropyl alcohol
Ketones
Manganese compounds
Mercury compounds
Methyl alcohol
Methyl bromide
Methyl chloride
Methylene chloride
Molybdenum compounds
Nitric acid
Nitroglycerin
Nitrophenols
Perchloroethylene
Phenylhydrazine
Phosphoric acid
Phthalic anhydride
Picric acid
Platinum and compounds
n-propyl alcohols
Pyridine
Radiation

Ionizing radiation
Microwaves
Ultraviolet radiation
Selenium
Silver and compounds
Sulfuric acid
Sulfur monochloride
Talc
Thallium and compounds
Toluene
Trichloroethylene
Turpentine
Xylene
Zirconium compounds
DRY CLEANERS
Amyl acetate
Benzene
Coal tar fractions (naphtha)
Carbon disulfide
Carbon tetrachloride
Cellosolves
Chlorinated benzenes
Dichloroethylene
Ethyl ether
Methyl alcohol
Methyl chloroform
Perchloroethylene
Propylene dichloride
Trichloroethylene
DYERS AND DYE MAKERS
Acetic acid
Acetic anhydride
Alkalies
Amines
Aniline
Antimony compounds
Benzene
Bismuth compounds
Calcium salts, calcium oxide
chloride of lime
Carbon dioxide
Cellosolves
Chlorinated benzenes

Chlorinated hydrocarbons
Chromates
Coal tar products (polycyclic
hydrocarbons)
Copper and compounds
Cresol
Ethylene
Formaldehyde
Formic acid
Gums
Hydrochloric acid
Hydrogen cyanide
Hydrogen peroxide
Hydrogen sulfide
Hydroquinone
Lead compounds
Manganese compounds
Mercaptans
Molybdenum compounds
Naphthalene
beta-naphthylamine
Nickel compounds
Nitric acid
Nitrophenols
Oxalic acids
Phenylhydrazine
Phosgene
Phosphoric acid
Phthalic anhydride
Picric acid
Pyridine (textile dyeing)
Solvents
Amyl acetate
Dimethyl formamide (DMF)
Dimethyl sulfate
Dioxane
Ethylene chlorohydrin
Ethylene glycol
Methyl alcohol
Sulfur monochloride (textile
dyeing)
Sulfuric acid
Thallium compounds
Tin compounds

Vanadium

Zinc

ELECTRICAL AND ELECTRONICS
WORKERS, INCLUDING APPLI-
ANCE AND SCIENTIFIC EQUIP-
MENT MAKERS

See also Lamp and light makers;
Phosphor makers; Vacuum
tube makers

Aluminum compounds

Asbestos

Beryllium and compounds

Bismuth compounds (for fuses)

Boron trifluoride (for nuclear
instruments)

Cadmium (in solder flux)

Chlorinated diphenyls and
naphthalenes

Coal tar and fractions

Germanium

Graphite

Ketones

Lead

Mercury and compounds

Naphthalene

Osmium and compounds

Plastics
Allyl resins
Diisocyanate resins (for refriger-
ators and freezers)
Epoxy resins
Fluorocarbons
Phenolic resins
Polyurethane (for refrigerators
and freezers)

Platinum and compounds

Radiation
Infrared
Ionizing (in radar tube manu-
facture)
Microwaves
Ultraviolet

Selenium (in rectifiers)

Silver

Tellurium

Thallium (in infrared instruments)

Thorium

Titanium and compounds

Trichloroethylene

Welding fumes

Xylene (for quartz crystal os-
cillators)

Zinc

ELECTRICAL AND ELECTRONICS
WORKERS: SEMI-CONDUCTOR
MAKERS

See also Electrical and electronics
workers

Arsenic

Bismuth and compounds

Carbon tetrachloride

Germanium and compounds

Phosphorus (white or yellow)

Selenium and compounds

Tellurium and compounds

EMBALMERS

Bacteria infections

Barium and compounds

Fluorides

Formaldehyde

Ionizing radiation

Methyl alcohol

EMULSIFYING AGENT MAKERS
AND WORKERS

Dioxane (diethylene ether)

n-butylamine

Ethylenediamine

Styrene

ENAMEL MAKERS AND
WORKERS

See also Pottery makers and workers

Amyl acetate

Arsenic

Barium and compounds

Benzene

Bismuth and compounds (in luminous enamel)
n-butyl acetate
Carbon disulfide
Cellosolves
Cerium (in vitreous enamel)
Chromium compounds
Cresol
Fluorides (in vitreous enamel)
Hydrogen chloride
Hydrogen fluoride (for enamel etching)
Lead
Manganese compounds
Nickel and compounds
Phthalic anhydride
Sodium and potassium hydroxide
Titanium
Toluene
Xylene
Zinc compounds
Zirconium compounds

ENGRAVERS
See also Metal Workers
Cadmium
Defective illumination
Lead (in steel engraving)
Phosphoric acid
Sodium and potassium hydroxide

ETCHERS
See also Metal Workers
Acids
Alkalies
Arsine
Chromium compounds
Hydrogen fluoride
Phenol
Silver compounds (for ivory etching)
Zinc compounds

EXPLOSIVES WORKERS, INCLUD-
ING DETONATORS, CLEANERS,
FILLERS AND PACKERS, SMOKE-
LESS POWDER MAKERS
Acetaldehyde
Acetic anhydride
Ammonia and salts
Amyl acetate (in smokeless powder)
Amyl alcohol
Aniline and derivatives
Barium compounds
Benzene
Carbon dioxide
Carbon disulfide
Cerium
Chromium compounds
Cresol
Dinitrobenzene
Dinitrophenol
Dinitrotoluene
Ethyl acetate (in smokeless powder)
Ethyl alcohol
Ethylene glycol
Ethyl ether
Graphite
Hexamethylenetetramine
Hydrazine
Ketones
Mercury and compounds
Methyl alcohol
Naphthalene (in smokeless powder)
Nitric acid
Nitrobenzene
Nitroglycerine
Nitrophenols
PETN
Phenol
Picric acid
Pyridine
Sulfuric acid
Trinitrotoluene (TNT)

Tetryl
Toluene
Zirconium

FARMERS AND AGRICULTURAL WORKERS

See also Fertilizer makers and users; Fungicide makers and users; Pesticide and insecticide makers and users

Ammonia (corn growing)
Arsenic
Asbestos
Bacteria infections
Calcium cyanamide (in fertilizer)
Calcium oxide
Coal tar and fractions (polycyclic hydrocarbons)
Cold
Detergents (synthetic)
Ethylene dibromide (cabbage growers)
Feeds
Fertilizers
Fluorides (vegetable growers)
Fruits (allergies)
Fungus infections
Heat
Kerosene
Lead
Lubricants
Mercury compounds
Oils
Parasitic infections
Pesticides
Poisonous plants
Ragweed
Solvents
Sunlight
Vegetables (allergies)
Virus and rickettsia infections

FARMING AND AGRICULTURAL WORK: SEED HANDLERS AND PROCESSORS

Carbon tetrachloride (in extraction of seed oils)
Ethylene dibromide (as insecticide)
Hexachlorobenzene (as seed disinfectant)
Mercury compounds (as insecticide)
Selenium and compounds (for germination testing)
Tetramethylthiuram disulfide (as disinfectant)
Zinc compounds (in seed treatment)

FARMING AND AGRICULTURAL WORK: SOIL TREATMENT

Carbon disulfide
Chlorobenzenes
Ethylene dibromide (as fumigant)
Fluorides
Methyl bromide (as fumigant)
Naphthalene
Tetrachloroethane

FEATHER WORKERS

Aniline and derivatives
Hydrogen peroxide
Methyl alcohol

FEED MAKERS (ANIMAL FEEDS)

Cobalt compounds (in mineral feeds)
Manganese compounds
Phosphoric acid
Zinc compounds (as additives)

FELT MAKERS AND WORKERS

Acids
Coal tar and fractions (polycyclic hydrocarbons)
Dyes
Hydrogen peroxide
Hydrogen sulfide
Mercuric nitrate

Methyl alcohol
Sodium carbonate

FERTILIZER MAKERS AND USERS:
AGRICULTURAL CHEMICALS
Ammonia and compounds
Arsine
Bacteria infections: anthrax
Calcium cyanamide
Calcium oxide
Carbon dioxide
Castor bean pomace
Cobalt and compounds
Hydrazine
Hydrogen chloride
Hydrogen cyanide
Hydrogen fluoride
Hydrogen sulfide
Manganese compounds
Molybdenum compounds
Nitrogen dioxide
Phosphorus
Phosphorus trichloride
Sulfuric acid

FIBROUS GLASS MAKERS
Coal tar and fractions (poly-
cyclic hydrocarbons)
Chromates
Furnace slag
Plastics and polyester resins

FIRE EXTINGUISHER MAKERS
AND USERS, FIREMEN
Bromine
Carbon dioxide
Carbon monoxide
Carbon tetrachloride
Ethylene dibromide
Ethylene glycol
Freon
Heat and cold
Methyl bromide
Methylene chloride
Phosgene
Sulfur dioxide

FISHERMEN
Coal tar and fractions (poly-
cyclic hydrocarbons)
Cold
Heat
Infections
Sunlight
Ultraviolet light

FLAMEPROOFERS
Antimony and compounds
Chlorinated diphenyls
Titanium compounds

FLOTATION AGENT MAKERS AND
WORKERS
Amyl alcohol
Carbon disulfide
Copper and compounds
Cresol
Ethylene dichloride
Phosphorus pentasulfide
Thallium and compounds

FLOUR MILL AND GRAIN
ELEVATOR WORKERS
Bleaches
Chlorine
Hydrogen peroxide
Nitrogen dioxide
Sulfur dioxide
Dust
Ethylene oxide
Fumigants
Acrylonitrile
Ethylene dibromide
Carbon tetrachloride
Methyl bromide
Methyl formate
Phosphine
Fungi: infections, allergic reac-
tions
Fungicides
Parasitic infections: grain
itch
Pesticides

FLYPAPER MAKERS
Chromium compounds

FOOD FLAVORING MAKERS AND
USERS
See also Food processors
  Acetaldehyde
  Acetic anhydride
  Amyl acetate
  Amyl alcohol
  n-butyl acetate
  Ethyl acetate
  Methylene chloride
  Nitrobenzene (in vanilla)
  Phosphoric acid (in syrups)
  Propyl acetate
FOOD PROCESSORS AND
HANDLERS, INCLUDING CAN-
NERS
See also Brewers and brewery
  workers; Flour mill and
  grain elevator workers; Food
  processors: fat processors;
  Food flavoring makers and
  users
  Acetic acid (as preservative)
  Allergies (vegetable material,
    molds and spores)
  Acrolein (coffee roasters, cooks)
  Ammonia (ice cream makers)
  Bleaches
    Hydrogen peroxide
    Nitrogen dioxide
    Sulfur dioxide
  Calcium oxide
  Carbon dioxide
  Citrus oil
  Detergents
  Ethyl acetate (confectioners)
  Ethylenediamine (in casein and
    albumin processing)
  Formic acid (as preservative)
  Fruit acids
  Fumigants
    Acrylonitrile
    Ethyl bromide
    Ethylene dibromide
  Ethylene oxide
  Formaldehyde
  Fungicides: mercurials
  Heat and cold
  Hydrogen chloride
  Infections
    Bacteria: anthrax
    Fungi
  Insecticides
  Lead
  Nitrous oxide (in aerosols)
  Ozone
  Phosphoric acid (gelatin mak-
    ers)
  Quinone (gelatin makers)
  Radiation
    Ionizing radiation
    Microwaves
    Ultraviolet radiation
  Resins
  Salts
  Soaps
  Trichloroethylene (in caffeine
    processing)
  Vegetable juices
  Zinc compounds (gelatin mak-
    ers)
FOOD PROCESSORS: FAT PROCES-
SORS, REFINERS, PURIFIERS, AND
FATTY OIL PROCESSORS
  Acetonitrile (animal oil proces-
    sors)
  Acrolein (acrylic aldhehyde)
  Amyl alcohol
  Barium and compounds
  Carbon disulfide
  Carbon tetrachloride
  Chromium compounds
  Cobalt and compounds (in hy-
    drogenation of oils)
  Cycloparaffins
  1,2-dichloroethylene
  Dichloroethyl ether
  Ethyl chloride

Ethylene dibromide
Ethylene dichloride
Ethyl ether
Hydrogen peroxide
Hydrogen sulfide
Hydroquinone
Isopropyl acetate
Infections: bacteria (anthrax)
Methylene chloride
Natural gas (in hydrogenation of oils)
Nickel (in hydrogenation of oils)
Nitroparaffins
Ozone (in bleaching)
Petroleum naphtha
Propylene dichloride
Sodium and potassium hydroxide (in vegetable oil processing)
Tetrachloroethane
Trichloroethylene

FOOD PROCESSORS: MEAT PACKERS
See Butchers, slaughterhouse workers, and meat handlers

FOOD PROCESSORS: SACCHARIN MAKERS
Phosphorus trichloride
Toluene

FOOD PROCESSORS: SUGAR PROCESSING AND REFINING
Ammonia
Bagasse (sugar cane)
Calcium oxide
Carbon dioxide
Chlorine
Hydrogen chloride
Hydrogen sulfide (in sugar-beet processing)
Methyl alcohol
Phosphoric acid
Sulfur dioxide

Sulfur monochloride
Sulfuric acid
Tin and compounds

FOOD PROCESSORS: VEGETABLE OIL EXTRACTION AND PURIFICATION
Acetonitrile
Barium and compounds
Chlorinated diphenyls and naphthalenes
Methyl bromide
n-propyl alcohol
Sodium and potassium hydroxide
Sulfur monochloride

FOOD PROCESSORS: YEAST MAKERS
Acetaldehyde
Carbon dioxide
Hydrogen fluoride
Phosphoric acid

FOUNDRY WORKERS
See Metal workers: foundry workers

FORMALDEHYDE MAKERS
Formaldehyde
Methyl alcohol
Natural gas

FREON MAKERS
See also Refrigerant makers and workers
Carbon tetrachloride
Chlorine
Freons
Hydrogen fluoride

FRIT MAKERS
Barium and compounds
Cobalt and compounds
Fluorides

FUEL: GASOLINE ADDITIVE MAKERS
Aniline and derivatives
Benzyl chloride

Boranes: pentaboranes, deca-
   boranes
Chlorine
Cobalt compounds
Ethyl chloride (for making tet-
   raethyl lead, TEL)
Ethylene dibromide (for mak-
   ing TEL)
Ethylene dichloride (for mak-
   ing TEL)
Hexamethylenetetramine
Hydrazine
Hydrochloric acid (for making
   TEL)
Lead (for making TEL)
Phosphorus oxychloride
Tetraethyl lead
Toluene

FUEL: MOTOR FUEL MAKERS
Aniline and derivatives
Benzyl chloride
Boranes: pentaborane, deca-
   borane
Chlorine
Cobalt compounds
Ethyl alcohol
Ethyl benzene
Ethylene dibromide
Ethylene oxide
Ethyl ether
Hexamethylenetetramine (anti-
   corrosive agent)
Hydroquinone
Hydrazine (anti-corrosive
   agent)
Isopropyl alcohol
Mercaptans
Methyl alcohol
Phosphorus oxychloride
Tetraethyl lead
Toluene
Tri-ortho-cresyl phosphate
   (TOCP)

FUEL: JET FUEL MAKERS AND
HANDLERS
Dimethyl hydrazine
Hydrazine
Kerosene
Methyl alcohol
Xylene

FUEL: ROCKET FUEL MAKERS
AND USERS
Ammonia
Aniline and derivatives
Boranes: diborane, pentabor-
   ane, decaborane
Butadiene
Carbon disulfide
Cerium
Dimethyl hydrazine
Ethyl alcohol
Ethylene oxide
Fluorine
Hydrogen fluoride
Freon
Hydrazine
Hydrogen peroxide
Isopropyl alcohol
Kerosene
Methyl alcohol
Nitric acid
Nitrogen oxides
   Nitrogen dioxide
   Nitrogen tetroxide
   Nitrous oxide
Nitroglycerin
Nitroparaffins

FUEL MAKERS: MISCELLANEOUS
Acetylene (motor boat)
Coal tar and fractions
Cycloparaffins (solid fuel for
   camp stoves)
Hexamethylenetetramine (fuel
   tablets)

FUMIGANT MAKERS AND USERS
Acrylonitrile

Benzene
Boron trifluoride
Carbon disulfide
Carbon tetrachloride
ortho-dichlorobenzene
Dioxane
Ethyl bromide
Ethylene dibromide
Ethylene dichloride
Ethylene oxide
Ethyl ether
Formic acid
Hydrogen cyanide
Mercaptans
Methyl bromide
Methyl formate
Methylene chloride
Naphthalene
Perchloroethylene
Propylene dichloride
Sulfur dioxide
Trichloroethylene

FUMIGATORS
Boron trifluoride
Ethyl ether
Formic acid
Hydrogen cyanide
Mercaptans
Methylene chloride
Propylene dichloride
Sulfur dioxide

FUNGICIDE MAKERS AND USERS
Barium compounds
Benzene
Calcium oxide
Chromates
Copper and compounds
Cycloparaffins
Dithiocarbamates
Ethylene oxide
Ferbam
Fluoride
Formaldehyde

Furfural (furfuraldehyde)
Hexachlorobenzene
Hexamethylenetetramine
Iron compounds
Mercury compounds
Naphthalene
Nitrophenols
Pentachlorophenol
Tetramethylthiuram disulfide
Tin compounds
Zinc compounds

FUR PROCESSORS
Alkalies
Alum
Bleaches: hydrogen peroxide,
    hydrogen sulfide
Carbon tetrachloride
Chromates
Dyes
Formaldehyde
Hydroquinone
Infections
    Bacteria: anthrax
    Fungi
Lime
Mercury compounds
Mordants (chrome, nickel)
Oils
Salts
Sulfuric acid

FURNITURE WORKERS
Amyl acetate (in polish)
Benzene
Chromium compounds
Dichloroethyl ether (in finish
    remover)
Formaldehyde
Lacquers
Methyl alcohol
Microwave radiation (in ve-
    neering)
Petroleum fractions, benzene,
    naphtha

Plastics: amino resins, tolylene
diisocyanate (TDI)
Pyridine
Rosin
Soaps
Sodium and potassium hydrox-
ide
Turpentine
Waxes
Woods
FURNITURE WORKERS:
UPHOLSTERERS
See also Cabinetmakers and car-
penters
Adhesives
Bacteria infections: anthrax
Glues
Lacquer solvents
Methyl alcohol
Stains
Tolylene diisocyanate (TDI)
Varnish
FURNACE WORKERS AND
REPAIRMEN
Asbestos
Carbon dioxide
Carbon monoxide
Coal tar fractions
Glass fibers
Heat
Radiation: infrared, ultraviolet
Sulfur dioxide
Zirconium
GALVANIZERS
See also Metal platers, eletroplat-
ers
Acrolein (acrylic aldehyde)
Ammonia and compounds
Arsine
Brass
Hydrogen chloride
Lead
Sulfuric acid

Trichloroethylene
Zinc and compounds
GARAGE WORKERS AND
MECHANICS
See also Auto workers and repair-
men
Asbestos
Carbon monoxide
Detergents (synthetic)
Diacetone alcohol
Ethylene glycol
Gasoline and additives
Glass fibers
Greases
Kerosene
Moisture
Oils
Paint
Paint removers
Plastics: polyester resins
Solvents
GASKET MAKERS
Asbestos
Glass fibers
Graphite
Plastics: fluorocarbons
Tetrachloroethane
GASOLINE ADDITIVES
See Fuel: gasoline additive mak-
ers
GAS PURIFIERS
Ammonia
Carbon monoxide
Hydrogen cyanide
Natural gas
Phenol
GAS: ILLUMINATING GAS
WORKERS
Ammonia
Arsine
Hydrogen cyanide
Phenol

Trichloroethylene
Zinc compounds

GAS: TEAR GAS MAKERS
See also Smoke bomb makers
Acrolein (aldehyde)
Coal tar fraction: anthracene

GEM MAKERS
Aluminum and compounds
Copper and compounds
Thallium and compounds
Titanium and compounds
Zirconium and compounds

GLASS CLEANER AND POLISH
MAKERS AND WORKERS
Ammonia
Cerium
Lead
Silver and compounds
Trichloroethylene

GLASSMAKERS
See also Etchers; Glass cleaner
and polish makers and work-
ers
Acrylonitrile (in safety glass)
Aluminum and compounds
Amyl acetate
Antimony and compounds
Arsenic
Barium and compounds (in
crystals for scientific instru-
ments)
Benzene
n-butyl acetate (in safety glass)
Cadmium
Calcium oxide
Carbon disulfide
Cerium
Chromium and compounds
Coal tar and fractions (glass
blowers)
Copper and compounds
Fluorides (in optical equip-
ment)

Formic acid (glass silverers)
Germanium compounds
Heat
Hydrogen chloride
Isopropyl alcohol
Lead
Manganese and compounds
Methyl alcohol (in safety glass)
Molybdenum compounds
Phosgene
Picric acid
Platinum
Radiation: infrared, ionizing,
ultraviolet
Selenium and compounds
Silver and compounds
Sulfur dioxide
Tellurium
Thallium (for glass with high
refraction)
Thorium
Titanium
Turpentine
Uranium and compounds
Vanadium and compounds
Zirconium and compounds

GLUE MAKERS AND USERS
See also Adhesive makers and
users; Airplane dope makers
and users; Gum makers;
Rubber cement makers and
users
Ammonia
Bacteria infections: anthrax,
folliculitis
Benzene
Carbon dioxide
Carbon disulfide
Chromium compounds
Copper and compounds
Cresols
Dioxane (diethylene ether)
Ethylene glycol

Hydrogen chloride
Hydrogen peroxide
Hydrogen sulfide
Nitric acid
Sulfur dioxide (as bleach)
Sulfuric acid
Trichloroethylene
Zinc compounds
GLYCERIN MAKERS
Allyl alcohol
Oxalic acid
GRAIN ELEVATOR WORKERS
See Flour mill and grain elevator
workers
GRINDING WHEEL MAKERS
Fluorides
Furfural (furfuraldehyde)
GUM MAKERS
Cellosolves
Chlorinated diphenyls and
naphthalenes
1,2 dichloroethylene
ortho-dichlorobenzene
Dichloroethyl ether
Epichlorohydrin
Ethyl ether
Isopropyl alcohol
Perchloroethylene
Propylene dichloride
Toluene
HEAT TRANSFER WORKERS
Chlorobenzene
ortho-dichlorobenzene
Chlorinated diphenyls and
naphthalenes
Freon
Perchloroethylene
Trichlorobenzenes
Trichloroethylene
HEAT TREATERS
Acrolein
Carbon monoxide
Coal tar fractions: anthracene

Hydrogen cyanide
Infrared radiation
HERBICIDE MAKERS AND USERS
Allyl alcohol
Ammonium sulfamate (Am-
mate)
Arsenic
Asbestos
Benzene
Calcium cyanamide
2,4,-D (2,4-dichlorophenoxyace-
tic acid)
Dinitrophenol
Pentachlorophenol
Phenol
Phenylmercuric acetate
Phthalic anhydride
2,4,5-T (2,4,5 trichlorophenoxy-
acetic acid)
HOSPITAL WORKERS, INCLUDING
NURSES AND DOCTORS
Anesthetics
Ethyl bromide
Ethyl chloride
Ethyl ether
Halothane
Methoxyfluorane
Nitrous oxide
Antibiotics
Antiseptics
Beryllium
Cobalt
Detergents (synthetic)
Disinfectants and germicides
Drugs
Fumigants
Infections
Bacteria
Viruses, especially hepatitis
Iodine
Isopropyl alcohol
Moisture
Radiation: ionizing, ultra-
violet, X-rays

Soaps
Talc
Tricresyl phosphate (in steril-
    izing surgical instruments)
HOSPITAL WORKERS: BACTERI-
OLOGISTS AND TECHNICIANS
    Infections: bacteria, virus
    Selenium and compounds
    Xylene
HOSPITAL WORKERS: HIS-
TOLOGY TECHNICIANS
    Benzene
    Chromium compounds
    Dioxane (diethylene ether)
    Ethyl alcohol
    Formaldehyde
    Mercuric chloride
    Picric acid
    Toluene
    Xylene
HOSPITAL WORKERS: MICRO-
SCOPISTS
    Benzidine
    Platinum and compounds
    Selenium and compounds
    Xylene
    Zinc compounds
HOSPITAL WORKERS: PATHOL-
OGISTS, AUTOPSY ATTENDANTS
    Epoxy resins
    Formaldehyde
    Infections: virus (hepatitis and
        others)
    Xylene
HYDRAULIC FLUID MAKERS
See Brake fluid makers and users
HYDROCHLORIC ACID MAKERS
    Arsine
    Chlorine
    Hydrochloric acid
    Hydrogen sulfide
HYDROGEN MAKERS
    Natural gas

HYDROGEN PEROXIDE MAKERS
    Hydrogen peroxide
    Ozone
    Quinone
HYDROQUINONE MAKERS
    Aniline and derivatives
    Hydroquinone
    Manganese compounds
    Quinone
INK MAKERS
See also Lithographers; Photoen-
    gravers; Printers; Typogra-
    phers, electrotypers and
    stenotypers
    Alkalies
    Aluminum and compounds
    Ammonia
    Aniline and derivatives
    Arsenic
    Barium and compounds
    Benzene
    Carbon black
    Carbon tetrachloride
    Cerium
    Chlorinated benzenes
    Chlorinated diphenyls and
        naphthalenes
    Chlorine
    Chromium compounds
    Cobalt and compounds (in sym-
        pathetic ink)
    Copper and compounds
    Cresols
    Diacetone alcohol (in quick-
        drying inks)
    Ethyl acetate
    Ethyl alcohol
    Ethylene glycol
    Formaldehyde
    Isopropyl alcohol
    Ketones
    Manganese compounds
    Mercury and compounds

Methyl alcohol
Nickel and compounds
Oxalic acid
Platinum and compounds (in indelible inks)
Selenium and compounds
Silver and compounds (in indelible inks)
Toluene
Turpentine
Vanadium
Xylene
Zinc compounds

INK REMOVER MAKERS AND USERS
See also Ink makers
Cresols
Oxalic acid

INSECTICIDE MAKERS AND USERS
See Farmers and agricultural workers; Pesticide and insecticide makers and users

JEWELRY MAKERS AND WORKERS, JEWELERS
Amyl acetate
Arsine
Chromium compounds
Hydrogen chloride
Hydrogen cyanide
Lead
Mercury and compounds
Nitric acid
Platinum
Silver
Sulfuric acid

LACQUER MAKERS AND USERS
Acetaldehyde
Ammonia
Amyl acetate
Amyl alcohol
Benzene

n-butyl acetate
n-butyl alcohol
Carbon disulfide
Carbon tetrachloride
Cellosolves
Chlorinated benzenes
Chlorinated diphenyls and naphthalenes
Cobalt and compounds
Cycloparaffins
Diacetone alcohol
Dichloroethyl ether
Dioxane (diethylene ether)
Epichlorohydrin
Ethyl acetate
Ethyl benzene
Ethylene chlorohydrin
Ethylene dichloride
Ethylene glycol
Ethyl ether
Formaldehyde
Infrared radiation
Isopropyl alcohol
Ketones
Methyl alcohol
Methylene chloride
Nitric acid
Plastics: alkyd resins; phenolic resins
Propyl acetate
n-propyl alcohol
Pyridine
Tetrachloroethane
Titanium and compounds
Toluene
Trichloroethylene
Turpentine
Xylene
Zirconium compounds

LAMPBLACK MAKERS
Naphthalene
Phenol
Polycyclic hydrocarbons

LAMP AND LIGHT MAKERS
  Amyl acetate
  Cadmium
  Cereum (in photographic illu-
    mination)
  Cobalt and compounds
  Fluorides (phosphorescent
    tubes)
  Hydrogen fluoride (for frost-
    ing)
  Mercury (in mercury lamps,
    fluorescent and neon lights)
  Methyl alcohol
  Osmium and compounds
  Ozone (from ultraviolet lamps)
  Thallium and compounds
  Thorium and compounds
  Titanium and compounds
  Zirconium compounds
LAUNDRY WORKERS
  Acetic acid
  Bacteriocides
  Bleaches: chloride of lime,
    chlorine
  Detergents (synthetic)
  Fluorides
  Formic acid
  Heat
  Hydrogen fluorides
  Oxalic acid
  Soaps
  Sodium and potassium hydrox-
    ides
LEAD BURNERS
  See also Pipefitters, pipe makers,
    plumbers
  Antimony and compounds
  Arsine
  Lead
  Polycyclic hydrocarbons: an-
    thracene
LEATHER WORKERS
  Acrylonitrile (in finishing)

Amyl acetate
Aniline and derivatives
Antimony and compounds:
  mordanters
Arsenic
Benzene
n-butyl acetate (in glue or
  dope)
Cellosolves (ethylene glycol de-
  rivatives)
Chromium compounds (in tan-
  ning)
Ethyl acetate
Ethylene glycol dyers
Formic acid
Methylene chloride (in fin-
  ishes)
Oxalic acid (as bleach)
Xylene
LEATHER AND TANNING
WORKERS
  Aluminum compounds
  Ammonia
  Amyl acetate
  Aniline and derivatives
  Bacteria infections: anthrax
  Barium and compounds
  Benzyl chloride (in making
    tannin)
  n-butylamine (in making tan-
    nin)
  Calcium oxide
  Carbon dioxide (in tannery
    pits)
  Chromates
  Copper and compounds
  ortho-dichlorobenzene
  Formaldehyde
  Formic acid
  Hydrochloric acid
  Hydrogen cyanide
  Hydrogen sulfide
  Lead

Mercury and compounds
Oxalic acid
Phthalic anhydride
Quinone
Sulfur dioxide
Titanium and compounds
Toluene
Zirconium compounds

LEATHER WORKERS (ARTIFI-
CIAL LEATHER)
Isopropyl acetate
Ketones
Plastics (vinyls)
Polyvinyl acetate
Polyvinyl alcohol
Polyvinyl chloride
Monomers

LIGHTER FLINT MAKERS
Cerium

LINEMEN
See also Cable splicers, coaters,
and workers
Carbon monoxide
Coal tar fractions (polycyclic
hydrocarbons)
Cold
Heat
Hydrogen disulfide
Infections: plague from rat
fleas, chigger bites
Methane
Sunlight

LINOLEUM MAKERS
Acrolein (acrylic aldehyde)
Amyl acetate
Asbestos
Asphalt
Barium and compounds
Benzene
Chromium
Dyes
Lead
Manganese compounds
Resins

Titanium compounds
Zinc compounds

LITHOGRAPHERS
See also Ink makers; Paper work-
ers; Photoengravers; Printers
Aluminum and compounds
Aniline and derivatives
Benzene
Cadmium
Chromium compounds
Copper and compounds
Formaldehyde
Hydrogen chloride
Hydrogen sulfide
Hydroquinone
Inks
Lead
Mercuric chlorine
Methyl alcohol
Nitric acid
Oxalic acid
Phosphoric acid
Photographic fluids
Sodium and potassium hydrox-
ide
Talc
Turpentine
Ultraviolet radiation
Xylene

LITHOPONE WORKERS
Barium and compounds
Cadmium
Hydrogen sulfide
Zinc compounds

MACHINISTS
Chlorinated diphenyls and
naphthalenes
Chromates
Cobalt soaps
Cresols
Cutting fluids
Formaldehyde
Germicides

Hydroquinone
Ionizing radiation
Lanolin
Lubricating oils
Metal fumes and dusts
Methyl chloride
Nitrobenzene
Phenolic amines
Phenylmercuric salts
Plastics: fluorocarbons, poly-
    urethane
Rosin
Rust inhibitors
Solvents
Tolylene diisocyanate (TDI)
MAGNET MAKERS
Bismuth and compounds
Cobalt and compounds
Nickel and compounds
MALEIC ACID MAKERS
Benzene
Cycloparaffins
MATCH MAKERS
Ammonium phosphate
Antimony and compounds
Barium and compounds
Carbon disulfide
Chromates
Dextrins
Dyes
Formaldehyde
Glues
Graphite
Gums
Lead
Manganese
Phosphorus pentachloride
Phosphorus sesquisulfide
Picric acid
Potassium chlorate
Sodium and potassium hydrox-
    ide
Thallium and compounds

Zinc compounds
MEAT WORKERS
See Butchers, slaughterhouse
    workers, and meat handlers
METAL BURNERS
See Lead burners
METAL CLEANERS, POLISHERS,
BURNISHERS, COATERS, AND
CONDITIONERS; MAKERS AND
USERS OF POLISHING SUB-
STANCES
See also Metal degreasers; Polish
    makers and users
Abrasive dusts: silica, silicates,
    corundum, etc.
Ammonia
Arsine
Chromium compounds
Detergents
Diethylenetriamine
Fluorides
Freons
Hydrogen chloride
Hydrogen cyanide
Hydrogen fluoride
Hydrogen peroxide
Kerosene
Metal dusts
Nitrogen dioxide
Petroleum naphtha
Oxalic acid
Phosphoric acid
Sulfuric acid
Solvents (for degreasing)
    Carbon tetrachloride
    Diacetone alcohol
    Dichlorobenzene
    Dioxane (diethylene ether)
    Ethylene glycol
    Ketones
    Methyl alcohol
    Nitrobenzene
Trichloroethane
Trichloroethylene

Triethanolamine
Waxes
Zinc compounds (in steel polish)
METAL DEGREASERS
Chlorinated benzenes
Chlorinated diphenyls and naphthalenes
Ethylene dichloride
Methyl chloroform
Naphtha (petroleum distillate)
Perchloroethylene
Propyl acetate
n-propyl alcohol
Trichloroethane
Trichloroethylene
METAL ETCHERS
See Etchers
METAL PLATERS, ELECTRO-PLATERS
See also Metal workers; Metalizers: metal platers, bronzers
Ammonia
Antimony and compounds (as metal bronzers)
Arsine
Barium and compounds
Bismuth and compounds
Cadmium
Calcium
Calcium oxide
Carbon disulfide
Chlorinated diphenyls and naphthalenes
Chromic acid, chromium compounds
Cobalt and compounds
Copper and compounds
Detergents
Fluorides
Formic acid
Graphite
Hydrogen chloride

Hydrogen cyanide
Hydrogen peroxide
Lead
Mercury and compounds
Molybdenum and compounds
Nickel compounds
Nickel carbonyl (gas platers)
Nitrogen dioxide
Platinum and compounds
Selenium and compounds
Soaps
Sodium and potassium hydroxides
Titanium
Waxes (synthetic)
Zinc and compounds
METAL WORKERS: ALLOY MAKERS
See also Coke oven workers; Ferralloy makers; Foundry workers
Acetylene (scarfing)
Aluminum
Ammonia
Arsenic
Asbestos
Barium (steel carburizers)
Cadmium
Calcium oxide
Carbon monoxide
Cerium
Chromium compounds
Cobalt
Copper dust and fumes
Gold
Graphite
Heat
Hydrogen sulfide
Iron
Lithium stearate
Manganese
Mercury
Metal dusts

Metal oxides
Molybdenum
Nickel
Osmium
Phosphine
Platinum
Radiation: infrared, ultraviolet
Selenium
Silver
Tellurium
Thallium
Thorium
Vanadium
Zinc, zinc oxide
Zirconium

METAL WORKERS: ALLOY MAKERS, BRITANNIA METAL WORKERS
Antimony
Copper
Tin

METAL WORKERS: ALLOY MAKERS, BRASS MAKERS
Arsenic

METAL WORKERS: ALLOY MAKERS, BRONZE MAKERS AND FOUNDERS
Arsenic
Copper
Phosphorus
Tin
Zinc

METAL WORKERS: BABBITT METAL MAKERS AND WORKERS
Arsenic
Antimony
Copper
Lead
Tin

METAL WORKERS: BERYLLIUM WORKERS, REFINERS, COMPOUND MAKERS, MACHINISTS, ETC.
Beryllium
Fluoride compounds (refiners)

METAL WORKERS: BLAST FURNACE WORKERS
Asbestos
Carbon dioxide
Carbon monoxide
Heat
Hydrogen cyanide
Hydrogen sulfide
Metal carbonyls

METAL WORKERS: COKE MAKERS
Carbon monoxide
Coal
Heat
Natural gas
Nitrogen dioxide
Radiation: infrared, ultraviolet
Sulfur dioxide

METAL WORKERS: COKE OVEN WORKERS
Ammonia
Asbestos
Benzene
Carbon monoxide
Coal
Coal tar and fractions
Cresol
Heat
Hydrogen cyanide
Hydrogen sulfide
Natural gas
Nitrogen dioxide
Phenol
Radiation: infrared, ultraviolet
Sulfur dioxide

METAL WORKERS: FERROALLOY MAKERS
Arsine
Asbestos

Barium and compounds (for carburizing steel)
Calcium cyanamide (steel case hardeners)
Carbon black
Carbon monoxide
Chromium
Coal tar and fractions
Cobalt (in high speed and steel tools)
Copper
Graphite
Hydrogen cyanide (for carburizing steel)
Iron
Manganese
Molybdenum
Nickel
Phosphine
Radiation: infrared, ultraviolet
Selenium
Sulfuric acid
Tellurium
Toluene
Vanadium
Zirconium
METAL WORKERS: FOUNDRY WORKERS
Acetylene (for scarfing)
Acrolein (acrylic aldehyde)
Aluminum
Arsenic
Barium (for carburizing steel)
Carbon dioxide
Carbon monoxide
Copper and compounds
Cresol
Ethyl silicate
Fluorides
Graphite
Heat
Hexamethylenetetramine

Hydrogen fluoride (in casting cleaners)
Iron
Lead
Mercury (in investment casings)
Methyl alcohol
Nickel carbonyl and other metal carbonyls
Phosphoric acid
Plastics: amino resins, phenolic resins
Radiation: infrared, ultraviolet
Silica
Sulfur dioxide
Tellurium
Tin
Titanium
Zinc (bronze founders)
Zirconium
METAL WORKERS: REFINERS AND SMELTERS
Acetylene
Amyl alcohol (ore upgraders)
Arsine
Bentonite
Brass (zinc founders and smelters, junk metal refiners)
Carbon monoxide
Cerium
Chlorine (for aluminum refining)
Dusts: rock, mineral, metal
Ethylene dichloride (ore upgraders)
Fluorides
Formic acid
Furfural (furfuraldehyde; rare-earth metal refining)
Heat
Hydrogen fluoride (ore dissolvers)

Hydrogen sulfide (pyrite burn-
ers)
Lead
Manganese compounds
Metal carbonyls
Nickel, nickel carbonyl
Nitric acid (ore flotation)
Nitrogen dioxide
Phosphine
Phosphorus: white or yellow
Radiation: infrared, ultraviolet
Selenium (pyrite roasters)
Silica dust
Taconite
Thallium (ore upgraders)
Thorium

METAL WORKERS: REFINERS,
SMELTERS OF COPPER (ELEC-
TROLYTIC PROCESS)
Arsenic (in electrolytic process)
Copper
Fluoride compounds
Selenium
Silver
Tellurium

METAL WORKERS: LEAD SCAV-
ENGERS
Ethylene dibromide
Ethylene dichloride
Tricresyl phosphate

METAL WORKERS: REFINERS,
GOLD EXTRACTORS, METAL
EXTRACTION
Arsenic
Arsine
Bromine
Chlorine
Diacetone alcohol (gold leaf)
Fluorides
Hydrogen cyanide
Hydrogen sulfide
Lead
Mercury

Silver
Sulfur monochloride
Tellurium

METAL WORKERS: REFINING
AND EXTRACTING SILVER
Arsenic
Chlorine
Mercury
Hydrogen cyanide
Fluorides
Tellurium

METAL WORKERS: REFINERS,
LEAD SMELTING
Arsenic
Fluoride
Lead
Selenium

METAL WORKERS: TIN SMELT-
ERS AND REFINERS
Arsenic
Bismuth
Hydrogen peroxide
Lead
Tin

METAL WORKERS: ZINC MILL-
ING, REFINING FOUNDERS AND
SMELTERS
Brass
Cadmium
Fluorides
Manganese
Selenium
Zinc

METAL WORKERS: RUST RE-
MOVERS, RUST PROOFERS
Abrasives: silica, corundum
Chromates
Decaborane (as rust inhibitor)
Oxalic acid (as rust remover)
Phosphoric acid (as rust inhibi-
tor)
Pyridine (as rust inhibitor)

Tetrachloroethane (as rust remover)

METAL WORKERS: SCRAP

METAL WORKERS
Chlorine
Lead
Metal oxides

METAL WORKERS: TEMPERERS OF STEEL
Hydrogen cyanide
Lead
Oils
Sodium carbonate
Sodium cyanide
Sodium dichromate

METALIZERS AND METAL SPRAYERS
Acetylene
Cadmium
Iron compounds
Zinc

METALIZERS: METAL PLATERS, BRONZERS
Acetone
Ammonia and salts
Amyl acetate
Antimony sulfide
Arsenic
Arsine
Benzene
Brass
Heat
Hydrogen chloride
Hydrogen cyanide
Lacquers
Lead
Mercury
Methyl alcohol
Petroleum hydrocarbons
Phosphorus
Resins
Sodium hydroxide
Sulfur dioxide

Turpentine
Varnishes

METHYLATION WORKERS
Dimethyl sulfate
Methyl alcohol
Methyl chloride

MILLINERY WORKERS
Aniline and derivatives
Benzene
Methyl alcohol

MINERS
Abnormal air pressure
Antimony and compounds
Carbon dioxide
Carbon monoxide
Coal tar and fractions
Dusts: rock, metal
Ethylene dinitrate
Explosives
Heat
Hydrogen sulfide
Lead
Mercury (mercury miners)
Radiation
Ultraviolet (strip and open-pit mining)
Ionizing (uranium miners)
Silica dust
Tolylene diisocyanate (TDI; mine tunnel coating)

MORDANTERS
See also Dyers and dye makers
Acids
Alkalies
Aluminum salts
Amyl alcohol
Antimony compounds
Arsenates
Barium (in textile dyeing)
Copper salts
Chromium salts
Fluorides (in textile dyeing)
Formaldehyde

Hydrogen cyanide
Iron salts
Lead salts
Phosphates
Silicates
Tin salts
Zinc chloride

NAIL ENAMEL AND POLISH MAKERS
See Cosmetics makers

NATURAL GAS MAKERS AND WORKERS
Coal tar fractions
Hydrogen sulfide
Natural gas

NITRIC ACID MAKERS
Ammonia
Natural gas
Nitric acid
Nitrogen dioxides

NITROCELLULOSE MAKERS AND WORKERS
See Textile workers: nitrocellulose makers and workers

NITROGEN COMPOUND MAKERS
Calcium cyanamide

NUCLEAR REACTOR WORKERS AND NUCLEAR TECHNOLOGISTS
Beryllium and compounds
Cobalt and compounds
Graphite
Ionizing radiation
Lead
Thorium and compounds
Uranium and compounds
Zirconium compounds

NYLON MAKERS
See Textile workers: nylon makers

OFFICE WORKERS
Air-conditioned air impurities
Asbestos
Fibrous glass
Fungal spores (causing allergy)
Adhesives
Duplicating-fluid removers
Duplicating materials
Ink removers
Inks
Rubber
Solvents

OIL: LUBRICATING OIL MAKERS
Aluminum compounds
Chlorinated benzenes
Dimethyl formamide
Ethylenediamine (as textile lubricant)
Furfural
Graphite
Hydroquinone
Ketones
Molybdenum
Naphthalene
Phenol
Phosphorus pentasulfide
Selenium and compounds
Tetramethylthiuram disulfide
Zirconium

OIL PROCESSORS
See also Food processors: vegetable oil extraction and purification; Petroleum refinery workers
Acetonitrile (animal oil processors, petroleum hydrocarbon purifiers)
Amyl alcohol
Barium compounds (oil additives)
Benzene
Carbon disulfide
Carbon tetrachloride
Cellosolves
Chloride of lime (as bleach)
Chlorinated diphenyls and

naphthalenes (mineral oil
processors)
Chromium compounds
Cobalt and compounds (in oil
dryers, pigments, hydroge-
nation)
Cresols
Cycloparaffins
Diacetone alcohol
1,2,-dichloroethylene
Dichloroethyl ether
Dimethyl formamide (for lu-
bricating oil extractors)
Dioxane (diethylene ether)
Ethyl chloride
Ethylenediamine (as oil neu-
tralizer)
Ethylene dibromide
Ethyl ether
Hydrogen peroxide
Isopropyl acetate
Ketones
Methylene chloride (oil extrac-
tors)
Naphtha (petroleum naphtha)
Natural gas (in hydrogenation
of oils)
Nickel (in hydrogenation of
oils)
Ozone (in oil bleaching)
Propylene dichloride
Sodium and potassium hydrox-
ide (in vegetable-oil proces-
sing)
Sulfur monochloride (in vul-
canization of oil)
Tetrachloroethane
Toluene
Trichloroethylene
Turpentine (oil additives)
OIL WELL WORKERS
Asbestos
Cement

Barium and compounds
Chromium compounds
Fluorides
Hydrogen chloride
Hydrogen fluoride
PAINT MAKERS AND USERS
See also Paint remover makers
and users; Pigment makers
and users
Alkalies
Aluminum and compounds
Ammonia (for water-base
paints)
Amyl acetate
Amyl alcohol
Aniline and derivatives (as anti-
fouling agents)
Antimony and compounds
Arsenic
Asbestos
Barium and compounds (in lu-
minous and flat paints)
Benzene
Bismuth compounds (in lumi-
nous paint)
Cadmium
Calcium oxide
Carbon disulfide
Cellosolves
Chlorinated benzenes
Chlorinated diphenyls and
naphthalenes
Chromium compounds
Coal tar and fractions (poly-
cyclic hydrocarbons)
Cobalt and compounds (pig-
ments in dryers)
Copper and compounds
Dichloroethyl ether
Dioxane (diethylene ether)
Epichlorohydrin
Ethylene glycol
Ethyl silicate (in silicate paints)

Fluorides
Graphite
Hydroquinone
Ketones
Lead
Manganese compounds
Mercury and compounds
Methyl alcohol
Nickel and compounds
Petroleum naphtha
Phenol
Plastics: epoxy resins
Pyridine
Selenium and compounds
Sulfuric acid
Talc
Tetrachloroethane
Titanium and compounds
Toluene (as thinner)
Trichloroethylene
Turpentine
Uranium (in uranium paints)
Xylene
Zinc compounds
Zirconium compounds

PAINT REMOVER MAKERS AND
USERS
Amyl acetate
Aniline and derivatives
Carbon disulfide
Cresol
Cycloparaffins
Diacetone alcohol
Dioxane (diethylene ether)
Ethylene dichloride
Furfural (furfuraldehyde)
Ketones
ortho-dichlorobenzene
Oxalic acid
Phenol
Sodium and potassium hydrox-
ide

Tetrachloroethane
Trichloroethylene

PAPER BOX MAKERS
Asbestos
Dyes
Paper dust
Plastics
Resins
Waxes

PAPER WORKERS
Acrylonitrile
Aluminum and compounds
Ammonia
Amyl acetate (for coated paper)
Arsine
Asbestos
Bagasse (sugar cane)
Barium and compounds
n-butyl acetate (for coated pa-
per)
Calcium oxide
Chloride of lime
Chlorinated diphenyls and
naphthalenes (for treated
paper, carbonless carbon
paper)
Chlorine
Chromium compounds (for pa-
per dyeing)
Diacetone alcohol
Formaldehyde
Formic acid
Graphite
Hydrogen sulfide (pulp work-
ers)
Oxalic acid
Plastics: amino resins for paper
treaters
Selenium compounds
Sodium and potassium hydrox-
ide
Sulfur dioxide
Sulfuric acid

Tin and compounds (in sensitized paper)
Titanium
Trichloroethylene (in paper cups)
Zinc compounds

PARAFFIN PROCESSORS
See also Wax makers and workers
Benzene
Carbon disulfide
Ethylene dichloride
Ketones
Perchloroethylene

PENCIL MAKERS
Adhesives and glues
Aniline and derivatives (colored pencils)
Benzene
Chromates (in colored pencils)
Graphite
Lacquer thinners
Methyl violet
Pyridine
Red cedar wood
Resins
Solvents
Waxes

PERCUSSION CAP MAKERS
Mercury compounds

PERFUME MAKERS
See also Cosmetics makers
Acetaldehyde
Acetic anhydride
Acetonitrile
Acrolein (acrylic aldehyde)
Ammonia
Amyl acetate
Amyl alcohol
Aniline and derivatives
Benzene
Benzyl chloride
Bismuth and compounds
n-butyl acetate

Cellosolves
Chlorinated benzenes
Chromium compounds
Cresol
Cycloparaffins
1,2-dichloroethylene
Dimethyl sulfate
Ethyl acetate
Ethyl chloride
Ethyl ether
Formic acid
Isopropyl acetate
Methyl alcohol
Methylene chloride
Phenol
Propyl acetate
Sodium and potassium hydroxide
Tin and compounds
Toluene
Trichloroethylene

PESTICIDE AND INSECTICIDE MAKERS AND USERS
Acetic acid
Arsenic
Asbestos
Barium and compounds
Benzene
n-butylamine
Calcium oxide
Carbon dioxide
Carbon tetrachloride
Cellosolves
Chlorinated benzenes
ortho-dichlorobenzene
para-dichlorobenzene
Trichlorobenzene
Chlorinated diphenyls and naphthalenes
Coal tar and fractions (polycyclic hydrocarbons)
Copper and compounds
Cresols

Ethylene chlorohydrin
Ethylene dibromide
Ethylene dichloride
Ethylene oxide
Fluorides
Formic acid
Hydrazine
Hydrogen cyanide
Kerosene
Lead
Mercury
Methyl bromide
Methyl formate
Naphtha (petroleum distillate)
Naphthalene
Phosphorus (white and yellow)
Phosphorus pentasulfide
Phthalic anhydride
Sulfur monochloride
Tetrachloroethane
Tetramethylthiuram disulfide
Thallium and compounds
Toluene
Turpentine
Xylene
Zinc compounds

PETROLEUM REFINERY
WORKERS
Aluminum and compounds
Ammonia
Aniline and derivatives
Arsenic
Arsine
Asbestos
Benzene
n-butylamine (for de-waxing)
Calcium oxide
Carbonyls of metals
Chlorinated diphenyls and
naphthalenes
Chlorine
Cold
Copper and compounds

Dimethyl formamide
Glass fibers
Fluorides
Furfural (furfuraldehyde)
Heat
Hydrogen bromide
Hydrogen fluoride
Hydrogen sulfide
Ketones
Methyl chloride
Molybdenum compounds
Naphtha (petroleum distillate)
Natural gas
Nickel
Nickel carbonyl
Nitrobenzene
Radiation: ionizing, ultraviolet
Sunlight
Sodium and potassium hydrox-
ide
Sulfuric acid

PHOSPHOR MAKERS
Cerium
Germanium
Selenium and compounds
Zirconium

PHOSPHORIC ACID MAKERS
Fluorides
Hydrogen cyanide
Phosphoric acid
Phosphorus (white or yellow)
Sulfuric acid

PHOSPHORUS PROCESSORS
Carbon disulfide
Ethyl chloride
Phosphorus compounds
Tetrachloroethane

PHOTOENGRAVERS
See also Ink makers; Lithographers; Paper workers; Photographic chemical makers and
users; Printers
Ammonia

Amyl acetate
Chromates
High-intensity light
Hydrogen cyanide
Hydrogen sulfide
Methyl alcohol
Nitric acid
Oxalic acid
Ozone
Phosphoric acid
Sodium and potassium hydroxide
Ultraviolet radiation

PHOTOGRAPHIC CHEMICAL MAKERS AND USERS
Acetic acid
Acetaldehyde
Ammonia (automatic film processing)
Amyl alcohol
Aniline and derivatives
Barium and compounds
Benzene
Benzyl chloride (in developers)
Bromine
Chlorine (in developers)
Chromates
Cresol (in developers)
Dimethyl hydrazine (in developers)
Dinitrophenol (in developers)
Hydrogen peroxide (in developers)
Hydroquinone (in developers)
Light: high intensity, photographing
Mercury compounds
Nitrophenols
Oxalic acid
Ozone (photographers)
Petroleum naphtha
Phenol
Quinone (in developers)

Selenium
Silver compounds
Sulfuric acid
Trichloroethylene (plate cleaners)
Trinitrotoluene (TNT)
Uranium and compounds
Vanadium

PHOTOGRAPHIC FILM MAKERS
Acetic anhydride
Ammonia
Amyl acetate
n-butyl alcohol
Cellosolves
Ethyl acetate
Formaldehyde
Ketones
Methyl alcohol
Methylene chloride
Phthalic anhydride
Silver and compounds
Tetrachloroethane

PHTHALIC ANHYDRIDE MAKERS
Cobalt and compounds
Naphthalene
Phthalic anhydride
Xylene

PICRIC ACID MAKERS
Benzene
Chlorobenzene
Phenol
Picric acid

PIGMENT MAKERS AND USERS
Antimony and compounds
Arsenic
Barium and compounds
Bismuth and compounds
Cadmium and compounds
Chromates
Copper and compounds
Graphite
Hydrogen chloride
Hydrogen cyanide

Molybdenum compounds
Oxalic acid
Selenium and compounds
Sulfuric acid
Thorium compounds (in luminous pigments)
Tin compounds
Titanium and compounds
Uranium
Zinc compounds (in pigments for steel)
Zirconium

PIPEFITTERS, PIPE MAKERS, PLUMBERS
Arsine
Asbestos
Coal tar and fractions (polycyclic hydrocarbons; pipe pressers)
Glass fibers
Graphite (in pipe-joint compound)
Lead
Nitrogen dioxide
Plastics
Polyester resins
Polyurethane
Tolylene diisocyanate (TDI)
Skin infections: parasitic eruptions
Welding fumes

PIPELINE WORKERS
See also Oil processors; Petroleum refinery workers
Asbestos
Coal tar and fractions (polycyclic hydrocarbons)
Chigger bites
Cold
Fibrous glass
Poisonous plants
Radiation: sunlight (ultraviolet light)

PITCH WORKERS

See Coal tar and pitch workers; Road workers and road builders

PLASMA TORCH OPERATORS
See Chapter Eight: Welding Hazards

PLASTIC MAKERS: RESIN MAKERS
See also Plastic makers; Plasticizer makers and users
Acetaldehyde
Acrolein (acrylic aldehyde)
Allyl alcohol
Ammonia
Benzene
Benzyl chloride
Carbon disulfide
Cellosolves
Chlorinated benzenes
Chlorinated diphenyls and naphthalenes
Cresol
Cycloparaffins
Decaborane
Diacetone alcohol
1,2,-dichloroethylene
Dichloroethyl ether
Dimethyl formamide
Dioxane (diethylene ether)
Ethyl acetate
Ethyl benzene
Ethyl chloride
Ethylene chloride
Ethylene chlorohydrin
Ethylene dichloride
Ethylene glycol
Formaldehyde
Furfural (furfuraldehyde)
Hexamethylenetetramine
Isopropyl acetate
Ketones
Methyl alcohol
Methylene chloride

Mineral oil
Naphthalene
Nitroparaffins
Oils: cashew nut oil, mineral
   oil
Phenol
Phosgene
Phthalic anhydride
Plastic resins
  Alkyd resins
  Allyl resins
  Amino resins
  Diisocyanate resins
  Epoxy resins
  Phenolic resins
  Polyester resins
Propyl acetate
n-propyl alcohol
Tetrachloroethane
Titanium and compounds
Toluene
Trichloroethylene
Turpentine
Xylene
PLASTIC MAKERS
Acetic acids
Acetic anhydride
Acetaldehyde (in phenolic res-
  ins)
Acetonitrile; plasticizers and
  polymethylacrylic resins
Acetone
Acrylic resins
Alkyd resins
Allyl alcohol
Allyl resins
Aluminum compounds
Amino resins
Ammonia
Amyl acetate
Amyl alcohol
Aniline and derivatives
Arsine
Asbestos

n-butyl alcohol (in plasticizers)
Cellulosics
Chlorinated diphenyls and
  naphthalenes
Cresol
Cycloparaffins (for molding
  plastic)
Cobalt and compounds (in
  polyester resins)
Diatomite
1,2-dichloroethylene
Diisocyanate resins
Dinitrobenzene
Dioxane (diethylene ether)
Epichlorohydrin
Epoxy resins
Ethyl acetate
Ethyl ether
Ethyl silicate (for protective
  coatings)
Ethylene dichloride
Ethylene oxide
Fibrous glass
Formaldehyde
Formic acid
Freon
Furfural (furfuraldehyde)
Glass fibers
Hydrogen chloride
Hexylmethylenetetramine
Hydrogen cyanide
Hydrogen fluoride
Hydrogen peroxide (in plastic
  foam)
Hydroquinone
Isopropyl acetate
Ketones
Lead
Methyl chloride (in polystyrene
  foam)
Mica
Nylon
Organic tin compounds

Phenol
Phosgene
Phthalic anyhydride
Polyester resins
Polyethylenes
Polystyrenes
Quartz
Selenium and compounds
Silica
Styrene
Tin compounds
Tolylene diisocyanate (TDI)
Tricresyl phosphate (in plasti-
cizers, polyvinyl chloride,
polystyrene)
Vinyl chloride
Vinyl plastics
Xylene (in polyester teraph-
thalate film)

PLASTICIZER MAKERS AND
USERS
Acrylonitrile
Allyl alcohol
n-butyl alcohol
Chlorinated diphenyls and
naphthalenes
Ethylene dichloride (in plasti-
cizer bath)
Phosphorus oxychloride
Phosphorus trichloride
Phthalic anhydride
Tricresyl phosphate

PLASTERERS
See also Construction workers;
Paint makers and users; Paint
remover makers and users
Asbestos
Bacteria infections: anthrax
Calcium oxide
Calcium sulfate (gypsum)
Glass fibers

POLISH MAKERS AND WORKERS

See also Metal Cleaners, Polish-
ers
Amyl acetate
Aniline and derivatives
ortho-dichlorobenzene
Dioxane (diethylene ether)
Graphite
Hydrogen fluoride
Hydrogen cyanide
Methyl alcohol
Nitrobenzene (in shoe polish)
Phosphoric acid
n-propyl alcohol
Titanium and compounds (in
white shoe polish)
Trichloroethylene
Turpentine (in store enamel
polishes)
Zirconium compounds

POTTERY MAKERS AND
WORKERS
See also Ceramic makers and
workers; Enamel makers and
workers
Aluminum and compounds
Carbon dioxide
Hydrogen chloride
Lead
Talc
Zirconium compounds

POSTAL WORKERS
Carbon monoxide
Cold
Dust
Improper illumination
Infections
Bacteria: dermatitis
Fungi: dermatitis
Lifting: back injuries

PRINTERS
See also Adhesive makers and
users; Glue makers and users;
Ink makers; Ink remover

makers and users; Lithographers; Photoengravers; Photographic chemical makers and users; Paper workers; Textile printers; Typographers, electrotypers, and stenotypers
Aniline and derivatives
Asbestos
Benzene
Carbon black
Carbon tetrachloride
Cellosolves
Chlorinated diphenyls and naphthalenes
Chromates
Cornstarch
Diacetone alcohol
Dioxane (diethylene ether)
Formaldehyde
Gums: acacia, arabic
Hexane (tulosol)
Hydrogen cyanide (in textile and art printing)
Ink mists
Isopropyl alcohol
Ketones
Lead (in wallpaper printing)
Methyl alcohol
Methyl chloride
Paper dust
Sodium and potassium hydroxide
Talc
Toluene
Trichloroethylene
Urea-formaldehyde resins
PROPELLENT AND AEROSOL MAKERS AND WORKERS
Amino resins: urea-formaldehyde and melamine-formaldehyde resins
Carbon tetrachloride

Chlorine
Carbon dioxide
Freon
Methyl chloride
Methylene chloride
PULP MAKERS
See Wood and paper pulp makers
RAILROAD SHOP WORKERS
Alkalies
Antiseptics
Asbestos
Chromates
Cutting oils
Detergents (synthetic)
Dichlorobenzene
Diesel fuel oil
Greases
Insecticides
Lacquers
Lubricating oils
Magnaflux
Paint
Paint stripper
Paint thinners
Solvents
Ultraviolet radiation
RAILROAD TRACK BUILDERS
See also Construction workers; Road builders
Copper and compounds (in tie preservers)
Cold
Heat
Increased pressure (tunnel diggers)
Sunlight
Zinc compounds (in tie preservers)
REFRIGERANT MAKERS AND WORKERS
See also Freon makers
Ammonia
Carbon dioxide

Carbon tetrachloride
Chlorine
Cold
Copper compounds
Ethyl bromide
Ethyl chloride
Ethyl ether
Formic acid
Freons
Hydrogen sulfite
Methyl bromide
Methyl chloride
Methylene chloride
Sulfur dioxide

RESIN MAKERS
See Plastic makers: Resin makers

ROAD WORKERS AND BUILDERS
See also Coal tar and pitch workers; Railroad track builders
Cement
Dust, clays, dirt
Coal tar
Cold
Furfural (furfuraldehyde)
Pressure (increased in tunnel work)
Plastics: epoxy resins
Plants: poison ivy, etc.
Sunlight
Vibrating tools

RODENTICIDE MAKERS AND USERS
See also Pesticide and insecticide makers and users
Arsenic
Barium and compounds
Fluorides
Strychnine
Tetramethylenethiuram disulfide
Thallium and compounds
Tin and compounds
Warfarin

ROOFERS
See Coal tar and pitch workers

ROPE MAKERS
Alkalies
Bleaches
Coal tar and fractions
Copper and compounds (in preservatives)
Dusts: cotton, hemp
Dyes
Oils
Pitch
Soaps

RUBBER CEMENT MAKERS AND USERS
Adhesives
Airplane dope
Ammonia
Amyl acetate
Benzene
n-butyl alcohol
Carbon disulfide
Gum
Ketones
Sulfur monochloride
Toluene
Trichloroethylene
Xylene

RUBBER: LATEX WORKERS
See also Rubber makers and workers
Ammonia
Fluorides
Phosphoric acid
Pyridine

RUBBER MAKERS AND WORKERS
Acetaldehyde
Acetic acid
Acetylene
Acrolein (acrylic aldehyde)
Acrylonitrile
Alkalies
Aluminum and compounds

Ammonia
Amyl acetate
Amyl alcohol
Aniline and derivatives
Antimony and compounds
Benzene
Benzidine
Benzyl chloride
Butadiene
n-butylamine
Calcium oxide
Carbon black
Carbon disulfide
Carbon tetrachloride
Chlorinated benzenes
Chlorinated diphenyls and
  naphthalenes
Chlorine
Chloroprene (chlorobutadine;
  in neoprene)
Chromates
Coal tar and fractions
Cobalt and compounds (as col-
  oring)
Copper and compounds
Cresol
Cycloparaffins
Decaborane
para-dichlorobenzene
1,2-dichloroethylene
Ethyl alcohol
Ethylenediamine
Formaldehyde
Formic acid
Freon (in sponge rubber)
Furfural (furfuraldehyde)
Graphite
Hexamethylenetetramine
Hydrochloric acid
Hydroquinone (in rubber coat-
  ing)
Ketones
Lead

Manganese compounds
Mercaptans
Methyl alcohol
Methyl chloride
Methylene chloride
Oxalic acid
Petroleum naphtha
Perchloroethylene
Phosphoric acid (in rubber la-
  tex)
Phosphorus pentasulfide
Phenol (in reclaiming rubber)
Propylene dichloride
Pyridine
Selenium
Styrene
Sulfuric acid
Talc
Tellurium
Tetrachloroethane
Tetramethylthiuram disulfide
  (in heat-resistant rubber)
Titanium
Tolylene diisocyanate (in abra-
  sion-resistant rubber)
Turpentine
Vinyl chloride
Xylene
Zinc compounds
RUBBER VULCANIZERS
Ammonia
Aniline and derivatives
Barium and compounds
Carbon dioxide
Hydrogen sulfide
Methyl alcohol
Sulfur monochloride
Tellurium
Tetramethylthiuram disulfide
SCIENTIFIC WORKERS AND
TECHNICIANS
See Hospital workers; Laboratory
  workers, chemical

SEWER WORKERS AND WATER
TREATERS
  Ammonia
  Bacteria infections
    Plague
    Leptospirosis
    Brucellosis
  Boron and compounds
  Bromine
  Calcium oxide
  Carbon dioxide
  Chloride of lime
  Chlorine
  Copper and compounds
  Fluorides
  Hydrazine
  Hydrogen peroxide
  Hydrogen sulfide
  Manganese compounds
  Ozone
  Phosphoric acid
  Silver compounds
SHIPYARD WORKERS
See Construction workers
SHELLAC MAKERS AND USERS
See also Cabinetmakers and Car-
    penters; Construction work-
    ers; Lacquer makers and
    users; Stain makers and us-
    ers; Varnish makers and us-
    ers; Wood preservative mak-
    ers and users
  Ammonia
  Amyl acetate
  Benzene
  n-butyl acetate
  1,2-dichloroethylene
  Ethyl alcohol
  Ethylenediamine
  Lead
  Methyl alcohol
SHOEMAKERS
See also Cobblers and cementers
    of rubber shoes; Glue makers

and users; Leather workers;
  Pigment makers and users;
  Polish makers and workers
Adhesives
Ammonia (in finishing)
Amyl acetate
Amyl alcohol
Dioxane (diethylene ether; in
    shoe creams)
Furfural (furfuraldehyde; in
    shoe dyes)
Glues
Ketones
Lead (in stains)
Methyl alcohol
Titanium and compounds (in
    shoe-whitening compounds)
Trichloroethylene
SMOKE BOMB MAKERS
See also Gas: tear gas makers
  Cadmium compounds
  Phosphorus
  Titanium compounds
  Zinc compounds
SOAP MAKERS
See also Bacteriocide makers and
    users; Detergent makers
  Acrolein (acrylic aldehyde)
  Alkalies
  Amyl acetate
  Barium and compounds
  Calcium oxide
  Cellosolves
  Chloride of lime (as bleach)
  Chromates
  Cobalt compounds
  Dichloroethyl ether
  Ethylene dichloride
  Hydrochloric acid
  Hydrogen peroxide (as bleach)
  Hydrogen sulfide
  Methyl alcohol
  Nitrobenzene

Perchloroethylene
Plastics: amino resins
n-propyl alcohol
Tetramethylthiuram disulfide
(as bacteriocide)
Trichloroethylene
Zinc compounds
SOLDER MAKERS; SOLDER FLUX
MAKERS AND USERS
See Chapter Eight: Welding Hazards
Acids
Antimony and compounds
Arsine
Bismuth and compounds
Cadmium and compounds
Copper and compounds
Epoxy resin
Hydrazine (in fluxes)
Infrared radiation
Lead
Silver and compounds
Tin and compounds
Zinc compounds (in fluxes)
SOLVENT MAKERS AND USERS
Benzene
Carbon tetrachloride
ortho-dichlorobenzene
Chlorinated diphenyls and
naphthalenes
Diacetone alcohol
1,2-dichloroethylene
Dichloroethyl ether
Dimethyl formamide
Dioxane (diethylene ether)
Epichlorohydrin
Ethyl acetate
Ethyl alcohol
Ethyl benzene
Ethyl bromide
Freon
Isopropyl acetate
Isopropyl alcohol

Ketones
Methyl alcohol
Methyl chloride (in low-temperature solvents)
Methylene chloride
Perchloroethylene
Petroleum naphtha
n-propyl alcohol
Propylene dichloride
Pyridine
Tetrachloroethene
Toluene
Trichloroethylene
Tricresyl phosphate
Turpentine
Xylene
STAIN MAKERS AND USERS
(FOR WOOD, ETC.)
See also Cabinetmakers and carpenters; Construction workers; Furniture workers; Lacquer makers and users; Shellac makers and users; Varnish makers and users; Wood preservative makers and users
Benzene
n-butyl acetate
Carbon tetrachloride
Cellosolves
Chlorinated benzenes
Chlorinated naphthalenes and
diphenyls
Chromates (in wood stains)
Cresol
Diacetone alcohol
Dioxane (diethylene ether)
Ethyl acetate
Ethyl alcohol
Ethylene glycol (in wood
stains)
Isopropyl alcohol
Ketones
Methyl alcohol (in wood stains)

Nitrobenzene
Nitroparaffins
Petroleum naphtha
Toluene
Trichloroethylene
Turpentine
Xylene

STAIN REMOVERS
See also Dry cleaners
Acetic acid
Amyl acetate
n-butyl acetate
Dichloroethyl ether
Ethylene dichloride
Methyl chloride
Methylene chloride
Oxalic acid
Propylene dichloride

SULFURIC ACID MAKERS AND
WORKERS
Ammonia
Arsine
Hydrogen sulfide
Nitric acid
Nitrogen dioxide
Sulfur dioxide
Sulfuric acid

SULFUR PROCESSORS
Carbon disulfide
ortho-dichlorobenzene
Ethyl chloride
Ethylenediamine
Tetrachloroethane

SURFACTANT MAKERS
Cresol
Ethylene oxide
Ethylenediamine
Phosphorus trichloride
Titanium compounds

TAR MAKERS, WORKERS, AND
REMOVERS
See Coal tar and pitch workers

TEXTILE PROCESSORS (BLEACH-
ING, CLEANING, TREATING, ETC.)
See also Dyers and dye makers;
Mordanters
Acetic anhydride
Acrolein (acrylic aldehyde; for
textile resins)
Acrylonitrile (in finishes)
Aluminum and compounds
Antimony and compounds (in
flameproofing)
Barium and compounds (as tex-
tile bleaches)
Cerium
Chloride of lime (as bleach)
Chlorinated diphenyls and
naphthalenes (in flame-
proofing)
Chlorine (as bleach)
Chromates
Copper and compounds
Cresol (in sizing)
Diacetone alcohol
Dichloroethyl ether (scourers)
Dioxane (in textile lubricants)
Ethylendiamine (in textile lu-
bricants)
Ethylene dichloride (in textile
cleaners)
Ethylene glycol
Ethylene oxide (as fumigant
and textile lubricant)
Formaldehyde (in waterproof-
ing)
Formic acid
Hexamethylenetetramine
Hydrogen peroxide (as bleach)
Hydroquinone (as textile coat-
ing)
Ketones
Lead
Manganese compounds (as tex-
tile fiber bleach)
Nitroparaffins

Ozone (as bleach)
Plastics: alkyd resins, amino
   resins
Selenium
Sodium and potassium hydrox-
   ide (as bleaches)
Sulfur dioxide (as bleach)
Sulfuric acid
Thallium compounds
Tin and compounds
Toluene
Trichloroethylene (in textile
   cleaners)
Zinc compounds
Zirconium compounds

TEXTILE PRINTERS
See also Dyers and dye makers;
   Mordanters; Printers
Acetic acid
Amyl acetate
Aniline and derivatives
Antimony and compounds
Arsenic
Barium and compounds
Cadmium and compounds
Cellosolves
Chloride of lime
Chromates
Ethylene chlorohydrin
Formaldehyde
Hydrogen cyanide
Hydrogen sulfide
Manganese compounds
Mercury compounds
Methyl alcohol
Oxalic acid
Phenol
Picric acid
Tin and compounds

TEXTILE WORKERS: COTTON,
THREAD, AND CLOTH WORKERS
(MILLING, BLEACHING, PREPAR-
ING, AND SIZING CLOTH)

Acids
Alkalies
Aluminum salts
Amino resins
Arsensic salts
Artificial fiber dusts and fumes
Cotton dust
Calcium
Calcium salts
Carbolic acid
Cellosolves
Chlorine
Detergents
Dicyanodiamide formaldehyde
Dyes
Formaldehyde
Fungicides
Magnesium salts
Melamine-formaldehyde resins
Moisture
Nitrogen dioxide
Potassium salts
Soaps
Sodium hydroxide
Sodium metasilicate
Sodium salts
Sodium silicate
Starch
Urea-formaldehyde resins
Zinc chloride

TEXTILE WORKERS: CELLULOS-
ICS MAKERS AND WORKERS
(CELLULOSE ACETATE, CEL-
LULOSE NITRATE, CELLU-
LOSE BUTYRATE)

Acetic acid
Acetic anhydride
Acetone
n-butyl acetate
Chlorinated benzenes
Chlorinated diphenyls and
   naphthalenes

Cobalt (in making ethyl acry-
late)
Diacetone alcohol
Dichloroethyl ether (in making
ethyl cellulose)
1,2-dichloroethylene
Dioxane
Copper and compounds
Ethyl chloride (in making ethyl
cellulose)
Ethylene chlorohydrin (in mak-
ing ethyl cellulose)
Furfural (furfuraldehyde)
Hydrogen cyanide
Ketones
Methylene chloride
Methyl alcohol
Methyl formate
Nitric acid
Nitroparaffin
Perchloroethylene (in cellulose
ester)
Phosphorus pentachloride (in
making acetyl cellulose)
Phthalic anhydride (as plasti-
cizer)
Sulfuric acid
Tetrachloroethane
Zinc chloride

TEXTILE WORKERS: NITROCEL-
LULOSE MAKERS AND WORKERS
Amyl acetate
Amyl alcohol
Arsine
Benzene
n-butyl alcohol
Cellosolves
Diacetone alcohol
Ethyl acetate
Furfural (furfuraldehyde)
Isopropyl acetate
Nitroparaffins: nitromethane,
nitroethane, nitropropane

Propyl acetate
n-propyl alcohol
Tricresyl phosphate

TEXTILE WORKERS: NYLON
MAKERS
Cycloparaffins
Furfural (furfuraldehyde)
Hydrogen cyanide
Polyamides

TEXTILE WORKERS: RAYON
VISCOSE MAKERS
See also Cellophane film makers
Acetonitrile
Ammonia
Arsine
Asbestos
Carbon dioxide
Carbon disulfide
Chlorinated diphenyls and
naphthalenes
Chlorine
Copper and compounds
Dimethyl amine
Ethylene oxide
Fluorides
Hydrogen cyanide
Hydrogen sulfide
Lead (in lead burning)
Oxalic acid (as bleach)
Plastics: fluorocarbons
Sodium and potassium hydrox-
ide
Sulfuric acid
Zinc compounds

TEXTILE WORKERS: SILK
PROCESSORS
Acids
Alkalies
Amyl acetate
Bromine (as bleach)
Chromates (in silk-screen mak-
ing)
Hydrogen peroxide (as bleach)

Hydrogen sulfide
Isopropyl acetate
Nitrogen dioxide (in raw-silk
bleaching)
Xylene (in silk finishing)
Zinc compounds
TEXTILE WORKERS: WOOL
PROCESSORS
Ammonia (scourers)
Bromine (in shrinkproofing)
Cresol (scourers)
ortho-dichlorobenzene (in de-
greasing)
Ethylene dichloride (as cleaner)
Ethylene dibromide (reclaim-
ers)
Hydrogen peroxide (wool
printers)
Methyl bromide
Petroleum naphtha
Infections
Bacteria: anthrax, rickettsia,
fever
Fungi: ringworm
Perchloroethylene (scourers)
Sulfur dioxide (as bleach)
Trichloroethylene (scourers)
TYPOGRAPHERS, ELECTROTYP-
ERS, AND STENOTYPERS (PRINT-
ING TRADES)
See also Engravers; Ink makers;
Lithographers; Paper work-
ers; Printers
Ammonia
Antimony
Arsenic
Graphite
Inks
Lead
Silver
Tin
Type-cleaning solvents

VACUUM TUBE MAKERS
See also Electrical and Electronics
workers; Lamp and light
makers
Carbon disulfide
Germanium
Ionizing radiation
Methyl alcohol
Molybdenum and compounds
Perchloroethylene
Thorium
Titanium
Trichloroethylene
Zirconium
VARNISH MAKERS AND USERS
See also Cabinetmakers and car-
penters; Construction work-
ers; Lacquer makers and us-
ers; Shellac makers and us-
ers; Stain makers and users
Acetaldehyde
Acids
Alkalies
Amyl acetate
Amyl alcohol
Aniline and derivatives
Barium and compounds
n-butyl alcohol
Carbon disulfide
Cellosolves
Chlorinated benzenes
Chlorinated diphenyls and
naphthalenes
Cobalt and compounds
Dichloroethyl ether
Ethyl acetate
Ethyl alcohol
Ethylene chlorohydrin
Ethylene dichloride
Furfural (furfuraldehyde)
Hydroquinone
Isopropyl alcohol

Ketones
Lead
Manganese
Methyl alcohol
Naphtha (petroleum distillate)
Nickel and compounds
Phenol
Plastics: phenolic resins
Propyl acetate
Styrene
Tetrachloroethane
Titanium and compounds
Toluene
Trichloroethylene
Turpentine
Zinc compounds
Zirconium compounds

VARNISH REMOVER MAKERS
AND USERS
See also Varnish makers and us-
    ers
Carbon disulfide
Carbon tetrachloride
Cellosolves
Chlorinated benzenes
Cresols
Epichlorohydrin
Ethylene dichloride
Ketones
Methylene chloride
Oxalic acid
Styrene

VETERINARIANS
See also Farmers and agricultural
    workers; Hospital workers,
    including nurses and doctors
Carbon disulfide
Cresol
Hexamethylenetetramine
Hydrogen chloride
Hydrogen peroxide
Infections
    Viruses: cat-scratch fever
    Bacteria: anthrax, brucellosis

Trichloroethylene
WATERPROOFING AND
WATERPROOFERS
Alum
Amyl acetate
Ethylene dibromide
Paraffin
Pitch
Residine
Resins: melamine-formalde-
    hyde resins
Rubber
Solvents
Titanium compounds
Tricresyl phosphate
Waxes

WATER TREATMENT
See Sewer workers and water
    treaters

WAX MAKERS AND USERS
See also Lacquer makers and us-
    ers; Shellac makers and us-
    ers; Stain makers and users;
    Varnish makers and users
Amyl alcohol
Barium and compounds
Benzene
Carbon disulfide
Carbon tetrachloride
Cellosolves
Chlorinated diphenyls and
    naphthalenes
Chromates (in bleaching)
Cycloparaffins
Diacetone alcohol
1,2-dichloroethylene
ortho-dichlorobenzene
Ethyl chloride
Ethylene dibromide
Ethylene glycol
Ethyl ether
Hydrogen peroxide (as bleach)
Isopropyl acetate

Methylene chloride (as wax re-
mover)
Nitroparaffins
Ozone
Perchloroethylene
Petroleum naphtha
n-propyl alcohol
Propylene dichloride
Tetrachloroethane
Toluene
Trichloroethylene
Turpentine
WELDERS
See Chapter Eight: Welding Haz-
ards
WETTING AGENT MAKERS AND
WORKERS
Benzyl chloride
Furfural (furfuraldehyde)
WOOD PRESERVATIVE MAKERS
AND USERS
See also Cabinetmakers and car-
penters; Furniture workers;
Lacquer makers and users;
Shellac makers and users;
Stain makers and users; Var-
nish makers and users
Arsenic
Chlorinated diphenyls and

naphthalenes
Chromates
Coal tar and fractions
Copper and compounds
Creosote
Diacetone alcohol
Dinitrophenol
Fluorides
Formaldehydes
Manganese compounds
Mercuric chloride and phenyl
mercuric compounds
Phenol
Resins
Sulfur monochloride (as hard-
ener)
Zinc chloride
Zinc sulfate
WOOD AND PAPER PULP
MAKERS
Ammonia
Chloride of lime
Chlorine
Hydrogen peroxide
Hydrogen sulfide
Sodium and potassium hydrox-
ide
Sulfur dioxide

*Bibliography
and Indexes*

Bibliography

THIS LIST is intended to allow the reader to find basic references, in addition to those mentioned in the text, footnotes, and chapter references. The books listed vary from highly technical textbooks, useful as reference sources, to works written for lay audiences.

All books published by the government are available from the Government Printing Office, Washington, D.C. 20402.

Physical Hazards

Margot Bennett. *Intelligent Woman's Guide to Atomic Radiation.* Baltimore: Penguin Books, 1964.

Otto G. Edholm. *Biology of Work.* World University Library Series. New York: McGraw Hill Book Co., 1967.

John W. Gofman and Arthur R. Tamplin. *Poisoned Power: The Case Against Nuclear Plants.* Emmaus, Pa.: Rodale Press, 1971.

Karl D. Kryter. *The Effects of Noise on Man.* Environmental Science Series. New York: Academic Press, 1970. This recent book on noise contains much detailed information, including internationally accepted standards, and is excellent for reference.

National Academy of Sciences, National Research Council. *The Effects on Populations of Exposure to Low Levels of Ionizing Radiation.* Washington, D.C.: Government Printing Office, 1972.

Noise in Industry. International Occupational Safety and Health Information Centre (CIS) Information Sheet no. 17. Geneva: International Labour Office, 1968.

Rupert Taylor. *Noise.* Baltimore: Penguin Books, 1971.

U.S. Department of Health, Education, and Welfare, National Institutes for Occupational Safety and Health (NIOSH). *Criteria for Recommended Standards: Occupational Exposure to Hot Environments.* HSM 73-10269. Washington, D.C.: Government Printing Office, 1972.

————. *Criteria for Recommended Standards: Occupational Ex-*

posure to Noise. HSM 73-11001. Washington, D.C.: Government Printing Office, 1972.

————. *Criteria for Recommended Standards: Occupational Exposure to Ultraviolet Radiation.* HSM 73-11009. Washington, D.C.: Government Printing Office, 1972.

Chemical Hazards

Paul Brodeur. *Asbestos and Enzymes.* New York: Ballantine Books, 1972.

Ethel Browning. *Toxicity and Metabolism of Industrial Solvents.* Amsterdam: Elsevier, 1965.

————. *Toxicity of Industrial Metals.* 2nd ed. London: Butterworth & Co., 1969.

Marion N. Gleason, R. E. Gosselin, and H. C. Hodge. *Clinical Toxicology of Commercial Products.* 3rd ed. Baltimore: Williams & Wilkins Co., 1969.

Alice Hamilton and H. L. Hardy. *Industrial Toxicology.* New York: Harper & Brothers, 1949.

The Merck Index of Chemicals and Drugs. 8th ed. Rahway, N.J.: Merck & Co., 1968.

Frank A. Patty, ed. *Industrial Hygiene and Toxicology.* 2nd ed. Vol. 2, *Toxicology.* New York: Interscience Publishers, 1963.

N. Irving Sax et al. *Dangerous Properties of Industrial Materials.* 3rd ed. New York: Reinhold Publishing Corp., 1968.

U.S. Department of Health, Education, and Welfare. *Occupational Diseases: A Guide to Their Recognition.* Ed. W. M. Gafafer. Public Health Service Publication no. 1097. Washington, D.C.: Government Printing Office, 1966.

————. National Institutes of Occupational Safety and Health. *Criteria for Recommended Standards: Occupational Exposure to Asbestos.* HSM 72-10267. Washington, D.C.: Government Printing Office, 1972.

————. *Criteria for Recommended Standards: Occupational Exposure to Beryllium.* HSM 72-10268. Washington, D.C.: Government Printing Office, 1972.

————. *Criteria for Recommended Standards: Occupational Exposure to Carbon Monoxide.* HSM 73-11000. Washington, D.C.: Government Printing Office, 1972.

————. *Criteria for Recommended Standards: Occupational Exposure to Inorganic Lead.* HSM 73-11010. Washington, D.C.: Government Printing Office, 1972.

General References

Barry Commoner. *The Closing Circle.* New York: Alfred A. Knopf, 1971.

Encyclopedia of Occupational Health and Safety. Geneva: International Labour Organization, 1971.

Harrison's Principles of Internal Medicine. Ed. M. M. Wintrobe et al. 6th ed. 2 vols. New York: McGraw-Hill Book Co., 1970.

Julian B. Olishifski and Frank E. McElroy, eds. *Fundamentals of Industrial Hygiene.* Chicago: National Safety Council, 1971.

Harry Rothman. *Murderous Providence: A Study of Pollution in Industrial Societies.* Indianapolis: Bobbs-Merrill Co., 1972.

General Index

Numbers in boldface indicate pages on which the disease is explained.

accidents, industrial 7, 91, 104, 190, 204, 208, 231. *See also* nervous system, central, abnormalities
hearing interference and, 113
heat and, 122, 124
acidosis, **41**
acute disease. *See* occupational disease, acute reactions
adrenaline, 29, 104. *See also* hormones
and irregular heartbeat, 204
AFL-CIO, 16
agitation. *See* nervous system, central, abnormalities
air conditioners, 175, 183
air-flow meter, 288, 336
air pollution, 14, 26, 158–9
asbestos, 175
monitoring of, 327–42
sulfur dioxide, 160
air sampling, 327–42
continuous monitoring, 336
equipment sources, 344
representative samples, 13, 329–30
alcoholic beverages, 90–1, 206. *See also* alcohols, in Substances Index
allergies, 58–9, 154. *See also* lung allergies and asthma; skin allergies and contact dermatitis
dithiocarbamates, 223
kidney involvement, 44
liver involvement, 38
organic acids, 211, 218
tolylene diisocyanate, 231
Allied Chemical Corporation, 12
alpha rays. *See* radiation, ionizing

American National Standards Institute (ANSI), 9
anemia, **35, 36**
anesthetics, 157, 195, 203, 205, 227. *See also* nervous system, central, abnormalities
anorexia. *See* appetite, loss of
antibodies, 34, 36
anxiety. *See* nervous system, central, abnormalities; stress
aorta, 29
appendicitis, 252
appetite, loss of (anorexia), 186, 206, 225, 230, 238, 252, 254, 258
arrhythmia. *See* heartbeat, irregular
arteries, 28–33. *See also* heart disease, hardening of arteries
arthritis, **71, 107, 108,** 242
asbestosis, 10, 172–3, 343
ascites, **38**
asphyxia, 157, 240. *See also* gases, in Substances Index; oxygen, content in blood, low; oxygen, lack of, to body tissues
asthma. *See* lung allergies and asthma
Atomic Energy Commission, 149
audiometer, 92, 118

back injuries and disease, **67–74**
arthritis, **71, 107**
lifting and carrying, **71–4**
ruptured disc, 68, 70
strain, **70–1**
bagassosis, 182–3
baritosis, 181, 246

Berman, Dan, 354
berylliosis, 180–1
beta rays. *See* radiation, ionizing
bile, 37
bilirubin, 37
black lung, 7, **169**
  Compensation Law, West Virginia, 7
bladder, urinary, **41**
  cancer of, **197, 198,** 199, 200
  irritation of, 199, 200
blast-furnace workers, 165–6
blindness
  alcohols and, 184
  acids and alkalies and, 239–40
  dimethyl sulfate and, 238
  osmium and, 257
  radiation and, 138, 143
blood, 17, 18, 28, **33–6**
  clotting, 33, 104
  hemoglobin, 34–5, 37, 43–4
  plasma, 34
  platelets, 33, 34, 36
  red cells, 33–6
  white cells, 33, 34, 35–6
blood cancer. *See* leukemia
blood disorders, 20, **154**
  amines and, 237
  anemia, **35, 36**
  antimony and, 244
  aromatic hydrocarbons and, **190–1**
  aromatic nitro compounds and, 196
  arsenic and, 245–6
  cadmium and, 248
  carboxyhemoglobinemia, 30, 154, 164–5
  clotting disorders, **36,** 191
  copper and, 251
  cyanosis, **196–7,** 237
  epoxides and, 227, 228, 230
  glycol ethers and, 188
  kidney disease and, 41
  lead and, 252
  methemoglobinemia, 154, 196–7, 200, 237
  nickel and, 256
  nitriles and, 166
  nitro and amino aromatic compounds and, 199, 200
  red blood cells, rupture of, **35, 43,** 237, 245, 251, 280
  zinc and, 260

blood oxygen. *See* oxygen, content in blood
blood pressure
  high, **44,** 104, 239, 252
  low, **45,** 162
blood vessels, 28–33
bone disease. *See* back injury and disease; musculo-skeletal system
bone marrow, 36, 190
bones. *See* musculo-skeletal system
boredom, 80, 81, 82, 83
brain, 48–50, 52
brass chills. *See* metal fume fever
brass foundry workers' ague. *See* metal fume fever
breathing. *See* respiratory system
bronchi, 22–3
bronchioles, 23
bronchitis, 159–64. *See also* lung disease, chronic bronchitis and emphysema
brown lung. *See* byssinosis
burns. *See* irritation of membranes; skin burns
byssinosis, 6–7, 181–3

cancer, 46, **62–4.** *See also* carcinogens
  anthracene and, 201
  arsenic and, 245
  asbestos and, 173–4, 350–1
  benzene and, 190
  bladder, **197, 198,** 199, 200
  bone, 148
  chromium and, 249
  cigarette smoking and, 31, 174, **348–9**
  kidney, **46**
  leukemia, **36,** 148, 190
  lung, 31, 173–5, 203, 245, 249, 256, 350–1
  nickel and, 256
  oils and, 203
  radiation and, 137–8, 145, 148, 149
  talc and, 180
capillaries, 28
carboxyhemoglobinemia, 29–30, 154, 164–5. *See also* carbon monoxide, in Substances Index
carcinogens, **154.** *See also* cancer
  DDT, 222–3
  dithiocarbamate, 228

epoxides, 227
nitrosamines, 235
ozone, 164
phenol, 192, 194
polyurethane, 232
cardiovascular system. *See* circulatory system
cartilage, 107
cataract. *See* eye disease
chemical hazards, exposure
standards, 155–7, **261–74**. *See also*
individual chemicals, in Substances Index
chest pain. *See* heart disease; lung
disease
cholesterol, 30–1, 32
chronic disease. *See* occupational
disease, chronic effects
cigarette smoking, effects of, 19,
25–6, 29–30, 31, 32, **157**, 158–9,
165, 169, 172, 174, **347–9**
cilia, 21, 23
circulatory system, **28–33, 52**, 105–6,
108–9. *See also* blood pressure,
high; heart disease
cirrhosis of liver, **38**, 91
clotting, of blood, 33, 104
disorders, **36**, 191
Coal Mine Safety Act (1969), 170
coal-mining accidents, 7
coal workers' pneumoconiosis. *See*
black lung
coke-oven workers, 46, 165–6
cold, 29, 56, **128–30**
protective devices, 130–1
collective bargaining, 259–65
communication, interference with
by noise, 106
compensation. *See* workmen's compensation
congenital abnormalities. *See* genetic damage
congestive heart failure, **32**
conjunctivitis. *See* eye disease
contact dermatitis. *See* skin allergies
and contact dermatitis
coordination, loss of. *See* nervous
system, central, abnormalities
cornea. *See* eye, cornea damage
coronary arteriosclerosis. *See* heart
disease, hardening of arteries
cough. *See* irritation of membranes;
lung disease

curie, 132
cyanosis, **196–7**, 237
cystitis, 198, 199–200

dB. *See* decibels
deafness. *See* hearing loss
decibels (dB), 93, 95–7
defense system, of human body, 17–
18, 20, 21, 24, 204, 172. *See
also* immune system
degreasing, hazards of, 281
demolition operations, 176–7
Department of Labor. *See* Labor,
U.S. Department of
depression, mental, 105, 239. *See
also* stress
depression, of central nervous system. *See* nervous system, central, abnormalities
dermatitis. *See* skin allergies and
contact dermatitis; skin disease, dermatitis
dermis, 54
deTreville, Robert, 6
diabetes, 86
digestive system, 21, 34
disease. *See also* liver disease
gall bladder stones, 37
pancreatitis, 91
ulcerative colitis, 79
ulcers, 79, 85, 91, 239
lead, effects of, 252
noise, effects of, 105
phenol, effects of, 194
vanadium, effects of, 260
dilution ventilation. *See* ventilation
discs, spinal, 68, 70, 71
disease. *See* occupational disease;
heart disease, lung disease, etc.
dizziness. *See* nervous system, central, abnormalities
drowsiness. *See* nervous system,
central, abnormalities
drug workers, radiation exposure,
146
drugs, relief of stress with, 90–1
drunkenness. *See* alcoholic beverages; nervous system, central,
abnormalities
ducts (ventilation), 300–1

ear, 98–100
hearing, loss, 9–10, 94–6, 116–19

ear (*cont.*)
  protection, 112–13
  ringing in. *See* nervous system,
    central, abnormalities
ecology, 14–15. *See also* air pollution
edema, 42. *See also* lung disease,
    pulmonary edema
electrical shock, 43
electrocardiogram, 33
electromagnetic radiation, 132, 134,
    135
electrostatic precipitator, 332–3
emphysema. *See* lung disease,
    chronic bronchitis and emphy-
    sema; lung disease, emphysema
environment, 14–15. *See also* air
    pollution
Environmental Protection Agency,
    9, 15
epidemiology, 344
epidermis, 54
euphoria. *See* nervous system, cen-
    tral, abnormalities
eye, **135–7, 275**
  cornea damage
    acids and, 215, 218
    ammonia and aliphatic amines
      and, 163, 236–7
    arsenic and, 245
    carbon tetrabromide and, 207
    chlorobutadiene and, 226
    copper compounds and, **250**
    dimethyl sulfate and, 238
    epichlorohydrin and, 229
    radiation and, **138**, 278
    welding and, 278
  damage and injury. *See also*
    blindness
    alcohols and, 185
    radiation and, 137–53
    welding and, 277–8
  irritation. *See* irritation of mem-
    branes
  protection, 138, 143, 229, 240,
    284–8
  swollen lids. *See* irritation of
    membranes
eye disease
  cataracts
    radiation and, **137, 139, 140,**
      250–1
    welding and, 278
  conjunctivitis, 137, **138,** 213, 278,
    282. *See also* irritation of mem-
      branes
exhaust ventilation. *See* ventilation
explosimeter, 288. *See also* air sam-
    pling
explosions, prevention of, 299–300.
    See also ventilation

fans, 302–3
farmer's lung. *See* bagassosis
farm workers, effects of heat on, 123
fatigue, 80–3, 122
  olfactory, 24
fever, 194, 195
fibrosis, lung. *See* lung disease, scar-
    ring
filter tufts. *See* glomeruli
fingers, abnormalities in, 108–9, 232.
    See also musculo-skeletal sys-
    tem, abnormalities
fire, spontaneous combustion, 215
fire fighters, 163
fluorosis, 241–2
folliculitis, **62**
Food and Drug Administration
    (FDA), 180
foot drop. *See* nervous system, pe-
    ripheral, abnormalities
foot-candles, 132
fractures, spontaneous, 242, 247–8.
    See also musculo-skeletal sys-
    tem, abnormalities
frequency (of sound), 93, 94
frostbite, 129
fulminant disease, 18

gall bladder, 37
gamma rays. *See* radiation, ionizing
gangrene, 226
garlic odor, of breath, 258
gas workers, 165–6
gastro-intestinal system. *See* diges-
    tive system
Geiger counter, 151, 343–4
genetic damage, 137, **141,** 145, 164,
    190, 223, 227
giddiness. *See* nervous system, cen-
    tral, abnormalities
glassblowers, hazards, 137
glass workers, 128
glomeruli, 40, 41, 44, 45

glomerulitis, 45–6
grievance procedure, 264–5, 365–6.
    See also collective bargaining;
    work groups, informal
Grospiron, A. F., 15
grippe. *See* influenza

half-life (radioactivity), 132
hallucinations. *See* nervous system,
    central, abnormalities
hardening of arteries. *See* heart dis-
    ease
headache. *See also* nervous system,
    central, abnormalities
  alcohols and, 184, 185
  arsenic and, 246
  asphyxiating gases and, 279, 281
  benzene and, 190
  carbon monoxide and, 165
  diborane and, 247
  glycols and, 188
  halogenated hydrocarbon solvents
    and, 206
  hydrogen cyanide and, 165, 166
  lead and, 252
  light, excessive, and, 277
  manganese and, 254
  metal fume fever and, 248, 251,
    283
  migraine, 79
  nickel carbonyl and, 268
  phenol and, 194
  pyridine and, 187
  respirators and, 313
hearing loss, 9–10, 94–6, 116–19
  medical diagnosis, 117–20
heart attacks. *See* heart disease,
    hardening of arteries
heart disease, 26, 28–33, 42
  aliphatic nitro compounds and,
    234
  antimony and, 244
  bromine compounds and, 162
  carbon monoxide and, 165
  chlorobutadiene and, 226
  congestive heart failure, 32
  enlargement of heart, 168
  hardening of arteries, 30–1, 32,
    104, 105, 239, 258
  kidney disease and, 41
  medical diagnosis, 33
  microwave radiation and, 140, 141

  in silicosis, 168
  stress and, 79
heartbeat, irregular
  adrenaline and, 204
  barium and, 246
  halogenated hydrocarbon solvents
    and, 203–4
  nitroolefins and, 235
  tetrachloroethylene and, 206
heat
  medical effects, 29, 44, 56, 123–8,
    140
  protection from, 128
hemoglobin, 34–5, 37, 43–4
  carboxyhemoglobin, 164–5
  methemoglobin, 196–7
hemorrhage, 36, 191
hepatitis, 37–8, 91, 206, 237
hertz, 93, 94
high blood pressure. *See* blood pres-
    sure, high
hives. *See* allergies; skin allergies
    and contact dermatitis
hoarseness. *See* irritation of mem-
    branes
hoods, 288, 294–300
hormones, 28–9, 38, 223. *See also*
    adrenaline
housekeeping, in industrial opera-
    tions (asbestos), 178
hypersensitivity. *See* allergies

IHF. *See* Industrial Health Founda-
    tion
illumination, recommended levels,
    144. *See also* radiation, visible
    light
immersion foot, 129–30
immune system, of body, 58–9, 60
impingers, 328, 331–2
Industrial   Health   Foundation,
    (IHF), 6–7
influenza, 230, 243. *See also* metal
    fume fever; polymer fume
    fever
informal work groups, 74–5, 80, 89–
    90
infrared radiation, 135, 137, 139–40,
    277–8
inorganic compounds, 154, 239–61
interstitial fibrosis. *See* lung disease,
    scarring

intoxication. *See* alcoholic beverages; nervous system, central, abnormalities
intravenous pyelogram, 47
ion, 132, 133
ionizing radiation. *See* radiation, ionizing
irritability. *See* nervous system, central, abnormalities
irritation of membranes, 20–1, **23–5**
  acetates and, 220–1
  acids, mineral, and, 240–1
  acids, organic, and, 212–18
  alcohols and, 185
  aldehydes and, 211
  amines and, 236
  aromatic hydrocarbons and, 190–5
  aromatic nitro compounds and, 198–200
  boron compounds and, 244
  cadmium and, 248
  chromium compounds and, 249
  coal tar and, 200, 201–2
  copper compounds and, 250
  cyanogen and, 166
  ethers and, 186–7
  germanium and, 251
  glycols and, 188
  halogenated hydrocarbon solvents and, 204–6
  irritating gases and, **157, 158–64**
  ketones and, 210
  lithium and, 253
  manganese and, 254
  molybdenum and, 256
  noise and, 106
  osmium and, 257
  plastics manufacture and, 225–8
  selenium and, 258
  tellurium and, 259
  vanadium and, 260
  welding operations and, 280–2
itching. *See* skin disease

jaundice, **35**, 37, **38–9**

keratin, 54, 55
kidney disease, 28, 34, **41–6**
  acidosis, **41**
  alcohols and, 184
  aromatics and, 191, 195
  arsine and, 245

beryllium and, 180
blood disease, as cause of, **35**
boron compounds and, 247
cadmium and, 248
cancer, **46**
chlorobutadiene and, 226
copper and, 251
N,N-dimethyl formamide and, 214
dioxane and, 187
diphenyls and, 192–3
freons and, 232
glycols and, 188–9
halogenated hydrocarbon solvents and, 203–8
hydrazine and, 237
ketones and, 209, 211
lead and, 252
lithium and, 253
medical diagnosis, 47
mercury, 255–6
nitroparaffins and, 235
phenol and, 194
phosphine and, 280
silver and, 258
stones, 248
uremia, **41, 43**
kidneys, 40–1

Labor, U. S. Department of, funding, 6
  Occupational Safety and Health Act, 9, 12
  occupational health statistics, 3, 13
lasers, 137, 142–3
leukemia, **36**, 148, 190
life expectancy decrease, 149
lifting and carrying, 71–3
  recommended weights, 73
ligaments. *See* musculo-skeletal system
light, sensitivity to. *See* skin disease, photosensitivity
liver disease, 21, **36–9**
  acrylamide and, 215
  alcohols and, 195
  allyls and, 225
  aromatics and diphenyls and, 193, 195
  arsine and, 245
  ascites, **38**

beryllium and, 180, 247
boron compounds and, 247
butadiene and, 226
cancer, 213, 222
cirrhosis, **38**, 91
copper compounds and, 251
N,N-dimethyl formamide and, 214
dioxane and, 187
ethyl benzenes and, 230
freons and, 232
glycols and, 188
halogenated hydrocarbon solvents
and, 203–8
hepatitis, **37–8**, 91, 206, 237
hydrazine and, 237
jaundice, 37, **38–9**
ketones and, 209–11
nitriles and, 166
nitroparaffins and, 235
phthalic anhydride and, 218
selenium and, 258
solvents and, 183
locker room, cleanliness, 179
lung allergies and asthma, **25**, 79
acids, organic, and, 218
chromic acid and, 249
cobalt and, 250
diborane and, 247
dust, organic, and, 181–3
epoxides and, 228
"ferric" compounds and, 251
formaldehyde and, 211
nickel and, 256
para-phenylenediamine and, 198
phthalate and, 213
platinum and, 257
toluene diisocyanate and, 231
lung cancer
asbestos and, 173–5, 350–1
arsenic and, 245
chromium and, 249
cigarette smoking and, 31, 174,
348–9
nickel and, 256
oils and, 203
lung disease
asbestos and, 10, 172–3, 345. *See
also* lung cancer
bagassosis, 182–3
berylliosis, 180–1
black lung, 7, **169**
bronchitis, acute, 159–64

byssinosis, 6–7, 181–2
infections, **26**, 86
irritation. *See* irritation of membranes
medical diagnosis, 27–8
pneumoconiosis, **154**, 169
carbon black and, 171
cobalt and, 249
progressive massive fibrosis, **168**
shaver's disease, 169, 243
silicosis, **167–9**, 171, 240
lung disease, chronic bronchitis and
emphysema, **19**, **25–6**
acids and alkalies and, 215, 240,
241, 242
acrolein and, 225
amines and, 236
chlorobutadiene and, 226
cold temperatures and, 130
cyanogens and, 163–4
diboranes and, 247
dusts and, 169, 182, 184
epoxides and, 228–9
ethers, 186–7
ethylene dibromide, 208
formaldehyde, 211
irritating gases and, **159–64**, 280–1
ketones and, 209
shift work and, 86
zinc compounds and, 260
lung disease, emphysema, **25–6**. *See
also* lung disease, chronic
bronchitis and emphysema
cadmium and, 248
platinum and, 257
pneumoconiosis and, 169, 171
lung disease, pneumonia, **25**
amines, aliphatic, and, 236
ethers and, 188, 228
cadmium and, 248
manganese and, 254
osmium and, 257
lung disease, pneumonia, chemical
acids and alkalies and, 240–1
antimony and, 244
fibrous glass and, 179
kerosene and, 203
uranium compounds and, 259
. zinc compounds and, 260
lung disease, pulmonary edema, **24,
42, 154**
acetates and, 221

lung disease, pulmonary edema, (*cont.*)
  acids and alkalies and, 241–2
  aliphatic amines and, 236
  bromoethane and, 207
  cadmium and, 248
  coal tar and, 201
  cyanogens and, 160–4, 166
  ethers and, 180, 228
  ethylene dibromide and, 208
  glycols and, 188
  irritating gases and, **159**, 275, 280–2
  mercaptans and, 237
  methylene chloride and, 204
  titanium and, 259
lung disease, scarring, **154**
  asbestos and, **26–7**, **172–3**
  beryllium and, 180–1
  fibrous glass and, 179
  organic chemicals and, 183
  platinum and, 257
  sulfuric acid and, 242
  tantulum and, 258
  titanium and, 259
  zinc and, 260
lung function tests, 27–8
lungs. *See* respiratory system

magnesium workers, 160
manganism, 254–5
measurements, of ventilation system, 305. *See also* air sampling
meningitis, 50
menstruation, irregular, 231. *See also* hormones
mental abnormalities, 105, 239. *See also* nervous system, central, abnormalities; stress
mental health, 81. *See also* stress
mesothelioma, 173–4, 180, 350–1
metabolism, body, **162–3**, **165–6**
  interruption of, 212
metal fume fever, **154**, **243**, 248, 250, 253, 258, 259, 260, 283
methemoglobinemia, 154, 196–7, 199–200, 237
microwaves, 137, 140–2, 146
migraine headache, 79
miners, 7, 123, 170
Monday fever. *See* byssinosis
monitoring equipment, 288. *See also* air sampling

monotony, 80, 81
mouth, sores. *See* irritation of membranes
mucous membranes, 22, 24. *See* irritation of membranes
muscles. *See* musculo-skeletal system
musculo-skeletal system, 67–74
  abnormalities, 232. *See also* back injuries and disease
    bone pain, 248
    finger bones, 232
    fluorosis, 241–2
    muscle contraction (tetany), 254
  diseases
    arthritis, **107**, **108**, 242
    cancer, 148
    tendonitis, **75**, **108**
    vibration effects, **106**, **108**
myoglobin, 43

Nader, Ralph, 43
narcotics, 157, 190, 191, 204, 206, **208**, 209. *See also* nervous system, central, abnormalities
nasal septum, perforation of, 241, 243, 249, 250
nausea
  alcohols and, 184
  aliphatic amines and, 236–7
  aminothiazole and, 238
  antimony and, 244–5
  arsine and, 246
  boron compounds and, 247
  ethers and, 186
  ethyl benzenes and, 230
  halogenated hydrocarbon solvents and, 206
  metal fume fever and, 250, 251, 279
  nitrogen oxides and, 160
  phenol and, 194
  plastics manufacture and, 225
  tellurium and, 258
  vanadium and, 260
nervous system, **47–54**
nervous system, autonomic, 50–2
nervous system, central, abnormalities, **49–50**
  acetates and, 220
  acetylene and, 158–280
  alcohols and, 90–1, 185
  aliphatic amines and, 234, 236–7

aliphatic nitro compounds and, 233–4, 235
acrylamide and, 141, 214
aromatic hydrocarbons and, 190–5
arsenic and, 246
boranes and, 247
benzene and, 190–1
carbon disulfide and, 238
carbon monoxide and, 165
chlorobutadiene and, 226
cyanide and, 166
epichlorohydrin and, 229
ethers and, 186
ethyl benzenes and, 230
glycols and, 186
halogenated hydrocarbon solvents and, 203–8
hydrogen sulfide and, 161
ketones and, 210–11
metals and, 12, 252, 254–6, 258, 260, 283
nicotine and, 223
oxygen, lack of, 308
pesticides and, 221–2, 223
phenol and, 192–4
phosphate plasticizers and, 233
pyridine and, 187
toluene and, 191
vibration and, 105
nervous system, central, stimulation and depression. *See* nervous system, central, abnormalities
nervous system, peripheral, abnormalities, **48–9**
carbon disulfide and, 239
lead and, 252
organophosphate insecticides and, 222
thallium and, 259
tri-ortho-cresyl phosphate and, 233
neuritis. *See* nervous system, peripheral, abnormalities
neuropathy. *See* nervous system, peripheral, abnormalities
neutrons. *See* radiation, ionizing
New York Times, 15
noise, 29, 52, 79, 81, 83, **92–120**
control, 109–12
ear protection, 112–13
exposure standard, 9–10, 116
hearing loss, 9–10, 94–6, 116–19
monitoring, 115, 340–41

speech interference, 106
typical industrial levels, 96–7
non-ionizing radiation, 132, 135
nose
irritation. *See* irritation of membranes
perforation of nasal septum, 241, 243, 249, 250
nosebleed. *See* irritation of membranes
nutrition, lack of, 81

OCAW. *See* Oil, Chemical and Atomic Workers International Union
occupational acrosteolysis, 232
occupational disease. *See also* heart disease, lung disease, etc.
acute reactions, 18–19, 343
to cold, 129–30
to gases, 158–9
to heat, 125–7
lungs, 24–5
chronic effects, 18–19, 345–6
to gases, 158–9
to heat, 127–8
lungs, 25–7
Occupational Safety and Health Act of 1970 (OSHA)
enforcement of, 8, 9, 11, 12
imminent danger clause, 12–13
legislative action, 8
passage of, 7–8
rights under, 7–8
standards, 9
use of, 360
odors. *See* smell, sense of
Oil, Chemical and Atomic Workers International Union (OCAW), 12, 15, 263–4, 363
olfactory fatigue, 24
optic nerve damage, 239. *See also* blindness
organic chemicals, 154, 183–239
OSHA. *See* Occupational Safety and Health Act of 1970
oxygen
content in blood, **26–7, 28,** 34, 35
high, **165–6**
low, 29, 42, **165, 172,** 180, 183, **196,** 204, 235, 237, 248, 256, 279, 280, 308

oxygen (*cont.*)
  lack of, to body tissues, **165–6**, 194–5

painters, 191
pancreatitis, 91
paralysis. *See* nervous system
Parkinson's disease, 254–5, 283. *See also* nervous system, central, abnormalities
patch tests, 61
photosensitization. *See* radiation, ultraviolet; skin diseases, photosensitivity
pink-eye. *See* eye disease, conjunctivitis
plasma, 34
platelets, 33, 34, 36
pleura, 27
plexuses, nerve, 48
pneumoconiosis. *See* lung disease, pneumoconiosis
pneumonia. *See* lung disease, pneumonia; lung disease, pneumonia, chemical
pollution, air. *See* air pollution
polymer fume fever, 230
portal system, 36
pregnancy, effect of exposure to chemicals, 190, 193
progressive massive fibrosis, **168**
protection, body system of. *See* defense system, of human body
protection, personal
  barrier lotions, 62
  clothing, 56, 179
  disadvantages, 106, 113, 313, 318–21
  ears, 113
  eyes, 138, 143, 229, 240, 284–8
  from heat, 128, 315
  respirators, 245, **306–26**. *See also* noise; radiation (all types); ventilation
pulmonary disease. *See* lung disease
pulmonary edema. *See* lung disease, pulmonary edema
pulmonary fibrosis. *See* lung disease, scarring
pulmonary function tests, 27–8

rad, 132
radar, 140
radiation, electromagnetic, 132, 134, 135
radiation, infrared, 135, 137, 139–40
  exposure in welding, 277–8
radiation, ionizing, 18, 36, **132**, 133, 134, **145–53**, 227
  exposure standard, 149, 153
  medical effects, 18, 36, 63–4, 137–8, 141, 145, 147–50
  monitoring, 343–4
  protective measures, 147, 150–1
radiation, lasers, 137, 142–3
radiation, microwaves, 137, 140–2, 146
  exposure standard, 142
  fire hazard, 142
radiation, non-ionizing, 132, 135–45
radiation, radiofrequency waves, 137, 142, 146
radiation, ultraviolet, 63, 135, 136–9
  exposure standard, 138
  protective measures, 138
  in welding, 139, 278
radiation, visible light, 143–5, 277–8
radioactivity. *See* radiation, ionizing
radiofrequency waves, 137, 142, 146
record keeping, medical, 350–1
red blood cells, 33–6
  rupture of, **35**, **43**, 237, 245, 251, 280
refinery workers, exposure to sulfur dioxide, 160–1
rem, 132. *See also* radiation
repetition, as source of stress, 79–80
respirators, 245, **306–26**
respiratory system, 22–8
  allergies. *See* lung allergies and asthma
  disease. *See* lung disease
  inflammation. *See* irritation of membranes
  pleura, 27
  protective mechanisms, 24
  upper airways, 22–4
road builders, 123
roentgen, 132, 134
Rubber Workers, International Union of, 362

sampling train, 328–9
saws, power, ventilation of, 177. *See
also* ventilation
scintillation counters, 151
seizures. *See* nervous system, central,
abnormalities
sensitizer. *See* allergies
shaver's disease, 169, 243
shift work, 83–7
shock, electrical, 43
siderosis, 181
silicosis, **167–9**, 171, 240
sinuses (nasal), cancer, 256
sinusitis. *See* irritation of membranes
skin, 20, **54–67**
blisters, 162, 201, 213, 236
cold damage, 128–30, 226
irritation, direct, 56–8. *See also*
irritation of membranes
skin alleries and contact dermatitis,
58–61
acids, organic, and, 211–13
aminothiazole and, 231
aniline derivatives and, 197–9
chromates and, 249
copper and, 250
epoxides and, 227
hexamethylenetetramine and, 226
nickel and, 256, 282
platinum and, 257
thallium and, 259
vinyl acetate and, 220
zinc and, 260
skin burns, **56**
acids, organic, and, 218
acids and alkalies, mineral, and,
240–2
amines, aliphatic, and, 236
halogens and, 161–2
lithium hydride and, 253
radiation and, 137, 139, 142
selenium and, 258
tin compounds and, 259
skin disease
acne, 67, 162, 192–3, 196
beryllium and, 180
cancer, 62–4
arsenic and, 245
oils and, 203
radiation and, 138

dermatitis, **56–61**
acids, organic, and, 213, 217
acrylics and, 225
alcohols and, 184–5
aminothiazole and, 238
antimony and, 243
aromatic hydrocarbons and,
191–5
chlorinated hydrocarbons and,
206, 208, 211, 222
coal tar and, 201
ethers and, 186
fibrous glass and, 179, 227, 230
osmium and, 257
phosgene and, 163
polyamides and, 230
selenium and, 256
solvents and, **183**
sulfur compounds and, 238–9
tellurium and, 258
zinc and, 260
eczema, 67
folliculitis, **62**
medical diagnosis, **60**
photosensitivity, 154
coal tar and, 201
radiation and, 132, 138
prevention, 62, 64–7
psoriasis, 67
sleep, disturbances of. *See* nervous
system, central, abnormalities;
stress
sleepiness. *See* fatigue; nervous system, central, abnormalities
smell, sense of (function), 23–4
smelters (exposure to sulfur dioxide), 160
Smoking. *See* cigarette smoking, effects of
sound-level meter, 115, 340–1
spinal cord, 48–9, 50, 52, 68
spine, 68–72
spleen, disease of, 209
sputum. *See* lung disease
staggering. *See* nervous system, central, abnormalities
standards, exposure to toxic conditions, 155–7, 261–74
stanosis, 181
steelworkers, 123, 137
stillbirths, 193. *See also* genetic damage

stress
  biological response to, 77–8, 127–8
  emotional, 87–8
  sources of, 51, 52, 79–89, 101,
    123–6. *See also* heat; noise
sunlight, 62, 63. *See also* radiation,
    ultraviolet; skin disease, pho-
    tosensitivity

Teamsters Union, Local 688, 356
teeth, corrosion of, 226, 241, 242.
    See also irritation of mem-
    branes
temperature. *See* cold, heat
  recommended for work, 122
tendonitis, **75, 108**
tendons, 75, 108
tetany, 254
textile workers, lung disease in, 6–7,
    181–2
tinnitis (ringing in ears). *See* ner-
    vous system, central, abnor-
    malities
throat irritation. *See* irritation of
    membranes
thyroid gland, abnormalities, 141,
    162–3, 238
trachea, 22–3
tremor. *See* nervous system, central,
    abnormalities
tubules, **40–1**, 43, 44, 45, 46
tunnel workers, 33, 168

ulcers, 79, 85, 91, 239
  of mouth. *See* irritation of mem-
    branes
ultraviolet radiation. *See* radiation,
    ultraviolet
Union Carbide Corporation, 15
United   Automobile   Workers
    (UAW), 266, 366
United Paper Workers International
    Union, 262, 362
universal tester, 288, 337–8

upper respiratory tract. *See* respira-
    tory system, upper airways
uranium workers, 148
uremia, **41, 43**
urinary bladder. *See* bladder, uri-
    nary

velometer, 288, 336
ventilation, 288, **292–306**
  of asbestos operations, 176–8
  assessing efficiency, 178, 303–6
  equipment
    ducts, 300–1
    fans, 302–3
    hoods, 288, 294–300
    industrial vacuum cleaner, 177,
      178
vertebrae, 68, 70, 74
vibration, 44, 46, 106–9, 232
  control, 92, 114
visible light radiation, 143–5, 277–8
vision, loss of. *See* blindness
vomiting. *See* nausea

weakness. *See* nervous system, cen-
    tral, abnormalities
welding, **285–6**
  chemical hazards, 163–4, 278–83
  control of hazards, 284–9
  physical hazards, 137–8, 146, 276,
    277
  safety, sample survey form, 287
white blood cells, 33, 34, 35–6
woodworkers, 137
work clothes, 179
work groups, informal, 74–5, 80,
    89–90
workmen's compensation, 7, 18, 19
wrist drop. *See* nervous system, pe-
    ripheral, abnormalities

X-rays, 27. *See also* radiation, ion-
    izing

Zinc chills. *See* metal fume fever

Index of Substances

acetaldehyde, 57, 211, 261
acetamide, 213
acetanilide, 35
acetates, 53, 220
acetic acid, 57, 215, 261
acetic anhydride, 57, 218, 261
acetone, 57, 209, 261
acetonitrile, 39, 57, 166, 261
acetophenone, 210
2-acetyl-aminofluorene, 197
acetylkentene. *See* acetophenone
acetylene, 158, 280
acetylene black. *See* carbon black
acetylene dichloride. *See* 1, 2-di-
  chloroethylene
acetylene tetrabromide, 261
acetylene trichloride. *See* trichloroethylene
acid anhydride, 212
acid gases, 336
acids, 24–5, 55, 56, 240–2
  organic, 211–19
acridine, 57, 201
acrolein, 57, 225, 261
acrylamide, 213, 261
acrylates, 212, 225
acrylic acid, 216
acrylic aldehyde. *See* acrolein
acrylonitrile, 39, 57, 166, 231, 261
acrylics, 225
adipic acid, 217
alcohols, 53, 184–6. *See also* alcholic
  beverages, in General Index
aldehydes, 210–12
aldrin, 221, 261
aliphatic compounds, 154

alkalies, 24–5, 56, 240–1, 336
alkyds, 225
allyl acetate, 226
allyl alcohol, 57, 185, 225, 261
allyl amine, 236
allyl bromide, 226
allyl chloride, 206, 226, 261
allyl formate, 226
allyl glycidyl ether, 228, 261
allyl phthalate, 226
allyl propyl disulfide, 261
allyl resins, 225
alpha- compounds. *See* listing by
  second word in name
aluminum, 169, 243
aluminum fluoride, 241–2
aluminum salts, 57
amines, 236
amines, aromatic, 59
4-aminodiphenyl, 197
2-aminoethanol. *See* ethanolamine
aminophenols, 199
2-aminopyridine, 187, 261
amino resins, 226
aminothiazole, 238
ammonia, 20, 24–5, 57, 163, 231, 250,
  261, 336
ammonium bichromate, 59
ammonium sulfamate (ammate), 261
amyl acetate, 57, 220, 261
amyl alcohol, 57, 185
amyl nitrate, 233–4
aniline and derivatives, 20, 35, 57,
  154, 196, 198, 199, 261
anisidine, 261
anthracene, 63, 201

anthracite. *See* coal dust
antimony and compounds, 39, 57, 243, 261
antioxidants, 59
ANTU (alpha-naphthyl thiourea), 261
argon, 158
aromatic acids, 214
aromatic compounds, 154
aromatic hydrocarbons, 189–200
arsenic and compounds, 44, 53, 55, 57, 63, 243, 244, 250, 261
arsine, 35, 39, 53, 57, 245–6, 261, 280
asbestos, 171–9, 224, 261, 347, 349–51
   contamination of air-conditioned air, 175
   exposure standard, 10–11, 172, 261
   medical effects, 10, 18, 21, 26–7, 172–4
   protective measures, 174–9
asphalt, 201–2
azelaic acid, 217
azinphos-methyl, 261
10-azoanthracene. *See* acridine

barium and compounds, 57, 181, 246, 261
bases. *See* alkalies
bauxite, 169. *See also* aluminum
benzaldehyde, 231
benzanthracenes, 202
benzanthrone, 196
benzene, 35, 36, 39, 57, 189–90, 272
benzene hexachloride, 221
benzenes, chlorinated, 39, 57
benzidine, 57, 197, 199
para-benzoquinone. *See* quinone
benzoyl peroxide, 262
benzyl alcohol, 185
benzyl chloride, 57, 195, 262
beryllium and compounds, 21, 39, 57, 180–1, 246, 272
beta- compounds. *See* listing by second word in name
bismuth subnitrate, 57
bisphenol. *See* diglycidyl ether
bismuth, 39, 44, 46, 246–7, 261
bithionol, 59
bitumen. *See* asphalt
boranes, 247
boron, 247
boron compounds, 57, 262

bromine, 57, 126, 262
bromoacetic acid, 216
bromoacetone, 210
brass, 282
bromoform, 39, 207, 262
bromomethane. *See* methyl bromide
burnt lime. *See* calcium oxide
butadiene, 57, 226, 262
butadiene oxide, 228
butane, 158
butanediols, 189
butanols, 185
2-butanone, 262
2-butoxyethanol, 262. *See also* cello-solves
butyl acetate, 57, 220, 262
n-butyl alcohol, 57, 262
butylamine, 236, 262
butyl cellosolve, 57
tert-butyl chromate, 262
butylene oxides, 228
n-butyl glycidyl ether (BGE), 228, 262
n-butyl mercaptan, 238, 262
2-butyl-3-one, 39
para-tert-butyltoluene, 262
3-butyn-2-one, 209
butyramide, 213
butyric acid, 215
butyric anhydride, 218
gamma-butyrolactone, 219

cadmium and compounds, 39, 44, 46, 57, 247–8, 272, 282
cadmium oxide, 158
calcium arsenate, 262
calcium cyanamide, 166
calcium cyanide, 165
calcium fluoride, 241
calcium hydroxide, 240
calcium oxide, 57, 240, 262
camphene, chlorinated. *See* toxaphene
camphor, 262
caproic acid, 215
epsilon-caprolactam, 230
caprylic acid, 215
carbamic acid, esters, 224
carbaryl, 262
carbolic acid. *See* phenol
carbon black, 63, 64, 170–1, 262

carbon dioxide, 29, 158, 240, 262, 279, 336
carbon disulfide, 30, 33, 44, 45, 53, 57, 238–9, 272, 280
carbon monoxide, 29–30, 31, 33, 34–5, 42, 44, 53, 164–5, 204, 231, 240, 262, 279–80, 336
carbon tetrabromide, 39, 207
carbon tetrachloride, 37, 39, 43, 44, 57, 163, 183, 205, 272
carbonic acid, 279
carbonyl chloride, 57
carboryl, 224
carboxylic acids, 211
catalysts, for plastics, 59
caustic chemicals, 24–5, 55. *See also* potassium hydroxides
cellulose and derivatives, 226
cellulose acetate, 226
cellosolves, 57, 187
cement, asbestos-containing, 177, 178, 274
cerium, 248
cesium, 248
cesium oxide, 248
channel black. *See* carbon black
chloral hydrate, 210
chlordane, 221, 262
chloride of lime, 162
chlorinated benzenes, 39, 57
chlorinated camphene, 262
chlorinated diphenyl oxide, 262
chlorinated diphenyls, 39
chlorinated hydrocarbons, 37, 53, 57, 62, 221
chlorinated lime, 57
chlorinated naphthalenes, 39, 196
chlorine, 158, 161–2, 262, 309, 336
medical effects, 19, 20, 24–5, 57
chlorine dioxide, 262
chlorine trifluoride, 262
chloroacetaldehyde, 262
chloroacetic acid, 216–17
alpha-chloroacetophenone, 262
chloroanilines, 199
chlorobenzene, 57, 195, 262
ortho-chlorobenzylidene malononitrile, 262
chlorobromomethane, 262
chlorobutadiene, 39, 226, 262
chlorodiphenyls, 193, 263

beta-chloroethanol. *See* ethylene chlorohydrin
beta-chloroethyl alcohol, 185
chloroform, 39, 57, 205, 263, 271
chloromethyl ether, 186
1-chloro-1-nitropropane, 57, 263
5-chloro-ortho-toluidine, 199
chloropicrin, 207, 263
chloroprene, 263
alpha-chloropropionic acid, 215
2-chloropyridine, 187
chlorosulfuric acid, 238
chlorotrifluoroethylene, 232
chlorpromazine, 38
chlorthion, 224
chromates, 249, 273. *See also* ammonium bichromate; potassium bichromate; sodium bichromate
chromic acid, 57, 249, 273
chromium and compounds, 44, 46, 57, 249, 263, 282
citraconic anhydride, 218
citric acid, 217
coal dust, 169–70, 274
coal tar and derivatives, 55, 57, 63, 201, 263
cobalt, 214, 250, 263
cobalt acetate, 250
cobalt carbonyls, 250
cobalt compounds, 57, 59
columbium, 250
contraceptives, oral, 39
copper and compounds, 57, 250, 263
copper cyanide, 165
copper oxide, 250
corundum, 169, 190
cotton dust, 6–7, 19, 181–2, 263. *See also* byssinosis, in General Index
Crag herbicide, 263
creosote, 63, 170, 201
cresol, 39, 57, 263
cristobalite, 131
crotonaldehyde, 263
crotonic acid, 216
crotonic anhydride, 218
cryolite, 241–2
cumene, 191, 263
cyanide and compounds, 165–6, 263
cyclohexamine, 236
cyclohexanol, 57, 185, 263

cyclohexanone, 57, 210, 263
cyclohexene, 57, 263

DDT, 39, 221, 263
DDVP, 224, 263, 264
decaborane, 247, 263
decahydronaphthalene. *See* Decalin
Decalin, 191–2
Demeton, 224, 263
diacetone alcohol, 263
diatomite, 169, 224
diazinon, 224
diazomethane, 264
dibenzanthracenes, 202
diborane, 247, 264
dibutylphthalate, 264
ortho-dichlorobenzene, 57, 264
para-dichlorobenzene, 195, 264
dichlorodifluoromethane, 264
1,3-dichloro-5,5-dimethyl hydantoin, 264
1,1-dichloroethane, 264
1,2-dichloroethylene, 57, 264
sym-dichloroethyl ether, 186–7, 264
dichloroisopropyl ether, 187
dichloromonofluoromethane, 264
1,1-dichloro-1-nitroethane, 57, 264
dichlorotetrafluoroethane, 264
dicyclopentadiene dioxide, 228
dieldrin, 221, 264
diesel oil, 63
diethyl acetic acid, 215
diethylamine, 264
diethylamine ethanol, 264
N,N-diethyl aniline, 199
N,N-diethyl formamide, 214
di-2-ethyl hexyl phthalate, 233
diethylene glycol, 188
diethylene ether. *See* dioxane
diethylene glycol ether, 189
difluorodibromomethane, 264
diglycidyl ether, 228, 229, 264
diisobutyl ketone, 20, 264
diisocyanates, 231
diisopropylamine, 264
diisopropyl ketone, 209
N,N-dimethyl acetamide, 214, 264
N,N-dimethyl acrylamide, 214
dimethylamine, 236, 264
N,N-dimethylaniline, 57, 199, 264
dimethyl benzanthracenes, 202

dimethyl-1,2-dibromo 2,2-dichloro-ethyl phosphate, 264
dimethyl formamide, 39, 57, 214, 264
N,N-dimethyl formamide, 214
dimethyl formamide, 57
1,1-dimethylhydrazine, 57, 265
dimethyl phthalate, 233, 265
dimethyl sulfate, 39, 57, 238, 265
dinitrobenzene, 57, 265
dinitro-ortho-cresol (DNOC), 194, 265
dinitrophenol, 39, 57, 195
dinitrotoluene, 57, 265
dioxane, 39, 57, 187, 265
diphenyl aniline, 199
diphenyl ether, 193
diphenylmethane, 193
diphenyls, 192, 265
   chlorinated, 39
dipterex, 224
dipropylene glycol methyl ether, 265
di-sec-octyl phthalate, 265
dithiocarbamates, 223
DMA. *See* dimethyl amine
DNOC. *See* dinitro-ortho-cresol
dodecyl mercaptan, 238
Dowanol EE, 188
Dowanol EM, 188
Dowtherm, 193
dusts, 21, 26–7, 167–83, 274, 295–9.
   See also air sampling, in General Index
dyes, 224, 298. *See also* aniline and derivatives

endrin, 221, 265
EP-207. *See* dicyclopentadiene dioxide
epichlorohydrin, 39, 57, 229, 265
EPN, 224, 265
EPON 562, 229
epoxides, 227
epoxy resins, 59
epsilon- compounds. *See* listing by second word in name
ethane, 158
ethanol, 184–5. *See also* ethyl alcohol
ethanolamine, 236, 265
ethers, 53, 186–9
ethion, 224

2-ethoxyethanol, 265
2-ethoxyethylacetate, 265
ethyl acetate, 57, 220, 265
ethyl acrylate, 265
ethyl alcohol, 39, 265. *See also* ethanol
ethylamine, 236, 265
ethyl-sec-amyl ketone, 209, 265
ethyl benzene, 57, 265. *See also* polystyrenes
etnyl bromide, 207, 265
ethyl butyl ketone, 209, 265
ethyl chloride, 57, 265
ethylene, 158
ethylene chlorohydrin, 39, 57, 265
ethylenediamine, 57, 236, 265
ethylene dibromide, 39, 57, 207–8, 266, 272
ethylene dichloride, 39, 57, 266, 272
ethylene glycol, 44, 188
ethylene glycol dinitrate, 233–4, 266
ethylene glycol ethers, 188
ethylene glycol monomethyl ether, 39, 266
ethyleneimine, 236, 266
ethylene oxide, 229, 266
ethyl ether, 186, 265
ethyl formate, 57, 265
2-ethylhexanoic acid, 215
ethylidine chloride. *See* 1,1-dichloroethane
ethylidene dichloride, 205
ethyl mercaptan, 238, 265
N-ethylmorpholine, 266
ethyl nitrite, 235
ethyl silicate, 39, 265

Ferbam, 223, 266
ferric chloride, 251
ferric sesquichloride, 251
ferricyanates, 166
ferrocyanates, 166
ferrous sulfate, 39
ferrovanadium dust, 266
fibrous glass, 175, 179, 227, 230
filler, 224. *See also* asbestos; diatomite; mica; sand
fluoroacetic acid, 217
fluoride, 266, 272
fluorine, 158, 161, 266, 309, 336
fluorine compounds, 57
fluorosilicic acid, 241–2

fluorotrichloromethane, 266
fluothane, 39
formaldehyde, 57, 211, 226, 230, 231, 272
formalin, 59
formamide, 213
formic acid, 57, 215, 266
freon, 57, 231
fuel oils. *See* amyl alcohol
fumaric acid, 217
furfural, 57, 211, 266
furfuraldehyde. *See* furfural
furfuryl alcohol, 266
furnace black. *See* carbon black

gallium, 251
gamma- compounds. *See* listing by second word in name
gases, 157–66
    air sampling, 335–42. *See also* ventilation, in General Index
gasoline, 57
    additives, 205. *See also* ethylene dichloride
germanium and compounds, 251
glycidol, 266
glycols, 187–8, 266
graphite, 171, 274
guaiacol, 187
guthion, 224, 266

hafnium, 266
HETP, 224
halogens, 154, 161–3. *See also* bromine; chlorine; fluorine; iodine
helium, 158
heptachlor, 221, 266
heptane, 266
heptanoic acid, 215
hexachloroethane, 266
hexachloronaphthalene, 266
hexachlorophene, 59
hexamethylene tetramine, 57, 226
hexane, 266
2-hexanone, 266
sec-hexyl acetate, 266
hydrazine and derivatives, 39, 57, 237, 266, 336
hydrocarbons, brominated, 53
hydrocarbons, chlorinated, 37, 53, 57, 62, 221

hydrocarbons, halogenated, 203–8, 336
hydrochloric acid, 240–1. *See also* acids; hydrogen chloride
hydrocyanic acid, 57. *See also* hydrogen cyanide
hydrofluoric acid, 282. *See also* hydrogen fluoride
hydrogen, 158
hydrogen bromide, 241, 266
hydrogen chloride, 57, 238, 266. *See also* hydrochloric acid
hydrogen cyanide, 53, 165–6, 231, 266, 336. *See also* hydrocyanic acid
hydrogen fluoride, 57, 230, 241, 272, 336. *See also* hydrofluoric acid
hydrogen peroxide, 57, 266
hydrogen selenide, 57, 166, 266
hydrogen sulfide, 26, 57, 161, 273, 280, 309
hydroquinone, 57, 194, 266
hydroxy- compounds (acids), 212
hydroxylamine, 237
hypnone. *See* acetophenone

insecticides. *See* pesticides
indium, 251
INH. *See* hydrazine and derivatives
iodine, 57, 162, 266
iodoacetic acid, 217
iron, 181, 251
iron carbonyl, 251
iron oxide, 251, 267
isoamyl acetate, 267
isoamyl alcohol, 267
isoamyl nitrate, 236
isobutyl acetate, 267
isobutyl alcohol, 267
isobutyric acid, 215
isobutyronitrile, 166
isocyanates, 166, 231
isolan, 224
isonitriles, 166
isophorone, 210, 267
isopropyl acetate, 220, 267
N-isopropyl acrylamide, 214
isopropyl alcohol, 267
isopropylamine, 267
isopropyl ether, 186, 267
isopropyl glycidyl ether (IGE), 267

isovaleric acid, 215
itaconic acid, 217

kerosenes, 57, 62, 203
ketals, 210–11
ketene, 267
keto-acids, 212
ketones, 53, 57, 161, 162, 227

lacquer, 185
lamp black, 63
lauramide, 213
lanthanum, 251
lead, 30, 35, 44, 45, 48, 53, 214, 252–3, 272, 283
lead arsenate, 223, 267
lead arsenite, 223
lignite, 63
lime. *See* calcium oxide
lindane, 221, 267
linoleic acid, 215
linseed oil. *See* linoleic acid
lithium, 253
lithium hydride, 253, 267
LPG (liquified petroleum gas), 267
lye, 284, 288

magnesium, 253
magnesium oxide, 253, 267
malathion, 224, 267
maleic acid, 217, 218
maleic anhydride, 218, 230, 267
malonic acid, 218
manganese, 39, 53, 214, 254, 267, 282
MEK. *See* methyl ethyl ketone
melamine, 226
mercaptans, 57, 237
mercury, 12, 44, 45, 48, 53, 255, 273, 336
mercury compounds, 57, 59, 255
mesityl oxide, 39, 209, 267
mesitylene, 191
meta- compounds. *See* listing under second word in name
metals, 39, 53, 242–61
methacrylic acid, 216
methacrylonitrile, 166
methane, 158
methanol, 183, 231. *See also* methyl alcohol
methoxychlor, 221, 267
methoxyfluorane, 39

methyl acetate, 220, 267
methyl acetylene, 267
methyl acrylamide, 214
methyl acrylate, 267
methylal, 267
methyl alcohol, 39, 267. *See also* methanol
methyl amine, 236, 267
methyl-n-amyl ketone, 209, 267
methyl benzanthracenes, 202
methyl bromide, 57, 207, 267
methyl-n-butyl ketone, 209, 267
methyl cellosolve, 57, 267. *See also* cellosolves
methyl cellosolve acetate, 57, 267. *See also* cellosolves
methyl chloride, 35, 39, 57, 204, 231, 272
methyl chloroform, 267
methyl cyanide. *See* acetonitrile
methylcyclohexane, 210, 267
methylcyclohexanol, 267
ortho-methylcyclohexanone, 268
methylene bisphenyl isocyanate, 268
methylene chloride, 204–5, 272
methyl ether, 186
methyl ethyl ketone, 209, 268
methyl formate, 57, 268
methyl iodide, 268
methyl isobutyl carbinol, 268
methyl isobutyl ketone, 209, 268
methyl isocyanote, 268
methyl isopropenyl ketone, 208–9
methyl mercaptan, 238, 268
methyl methacrylate, 230, 268
methylmethacrylate, 225, 268
methyl nitrite, 234
methyl parathion, 224
methyl-n-propyl ketone, 209, 268
alpha-methyl styrene, 268
MIBK. *See* methyl isobutyl ketone
mica, 224, 274. *See also* silicates
mineral oil, 63–4, 203, 268
mineral pitch. *See* asphalt
mineral spirits. *See* kerosene
mineral wool, 179
molybdenum and compounds, 256, 268
monobromoethane. *See* ethyl bromide
monomethyl aniline, 268
monomethyl hydrazine, 268

morpholine, 236, 268
muriatic acid. *See* hydrochloric acid

nabam, 223
naneb, 223
naphtha (coal tar), 57, 268
naphtha, petroleum, 57, 269
naphthalene and derivatives, 39, 63, 191, 268
naphthalenes, chlorinated, 39, 196
naphthenic acids, 214
1-naphtholamine. *See* alpha-naphthylamine
2-naphtholamine, 200. *See also* beta-naphthylamine
beta-naphthylamine, 199
alpha-naphthylamine (1-naphthylamine), 200
beta-naphthylamine, 200
    effect on skin, 57
neon, 158
nickel carbonyl, 268, 336
nickel and compounds, 57, 59, 250, 256, 268, 282
nicotine, 29, 223, 268. *See also* cigarette smoking, effects of, in General Index
nilovar, 39
niobium. *See* columbium
nitric acid, 57, 159–60, 242, 268
nitric oxide, 268
nitro acids, 212
nitro compounds, 234
para-nitroanilines, 200, 268
nitrobenzene and derivatives, 39, 57, 196, 268
nitrobutanes, 235
nitroethane, 235, 268
nitrogen, 158
nitrogen dioxide, 19, 26, 158, 159–60, 231, 268
nitrogen oxides, 24–5, 159–60, 280, 336
nitrogen trifluoride, 268
nitroglycerine, 57, 233–4, 268
nitrohexene, 235
nitromethane, 57, 235, 268
nitrononene, 235
nitropropanes, 235, 268
nitrosamines, 235
para-nitroso-N,N-dimethyl aniline, 200

nitrotoluene, 268
nitrous oxide, 158
nylons. *See* polyamides

octachloronaphthalene, 268
octane, 268
oil, mineral, 63–4, 203, 268
oils, 56, 62, 63, 64, 203
oleamide, 213
oleum. *See* sulfuric acid
OMPA, 224
organic acids, 211–19
organo (alkyl) mercury, 273
organophosphate pesticides. *See*
    pesticides, organophosphate
organophosphate plasticizers. *See*
    plasticizers, organophosphate
ortho- compounds. *See* listing by
    second word in name
osmium and compounds, 257, 269
oxalic acid, 44, 57, 269
oxygen, 308, 336. *See also* oxygen,
    in General Index
oxygen difluoride, 269
ozone, 164, 269, 280–1, 336

palladium, 257
palmitamide, 213
para- compounds. *See* listing by sec-
    ond word in name
paraldehyde, 210
paraquat, 269
parathion, 224, 269
pentaborane, 247, 269
pentachloronaphthalene, 269
pentachlorophenol, 57, 269
pentane, 269
pentanone, 268
pentasol. *See* amyl alcohol
3-pentyn-2-one, 210
perchloroethylene, 57, 183. *See also*
    tetrachloroethylene
perchloromethyl mercaptan, 238,
    269
perchloryl fluoride, 269
peroxides, 55, 226
pesticides, 48, 53, 221–3
    organophosphate, 53, 222–3
petroleum, 55, 63, 269
phenol, 20, 39, 57, 183, 192
phenolic resins, 230

phenyl carbinol. *See* benzyl alcohol
para-phenylene diamine, 59, 198,
    200, 269
phenyl ether, 269
phenyl glycidyl ether, 269
phenylhydrazine, 39, 57, 237, 269
phenyl mercuric acetate, 255
phenyl methyl ketone. *See* aceto-
    phenone
phosdrin, 224, 269
phosgene, 24–5, 57, 163, 205, 269,
    281, 309, 336
phosphates, 233
phosphine, 57, 269, 280
phosphoric acid, 269
phosphorus, yellow, 39, 269
phosphorus pentachloride, 269
phosphorus pentasulfide, 269
phosphorus trichloride, 52, 57, 269
phthalates, 213, 233
phthalic anhydride, 57, 213, 218,
    230, 269
picoline, 187
picric acid, 57, 269
piperidine, 187
pitch, 63
pival, 269
plasticizers, 53, 224–5, 227, 233
    organophosphate, 53
plastics, 59, 223–33
platinum, 257
platinum salts, 57, 269
plumbago. *See* graphite
poison ivy, 59, 60
poison oak, 59
polyamides, 230
polybutylene glycols, 189
polyesters, 230
polystyrenes, 166, 204, 230
polytetrafluoroethylene, 227
polyurethane, 175, 231
polyvinyl alcohol, 232
polyvinyl acetate, 232
polyvinyl chloride, 232
potassium bichromate, 59
potassium cyanide, 165
potassium hydroxide, 58, 239–40
propane, 158, 269
propargyl alcohol, 269
beta-propiolactone, 219
propionamide, 213
propionic acid, 216

propionic anhydride, 218
propionitrile, 166
n-propyl acetate, 57, 220, 269
propyl alcohol, 269
propylene glycols, 189
propylamine, 236
propylene dichloride, 39, 57, 206, 269
propyleneimine, 236, 269
propylene oxide, 229, 269
n-propyl nitrate, 269
PTFE. *See* polytetrafluoroethylene
PVC. *See* polyvinylchloride
pyrazoxon, 224
pyrethrum, 223, 269
pyridine, 39, 57, 187, 269
pyrocatechol, 194
pyrolon, 224

quartz. *See* silica
quicklime. *See* calcium oxide
quinoline, 187
quinone, 57, 269

radium, 64
radon, 148
rayon viscose, 33. *See also* carbon
    disulfide
RDX, 270
resorcinol, 194
rhenium, 257
rhodium, 257, 270
rock wool, 175, 179
ronnel, 270
rotenone, 223, 270
rubber, 59
rubidium and compounds, 257
ruthenium and compounds, 258

salicylic acid, 216
sand, 224
selenium, 39, 57, 258, 270
selenium hexafluoride, 270
shale oil and wax, 63
silica, 21, 167–9, 240, 274
silicates, 175, 274
silver and compounds, 57, 258, 270
slaked lime. *See* calcium hydroxide
soap, 168
soapstone, 274
sodium bichromate, 59
sodium carbonate, 240
sodium cyanide, 165

sodium fluoroacetate, 217, 270
sodium hexafluorosilicate, 241–2
sodium hydroxide, 58, 239–40, 270
sodium peroxide, 240
sodium silicate, 240
solvents, 55, 183, 203, 227, 231
soot. *See* carbon black
stearamide, 213
stearic acid, 216
steroids, 39
stilkene, 53, 243, 270
Stoddard solvent, 58, 270
strontium, 148, 258
strychnine, 270
styrene, 39, 58, 230, 231, 273
succinic acid, 218
sulfonic acids, 214
sulfur dichloride, 238
sulfur dioxide, 24–5, 58, 158, 160,
    161, 238, 250, 270, 336
sulfur hexafluoride, 270
sulfuric acid, 58, 242, 250, 270. *See
    also* acids
sulfur monochloride, 58, 238, 270
sulfur pentafluoride, 270
sulfur chloride, 238
sulfuryl fluoride, 270
sumac, 59
systox, 224, 270

talc, 179, 274
tantalum, 258, 270
tar and tar oils, 62, 63, 170. *See also*
    coal tar
tartaric acid, 218
tartrates, 44
TDI. *See* tolylene diisocyanate
tear gas, 210
TEDP, 224, 270
Teflon. *See* polytetrafluoroethylene
tellurium, 258, 270
tellurium hexafluoride, 270
TEPP, 224
terpene polychlorinate, 221
terphenyls, 270
testosterone, 38. *See also* hormones
tetrachloroethane, 37, 39, 44, 58,
    205–6, 270
tetrachloroethylene, 39, 58, 206, 270,
    273
tetrachloronaphthalene, 270
tetracyclines, 39

tetraethyl lead, 252–3, 270, 336
tetrafluoroalkanes, 232
tetrafluoroethylene, 232
tetrahydrofuran, 270
tetralin, 191
tetramethyl lead, 270
tetramethyl succinonitrile, 270
tetramethylthiuram disulfide, 58
tetranitromethane, 271
tetryl, 270
thallium and compounds, 58, 259, 270
thimet, 224
thioacetamide, 213
thiocyanates, 166
thionyl chloride, 238
thiram, 271
thorium compounds, 58
thorazene, 38
tin and compounds, 58, 181, 259, 271
titanium and compounds, 259, 271, 282
titanium tetrachloride, 58
TOCP. *See* tricresyl phosphates
toluene, 35, 39, 58, 190, 273
tolylene diisocyanate (TDI), 58, 231, 271
toluidines, 198, 200
toxaphene, 221, 271
TPP. *See* tricresyl phosphates
tribromomethane. *See* bromoform
tri-butyl phosphate, 233, 271
trichloroacetic acid, 216, 217
trichloroethane, 39, 271
trichloroethylene, 39, 58, 206, 273
trichloromethane. *See* chloroform
trichloronaphthalene, 271
1,2,3-trichloropropane, 271
1,1,2-trichloro-1,2,2-trifluoroethane, 271
tridymite. *See* silica
triethylamine, 271
trifluoroacetic acid, 20, 217
trifluoromonobromomethane, 271
tri-isobutyl phosphate, 233

trimethyl benzanthracenes, 202
trimethyl benzene. *See* mesitylene
trinitrotoluene, 35, 271
tricresyl phosphates, 58, 233, 271
triphenyl phosphate, 271
trisodium phosphate, 240
trithion, 224
tungsten, 259–60
turpentine, 58, 271

uranimum and compounds, 44, 46, 148, 260, 271

valeramide, 213
valeric acid, 216
gamma-valerolactone, 219
vanadium, 260, 271
vanadium tetrachloride, 58
vapors, ventilation of. *See* ventilation, in General Index
vinyl acetate, 220
vinyl carbinol. *See* allyl alcohol
vinyl chloride, 58, 232, 271
vinyl ether, 186
vinyl pyridine, 187
vinyls, 232
vinyl toluene, 271
vitriol, oil of. *See* sulfuric acid

warfarin, 271
waxes, 62, 63

xylene, 58, 191, 271
xylidine, 200
xylol. *See* xylene

yttrium, 271

zectran, 224
zinc and compounds, 58, 260–1, 271, 283
zineb, 224
zirconium and compounds, 58, 261, 271
zuram, 223

About the Authors

Jeanne M. Stellman earned a Ph.D. in physical chemistry at the City University of New York. She is Assistant for Health and Safety to the president of the Oil, Chemical and Atomic Workers International Union. She has taught and developed several training courses for workers at the Rutgers University Labor Education Center and is a labor representative on several advisory boards for setting standards.

Susan M. Daum is an internist affiliated with the Environmental Science Laboratory at Mount Sinai School of Medicine. She is a graduate of Cornell Medical College and served her internship and residency in internal medicine at Mount Sinai Hospital. She has worked closely with Jeanne Stellman in the program of teaching industrial health hazards, and has also served as a consultant to the government, representing labor, on scientific advisory committees.